# Fiber Bragg Grating Sensors: Recent Advances and Future Perspectives

# Fiber Bragg Grating Sensors: Recent Advances and Future Perspectives

Editor

**Oleg G. Morozov**

MDPI

Basel • Beijing • Wuhan • Barcelona • Belgrade • Novi Sad • Cluj • Manchester

Oleg G. Morozov
Radio Photonics and
Microwave Technologies
Kazan National Research
Technical University
n.a. A.N. Tupolev-KAI
Kazan
Russia

*Editorial Office*
MDPI AG
Grosspeteranlage 5
4052 Basel, Switzerland

This is a reprint of the Special Issue, published open access by the journal *Sensors* (ISSN 1424-8220), freely accessible at: www.mdpi.com/journal/sensors/special_issues/Fiber_Bragg_Grating_Sensors.

For citation purposes, cite each article independently as indicated on the article page online and using the guide below:

| Lastname, A.A.; Lastname, B.B. Article Title. *Journal Name* **Year**, *Volume Number*, Page Range. |
|---|

**ISBN 978-3-7258-1640-8 (Hbk)**
**ISBN 978-3-7258-1639-2 (PDF)**
**https://doi.org/10.3390/books978-3-7258-1639-2**

# Contents

# About the Editor

**Oleg G. Morozov**

O.G. Morozov was born in Kazan, Tatarstan, Russia, USSR, on October 30, 1960. He became an engineer in radiotechnics in 1983 and obtained his PhD degree in 1987 at the Tupolev Aviation Institute, Kazan. Dr. Morozov became a Doctor of Technology at the Kazan National Research Technical University-Tupolev Aviation Institute, Kazan, in 2004. In this university, he was the head of the Quantum Electronics and Laser Technology R&D Lab. from 1989 to 1993, the head of the TV and Multimedia System Department, from 2005 to 2014, and the head of the Radiophotonics and Microwave Tech. Department from 2014 to 2023. From 2012 to 2021, he was the Director of the R&D Institute of Applied Electrodynamics, Photonics, and Live Systems at the same University. Currently, he is a full professor in the Radiophotonics and Microwave Tech. Department. His research areas comprise microwave photonics in radar systems, fiber optic sensory, and optical metrology. Previously, he obtained the S.E. Frisch Medal of the Russian Optical Society, n.a. D.S. Rozhdestvensky. Moreover, he is a senior member of the SPIE, a senior member of the OSA, and a senior member of the IEEE, so as EOS member, and ROS member.

# Preface

This reprint wishes to explore one of the most popular trends of relevance to sensors, which represents an influential platform for the exchange of scientific knowledge in the field of fiber Bragg gratings (FBGs) from the point of view of recent advances and future perspectives. The 35 years' history of FBG development has already transformed them from exotic novelties to classical sensors, without which it is impossible to imagine any of the technical and cyber-physical systems rounding the man. We are opening a little, but new, page in their history and giving the floor to three different groups of scientists. The first group consists of those who are just discovering FBGs and admire their capabilities in comparison with electronic and other types of optical sensors. The second comprises those who are masters of using FBGs and are trying to expand the scope of their applications toward infinity. Still others are those who, in spite of the conservatism of the classical mathematical models describing FBG structure, break it with new decisions or turn FBG interrogation systems on their heads, making them faster, more accurate, and, most importantly, cheaper. The ultimate goal of this reprint is to find ways to build smart FBG systems and identify their future applications, put together by authors representing the best scientific schools in the world. Modernization of FBG basics and its performances in various conditions are presented by H. Li et al. for line-by-line inscriptions of tilted FBGs, A. Morana et al. for studies of radiation effects on FBGs, and H.V. Tran and S.-Y. Seo for the short-term creep effect on strain transfer from fiber-reinforced polymer strips. New approaches are proposed for interrogation methods. V. Anfinogetov et al. consider in detail the algorithm of FBG spectrum distortion correction for classical OSA with CCD elements. U. Nordmeyer et al. perform noise analysis for new microwave photonic interrogation of fiber-based sensors. The next section of this reprint reflects the results of using FBGs in aviation systems: R. Nigmatullin et al. continue the topic of noise processing, namely noise cancellation of helicopter blade deformation measurement; H. Wu et al. design a high-precision shape sensor for wing shape reconstruction; and Z. Li et al. are developing a medium-frequency accelerometer based on flexible hinges. Separately, we discuss the transfer of FBG technologies to new devices based on integral photonics and new mediums in which FBGs can gain new life: G. Voronkov et al. are modeling a fully integrated micro-ring-based photonic system for liquid refractometry; I. Stepanov et al. are designing wavelength-tunable vortex beam emitters based on silicon micro-ring, twisted few-mode optical fiber with improved height of quasi-step refractive index profile, as proposed by A. Bourdine et al. The editor thanks all authors for their contributions to this reprint.

**Oleg G. Morozov**
*Editor*

sensors

MDPI

*Communication*

# Fabrication and Characterization of Line-by-Line Inscribed Tilted Fiber Bragg Gratings Using Femtosecond Laser

**Hongye Li [1,2,†], Xiaofan Zhao [1,3,†], Binyu Rao [1,3], Meng Wang [1,2,3], Baiyi Wu [1,2,3] and Zefeng Wang [1,2,3,*]**

1 College of Advanced Interdisciplinary Studies, National University of Defense Technology, Changsha 410073, China; lihongye@nudt.edu.cn (H.L.); zhaoxiaofan_zxf@nudt.edu.cn (X.Z.); raobinyu@nudt.edu.cn (B.R.); wangmeng@nudt.edu.cn (M.W.); wubaiyi@nudt.edu.cn (B.W.)
2 State Key Laboratory of Pulsed Power Laser Technology, Changsha 410073, China
3 Hunan Provincial Key Laboratory of High Energy Laser Technology, Changsha 410073, China
* Correspondence: zefengwang@nudt.edu.cn
† These authors contributed equally to this work.

**Abstract:** In this paper, we studied the basic characteristics of tilted fiber Bragg gratings (TFBGs), inscribed line-by-line. Experimental results showed that if the TFBGs were located within different planes parallel to the fiber axis, the spectra performed differently. For 2°TFBG, if it was located near the central plane, the Bragg resonance was stronger than ghost mode resonance, and the order reversed if it was located near the boundary between core and cladding. As the tilted angle increased, the range of cladding mode resonance increased. When the tilted angle was larger than 12°, the birefringence effect was observed. Based on the birefringence phenomenon, torsion characteristics were experimentally studied; the sensitivity was about 0.025 dB/degree in the linear variation range. The harmonic order of TFBGs also affected the transmission spectrum. Leaky mode resonance was observed in the 8th order TFBG, and torsion (or polarization) influenced the spectrum of the 8th order TFBG. Our research represented the theory of line-by-line inscribed TFBGs and provided an inscription guidance for TFBGs.

**Keywords:** femtosecond laser; direct writing; tilted fiber Bragg gratings



**Citation:** Li, H.; Zhao, X.; Rao, B.; Wang, M.; Wu, B.; Wang, Z. Fabrication and Characterization of Line-by-Line Inscribed Tilted Fiber Bragg Gratings Using Femtosecond Laser. *Sensors* **2021**, *21*, 6237. https://doi.org/10.3390/s21186237

Academic Editor: Oleg G. Morozov

Received: 20 August 2021
Accepted: 14 September 2021
Published: 17 September 2021



## 1. Introduction

TFBGs can realize the coupling between core mode and cladding modes [1,2]. Owing to the characteristics of cladding mode, TFBGs play an important role in fiber sensing. For example, small angle (<23°) TFBGs, which couple the forward-propagating core mode to backward-propagating cladding modes [1], are largely used in temperature sensing [3] and refractive index sensing [4]. If coated with metal, TFBG-assisted surface plasmon resonance can be excited, with which high sensitivity sensing can be carried out, such as gas/acoustic sensing [5], biological sensing [6], electrochemical activity in supercapacitors [7], and anemometers [8]. In excessively tilted fiber grating (Ex-TFG) (tilted angle > ~70°), the coupling occurs between core mode and cladding modes transmitted in the same direction, which behaves like long period fiber grating. Birefringence characteristics also presents in Ex-TFG. Thus, Ex-TFG is a good candidate for polarization-dependent sensing [9,10]. 45° TFBGs experience high polarization-dependent loss, which provides a method to realize an all-fiber polarizer [11]. In addition to their application in fiber sensing, TFBGs have been applied in fiber lasers. Small angle TFBGs were utilized to suppress the stimulated Brillouin scattering effect in high power fiber lasers [12]. 45° TFBG could play the role of saturable absorber in mode-locked fiber lasers [13,14].

Different methods have been reported to inscribe TFBGs. Ultraviolet exposure is the most common method used [15]. This method can inscribe TFBGs with low insertion loss and high stability. However, the inscription process is complex. Hydrogen loading and annealing are unavoidable during inscription. In the present study, the inscription of Ex-TFG was achieved by adjusting the optical system. Additionally, the resonant wavelength was limited by phase mask (PM). TFBGs could also be made by PM and femtosecond laser radiation [3,16]. Benefits from the femtosecond laser, hydrogen loading, and annealing were avoided, but limitations in the scale and depth of the femtosecond laser focus created difficulties in fiber alignment. To expand the area of refractive index modulation (RIM) and stretch the length of the grating, a complex electrical translation stage was necessary. The resonant wavelength was also limited by PM in this condition. Femtosecond laser plane-by-plane direct writing was an effective method to inscribe TFBG. The tilted angle was controlled by electrical translation stage [17] or the rotating angle of a cylindrical lens in the optical system [18]. The resonance could occur at any wavelength, triggered by adjusting the period of gratings. The birefringence decreased because the RIM was more uniform than TFBGs inscribed with PM [17], but insertion loss increased in the direct writing process. These methods could only fabricate non-localized TFBGs, in which the transmission spectrum was only influenced by the tilted angle. The RIM in the non-localized condition covered the whole core region and the coupling process was relatively stationary [19,20].

Another femtosecond laser direct writing method—namely, line-by-line inscription [21,22]—was also a potential method of fabricating TFBGs. In the year of this study's writing, Liu et al. proposed TFBGs inscribed line-by-line using femtosecond lasers [23]. Comparing their method with the aforementioned three methods, the TFBGs inscribed line-by-line were highly localized. A benefit of line-by-line inscription was that the insertion loss of TFBGs was low. Additionally, the range of cladding mode resonance was wider than TFBGs inscribed with other methods. Ref. [23] provided preliminary guidance for line-by-line inscription of TFBGs. Further investigation should be carried out to improve this method and the relevant theory of TFBGs.

In this paper, we investigated the fundamental characteristics of TFBGs inscribed line-by-line. The influence of the tilted angle, the line position, and the harmonic order of grating on the transmission spectrum of TFBG were experimentally studied. Some torsional experiments were also carried out to investigate the polarization-dependent effect of TFBGs inscribed line-by-line. Our research works improved the theory of line-by-line inscribed TFBGs and presented guidance for a femtosecond laser direct writing technique.

## 2. Inscription Method

Figure 1 shows the schematic of line-by-line inscription of TFBGs. In our experiment, an oil-immersion lens with a magnification of 100× was used to focus the femtosecond laser. The wavelength of the femtosecond laser was 515 nm, the repetition rate was 1 kHz, and the pulse energy after oil-immersion lens was about 36 nJ. The period of grating ($\Lambda$) and the tilted angle ($\theta$) were controlled by an electrical 3D translation stage. During inscription, the focus of the femtosecond laser translated from point A to point B. Then, the femtosecond laser was shut off, and the focus of the femtosecond laser moved to the beginning of the next line (point C in Figure 1). After the focus reached point C, the femtosecond laser switched on and began to generate the next line.

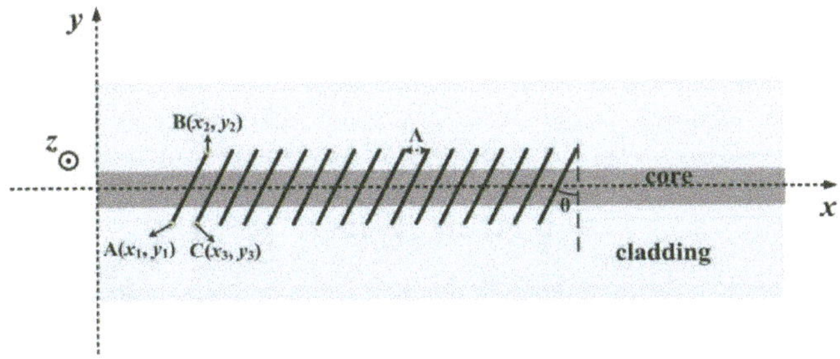


**Figure 1.** Phase matching condition of the first order harmonic resonance.

The relationship between these three points (A($x_1$, $y_1$), B($x_2$, $y_2$) and C($x_3$, $y_3$)) is as follows:

$$\begin{cases} y_1 = -y_2 = y_3 \\ x_2 = 2|y_1|\tan\theta + x_1 \\ x_3 = x_1 + \Lambda \end{cases} . \quad (1)$$

In order to cover the core diameter, the value of $|y_1|$ is 16 μm. By varying the period of grating (Λ), the resonant wavelength could be tuned. Cladding mode coupling was affected by tilted angle (θ). In our experiment, TFBGs were inscribed on single mode fibers (Corning, SMF28). The number of the period was 3000. A sweeping wavelength laser (resolution: 6 pm, and wavelength range: 1503.4–1620nm) was used to record the transmission spectra.

### 3. Effect of TFBG Plane Position on Transmission Spectrum

Figure 2 demonstrates the spectra and microscopies of two 2°TFBGs inscribed on two different planes parallel to the fiber axis. The period of grating was 1.11 μm. Comparing with Figure 2d,f, 2°TFBG-1 was located on a plane near the boundary between core and cladding (off-center inscription in z-direction), and 2°TFBG-2 was located at the central part of the core. That is to say, the overlap between core (or fundamental mode) and RIM of 2°TFBG-2 was stronger than that of 2°TFBG-1. As shown in Figure 2a,b, both the Bragg resonance and the cladding mode resonance of 2°TFBG-1 were weaker than that of 2°TFBG-2 as a result of lower overlap between fundamental mode and RIM. Additionally, the ghost mode resonance of 2°TFBG-1 was stronger than the Bragg resonance; however, the relationship between these two resonances was reversed in 2°TFBG-2. For non-localized inscriptions like the ultraviolet exposure method, the relationship between the Bragg resonance and the ghost mode resonance in TFBG is mainly influenced by tilted angle. As for highly localized inscriptions like femtosecond laser line-by-line inscription, the relationship is also affected by the grating position, and the overall resonant intensity will decrease if the off-center inscription is carried out—which is different from what occurs in non-localized conditions [19,20].

Figure 3 reveals the calculated results of the overlap integrals under different grating planes. As shown in Figure 3a, the offset distance between the fiber axis and the grating plane is defined as $d$. The overlap integral is read as

$$A = \int_{-\infty}^{+\infty} dx(\iint_S \Delta\tilde{n} \cdot \vec{E}_{01} \cdot \vec{E}^{*}_{coupled} \cdot circ(r \le r_{core})drd\varphi) \cdot e^{i\cdot\frac{4\pi}{\Lambda}\cdot x} \quad (2)$$

$$\Delta\tilde{n} = \Delta n \cdot \delta(z-d) \cdot \delta(y - \cot\theta(x + m \cdot \Lambda)) \quad (3)$$

Here, $E_{01}$ is the normalized electric field of fundamental mode, $E_{coupled}$ is the normalized electric field of the mode that fundamental mode couples to, and $\Delta n$ is the RIM. Additionally, $r_{core}$ is the radius of the core (in our simulation, this value was 4 μm). The value of θ was 2°. The integral along $x$ axis was used to extract the Fourier-expansion coefficient of the second order grating. Additionally, n (an integer) represents different lines causing effects on one specific cross section of the fiber core. Although the form of the function of RIM distribution is expressed as impulse function ($\delta$) along the $z$ axis, we set the modulation region to have a 0.6 μm length along the z axis in simulation. This was done in order to press close to the real inscription in line-by-line condition.

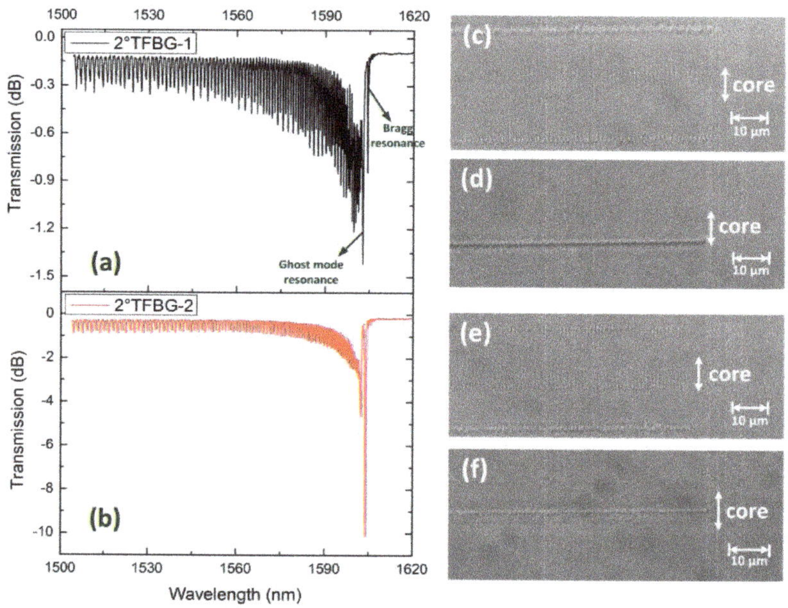


**Figure 2.** Transmission spectrum of (**a**) 2°TFBG-1 and (**b**) 2°TFBG-2. Microscopy image (100×) of 2°TFBG-1: (**c**) top view and (**d**) side view. Microscopy image (100×) of 2°TFBG-2: (**e**) top view and (**f**) side view.

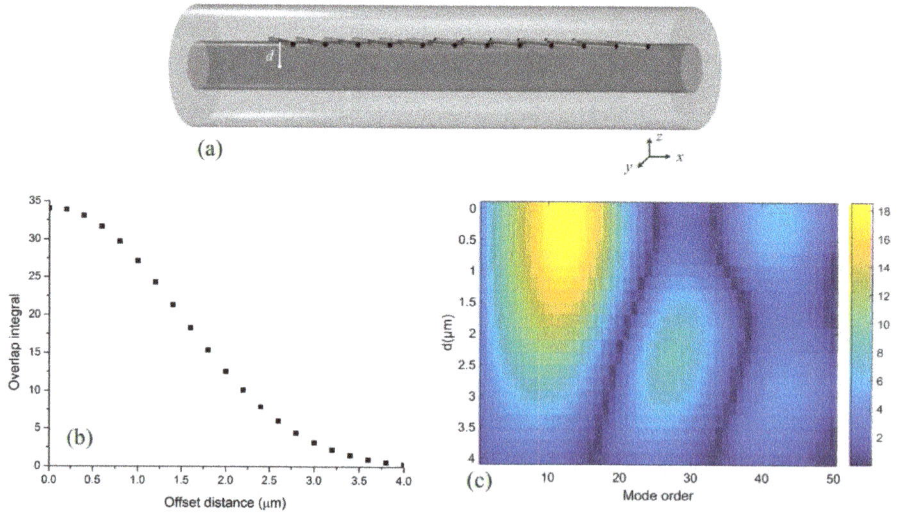


**Figure 3.** (**a**) Schematic of TFBG with offset distance *d*. (**b**) Overlap integral of fundamental mode versus offset distance. (**c**) Overlap integrals of the first 50 cladding modes under different offset distance.

Here, $E_{01}$ is the normalized electric field of fundamental mode, $E_{coupled}$ is the normalized electric field of the mode that fundamental mode couples to, and $\Delta n$ is the RIM. Additionally, $r_{core}$ is the radius of the core (in our simulation, this value was 4 μm). The value of $\theta$ was 2°. The integral along $x$ axis was used to extract the Fourier-expansion coefficient of the second order grating. Additionally, n (an integer) represents different lines causing effects on one specific cross section of the fiber core. Although the form of the function of RIM distribution is expressed as impulse function ($\delta$) along the $z$ axis, we set the modulation region to have a 0.6 μm length along the z axis in simulation. This was done in order to press close to the real inscription in line-by-line condition.

Figure 3b demonstrates the overlap integral of fundamental mode, which decreased with increases in $d$. Figure 3c shows the overlap between fundamental mode and the first 50 cladding modes under different offset distances ($d$). These overlap integrals showed decreasing trends when $d$ increased, especially for lower order cladding modes. These phenomena agreed well with what we observed in our experiment.

## 4. Effect of Tilted Angle on Transmission Spectrum

Figure 4 illustrates the transmission spectra and the microscopy images of TFBG with different tilted angles (2°, 4° and 6°). The period of grating was 1.11 μm, which guaranteed the 2nd order resonance at C + L band and the Bragg resonance near 1605nm. Moreover, these three TFBGs were located near the center of the core to avoid weak coupling. As the tilted angle increased, both the Bragg resonance and the ghost mode resonance showed a weakening tendency. Additionally, the envelope of the cladding mode coupling showed an expanding trend and shifted to a shorter wavelength. In other words, a larger tilted angle excited a number of higher order cladding modes and suppressed the coupling of lower order cladding modes.

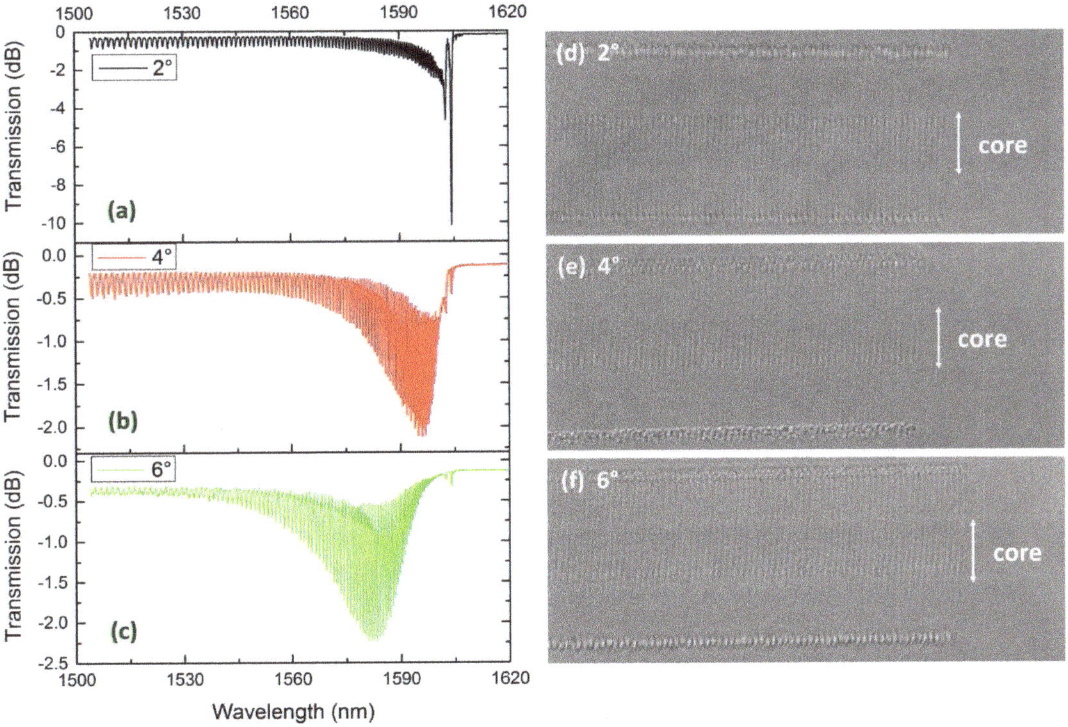


**Figure 4.** Transmission spectrum of (**a**) 2°TFBG, (**b**) 4°TFBG and (**c**) 6°TFBG, and microscopy image (100×) of (**d**) 2°TFBG, (**e**) 4°TFBG and (**f**) 6°TFBG.

Figure 5 performs the transmission spectrum and microscopy image of 12°TFBG, 15°TFBG and 18°TFBG. Considering the limited wavelength range of sweeping wavelength laser, the period of grating is set as 1.13 μm. Except the similar phenomena observed in Figure 3, birefringence is clearly seen on the spectra. Dual dips phenomenon occurs within several resonance wavelengths. Comparing with Figure 4a–c, the envelop of cladding mode coupling becomes blurred. These phenomena are polarization-dependent.

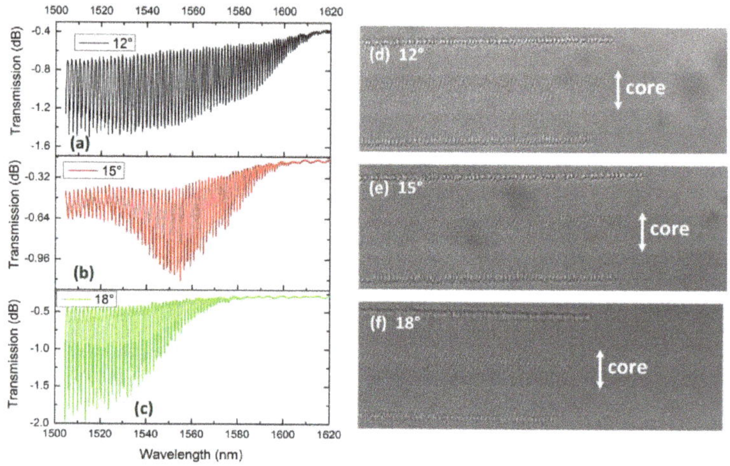


**Figure 5.** Transmission spectra of (**a**) 12°TFBG, (**b**) 15°TFBG and (**c**) 18°TFBG, and microscopy image (100×) of (**d**) 12°TFBG, (**e**) 15°TFBG and (**f**) 18°TFBG.

To test the polarization-dependent characteristics, the spectra of 18°TFBG (under different polarization states) were measured. The schematic of spectrum measurement system is illustrated in Figure 6a. After passing through a polarizer, light output from C + L ASE source became linearly polarized. Then, a polarization controller was utilized to adjust the polarization state. Finally, light with different polarization states passed through the TFBG and arrived at an optical spectrum analyzer (OSA). Figure 6b shows the spectra of 18°TFBG under different polarization states. It was obvious that the polarization state had a large impact on the transmission spectrum. Either of the resonant dips could be totally suppressed by tuning the polarization state of the input light.

We also investigated the torsional effects of 18°TFBG. The torsional detection setup is illustrated in Figure 7. The polarized laser output from a sweeping wavelength laser system, and then passed through a TFBG. After that, a power meter (integrated in the sweeping wavelength laser system) was used to record the spectra of TFBG. The TFBG was fixed between a fiber holder and a fiber rotator. By adjusting the rotating angle of the fiber rotator, different torsional stresses were exerted on the TFBG.

Figure 8 shows the spectra of 18°TFBG under different twist angles. For the purposes of our investigation, we only presented the spectra from 1508 nm to 1516 nm. Obviously, twist angles influence the spectrum of 18°TFBG. To quantify our study, we chose two dips (dip A and dip B in Figure 8) near 1510 nm to characterize torsional effects. The resonant wavelengths varied with the twist angle, which could have been induced by stress along the fiber axis in the process of twisting. However, the wavelength shift was extremely small and presented no regular change. The twist-related depth variations of dip A and dip B were much more obvious when compared to wavelength shifts. If one of them deepened, another would weaken.

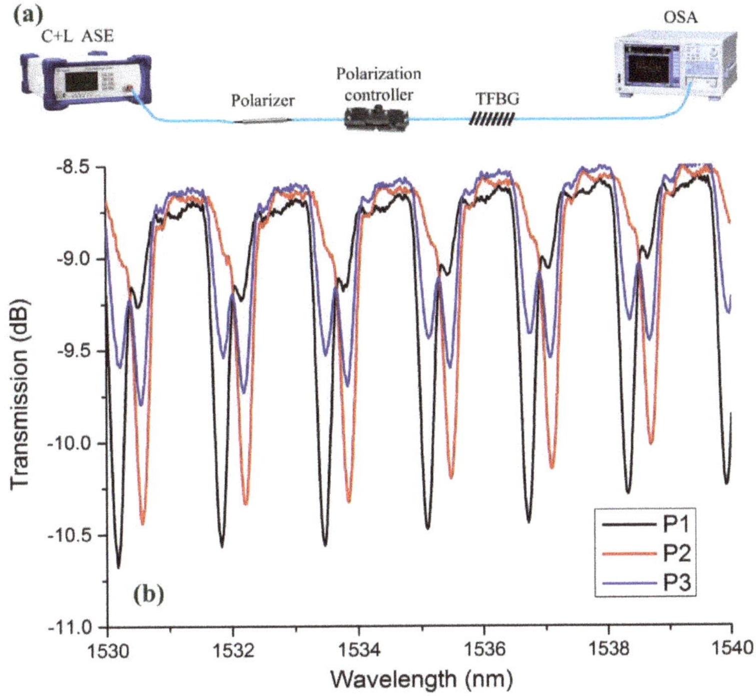


**Figure 6.** (**a**) Schematic of polarization-dependent spectrum measurement system. (**b**) Spectra of 18°TFBG under different polarization states.

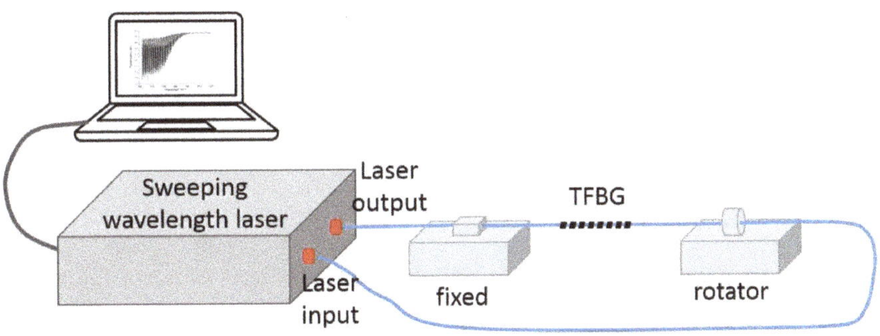


**Figure 7.** Schematic setup of torsional detection.

Figure 9 shows the depth variation of dip A and dip B as the twist angle increased with steps of 30 nm. The difference between dip A and dip B was also performed to magnify the twist-related variation. As shown in Figure 9, all the curves took on sine-like shapes. During twist angle increases, the polarization state of input light also rotated; thus, the variation of dip A and dip B also presented a periodical shape. As the blue line shows, the slope (sensitivity) between 120° and 270° was about 0.025 dB/degree. TFBGs with a tilted angle greater than 12° inscribed line-by-line are good candidates in torsional sensing applications. Compared with other kind of torsion sensors, the proposed torsion sensor in our manuscript was highly localized. This provided for the possibility of further integration and would be suitable for narrow area torsion sensing. Moreover, among the benefits of femtosecond laser direct writing, the cladding layer of our sensing head remained intact, a feature which could extend the service life of torsion sensors.

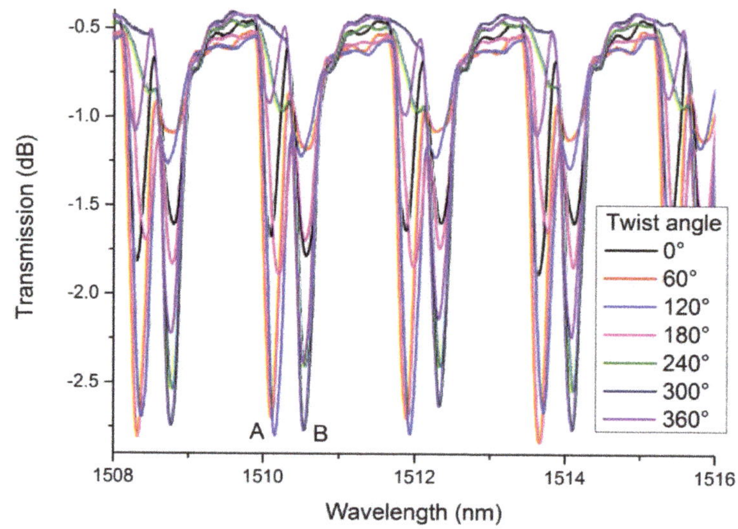


**Figure 8.** Spectra of 18°TFBG under different twist angles.

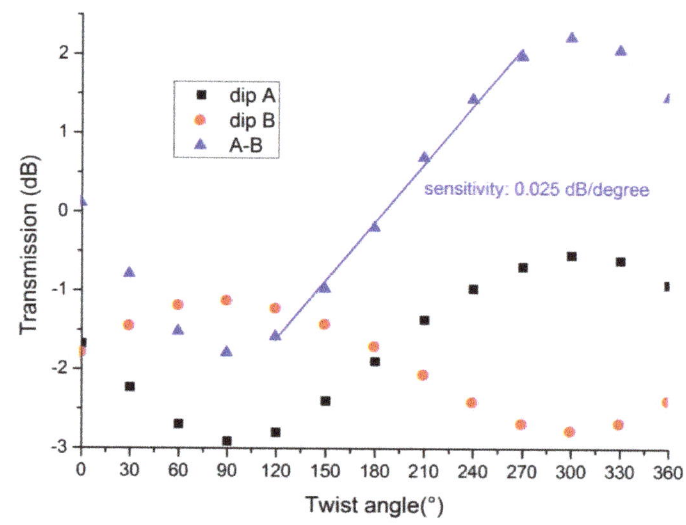


**Figure 9.** Depth variations of dip A and dip B under different twist angles.

## 5. Effect of Harmonic Order on Transmission Spectrum

In addition to the tilted angle, the harmonic order was another factor that influenced the shape of TFBGs. Figure 10 demonstrates the spectra and microscopy images of 6°TFBGs with different harmonic orders. Figure 10a shows the spectrum of the 4th harmonic order 6°TFBG ($\Lambda = 2.22$ μm). The characteristics of the 4th harmonic order 6°TFBG were similar to that of the 2nd order (see Figure 4c). However, when the harmonic order increased, the spectrum deteriorated. Figure 10b demonstrates the spectrum of the 6th order 6°TFBG. We observed some tiny fluctuations occurring on the spectrum, which were not present in the 2nd or 4th order 6°TFBG. The spectrum of 8th order 6°TFBG (Figure 10d) was chaotic. Except for the coupling between forward-propagating fundamental modes and backward-propagating cladding modes (the basic characteristics of TFBG), the coupling between the fundamental mode and higher order cladding mode transmitted in the same direction, occurred in the spectrum (the characteristics of LPFGs). Similar phenomena were reported in TFBGs inscribed plane-by-plane, and the authors described this phenomenon as leaky mode resonance (LMR) [24,25]. Additionally, the Bragg resonance and ghost mode resonance disappeared, as shown in Figure 10d. To confirm the existence of the Bragg resonance and ghost mode resonance, we measured the spectrum of 6°TFBG immersed in refractive index matching oil (Figure 10c). Cladding modes were stripped out of the cladding layer, and the Bragg resonance and ghost mode resonance existed in the 8th order 6°TFBG.

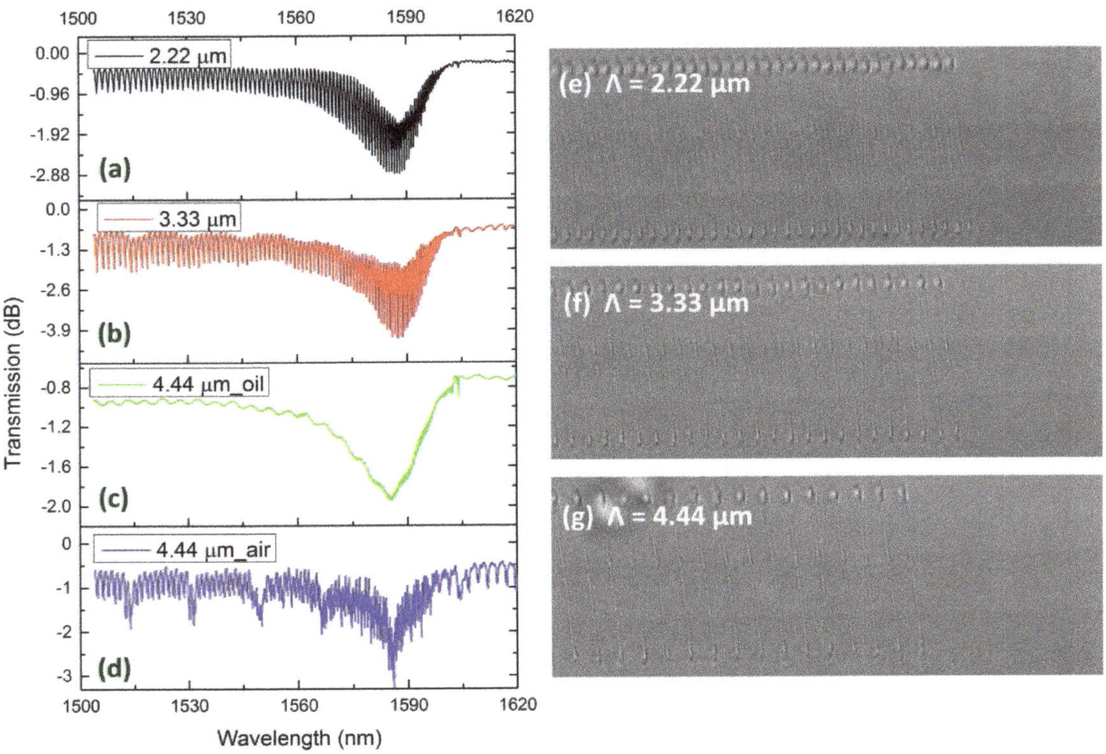


**Figure 10.** Spectra of 6°TFBGs with different harmonic order: (**a**) Λ = 2.22 μm, (**b**) Λ = 3.33 μm, (**c**) Λ = 4.44 μm (in oil), and (**d**) Λ = 4.44 μm (in air). Microscopy image (100×): (**e**) Λ = 2.22 μm, (**f**) Λ = 3.33 μm, and (**g**) Λ = 4.44 μm.

The torsional effect of the 8th order 6°TFBG was also investigated. The experimental setup was the same as that depicted in Figure 7. Figure 11 perform the spectra under different twist angle. Obviously, the intensity of LMR varied with twist angle, and the resonant wavelength also shifted. To exemplify that, the LMR was relatively weaker when the twist angle was 0° or 360°, the dips reached their deepest points when the twist angle was 180°, and the resonant wavelengths also shifted to longer wavelengths. However, with the impact of the coupling between forward-propagating fundamental mode and backward-propagating cladding modes, we could not simply describe this process by tracking the variation of the resonant wavelengths or the depth of LMR. A more appropriate methodology should be used to study this process.

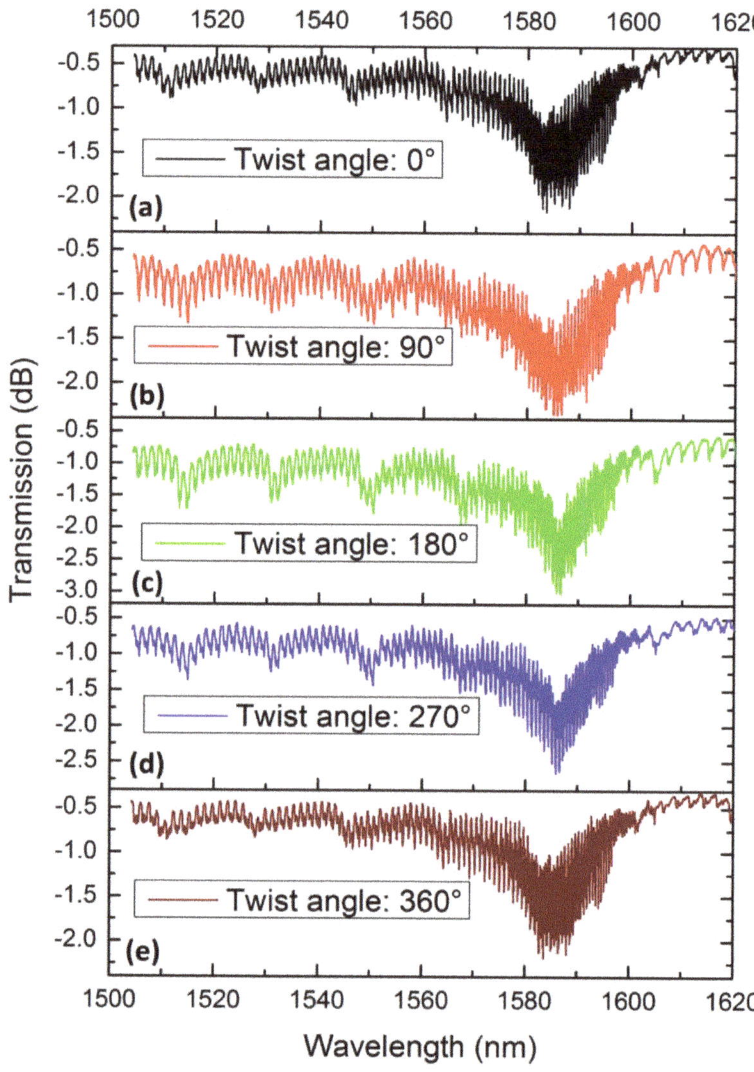


**Figure 11.** Spectra of the 8th order 6°TFBG under different twist angles: (**a**) 0°, (**b**) 90°, (**c**) 180°, (**d**) 270° and (**e**) 360°.

Correlation is a proper method to consider all phenomena simultaneously. The correlation can be calculated as

$$C_\alpha = \frac{\sum_{i=1}^{n} (A_0[i] \cdot A_\alpha[i])}{\sqrt{\sum_{i=1}^{n} A_0^2[i]} \cdot \sqrt{\sum_{i=1}^{n} A_\alpha^2[i]}} \tag{4}$$

$$A_\alpha[i] = S_\alpha[i] - \left( \sum_{i=1}^{n} S_\alpha[i] \right) / n \tag{5}$$

Here, $S_\alpha[i]$ represents the $i$th point on the spectrum of the 8th order 6°TFBG when the twist angle is $\alpha$. The total number of points on the spectrum was defined as n. Figure 12 illustrates the correlation function verses the twist angle. The whole curve took on a sinusoidal tendency, which indicates that the 8th order 6°TFBG was a polarization-dependent device. The slope (sensitivity) between 30° and 150° was −0.001/degree, and the slope (sensitivity) between 240° and 330° was 0.001/degree. These two ranges were deemed suitable for twist sensing.

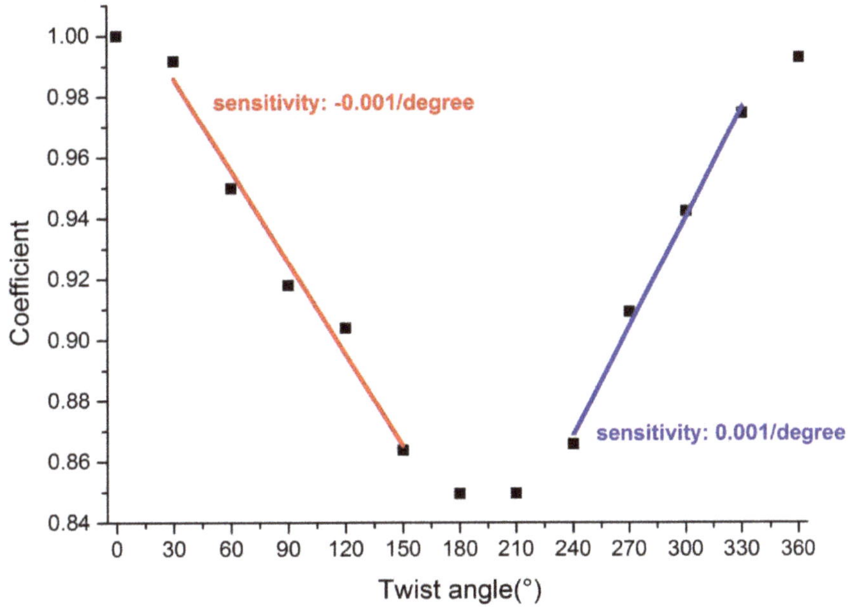


**Figure 12.** Correlation of spectra under different twist angles.

## 6. Conclusions

In conclusion, we experimentally studied the characteristics of line-by-line inscribed TFBGs. Studies show that, if the TFBG is located at a different plane parallel to the fiber axis, the transmission spectrum will show large differences. By increasing the tilted angle, the range of cladding mode resonance increased and the minimum value of cladding mode resonance shifted to a shorter wavelength. When the tilted angle was larger than 12°, birefringence effects became obvious. The dual resonant dips of 18°TFBG varied periodically with twist angle and the slope (sensitivity) of the linear variation range was about 0.025 dB/degree, which made these kinds of TFBG a good candidate for twist sensing. Clearly, LMR was observed in the 8th order TFBG, and twist (or polarization state) impacted the spectrum of the 8th order TFBG. By calculating the correlation function of the transmission spectra of the 8th order TFBG under different twist angles, we were able to demodulate the variation tendency, such that we realized twist sensing. TFBG is also a good candidate for refractive index sensing, e.g., biological sensing and chemistry sensing. Outside of their applications in sensing and communication, FBGs with relatively small angles can be applied in stimulated Brillouin scattering suppression in high power fiber lasers. Our research results provide guidance for highly-localized TFBG inscription, which could be meaningful in integrated device fabrication, multichannel sensing, and control of polarization states.

**Author Contributions:** H.L. investigation, data curation, writing—original draft preparation; X.Z. investigation, data curation, writing—review and editing; B.R. software; M.W. methodology, funding acquisition; B.W. validation; Z.W. supervision, project administration. All authors have read and agreed to the published version of the manuscript.

**Funding:** This work was supported by Outstanding Youth Science Fund Project of Hunan Province Natural Science Foundation (2019JJ20023), National Natural Science Foundation of China (NSFC) (11974427, 12004431), State Key Laboratory of Pulsed Power Laser (SKL-2020-ZR05, SKL2021ZR01), and Postgraduate Scientific Research Innovation Project of Hunan Province (CX20200046, CX20190027).

**Conflicts of Interest:** The authors declare no conflict of interest.

## References

1. Albert, J.; Shao, L.Y.; Caucheteur, C. Tilted fiber Bragg grating sensors. *Laser Photonics Rev.* **2013**, *7*, 83–108. [CrossRef]
2. Dong, X.; Hao, Z.; Liu, B.; Miao, Y. Tilted fiber Bragg gratings: Principle and sensing applications. *Photonic Sens.* **2011**, *1*, 6–30. [CrossRef]
3. Wang, R.; Si, J.; Chen, T. Fabrication of high-temperature tilted fiber Bragg gratings using a femtosecond laser. *Opt. Express* **2017**, *25*, 23684–23689. [CrossRef]
4. Gang, T.; Liu, F.; Hu, M.; Albert, J. Integrated Differential Area Method for Variable Sensitivity Interrogation of Tilted Fiber Bragg Grating Sensors. *J. Lightwave Technol.* **2019**, *37*, 4531–4536. [CrossRef]
5. Caucheteur, C.; Guo, T.; Liu, F.; Guan, B.; Albert, J. Ultrasensitive plasmonic sensing in air using optical fibre spectral combs. *Nat. Commun.* **2016**, *7*, 13371. [CrossRef]
6. Lao, J.; Han, L.; Wu, Z.; Zhang, X.; Huang, Y.; Tang, Y.; Guo, T. Gold Nanoparticle-Functionalized Surface Plasmon Resonance Optical Fiber Biosensor: In Situ Detection of Thrombin with 1 nM Detection Limit. *J. Lightwave Technol.* **2018**, *37*, 2748–2755. [CrossRef]
7. Lao, J.; Sun, P.; Liu, F.; Zhang, X.; Zhao, C.; Mai, W.; Guo, T.; Xiao, G.; Albert, J. In situ plasmonic optical fiber detection of the state of charge of supercapacitors for renewable energy storage. *Light Sci. Appl.* **2018**, *7*, 34. [CrossRef]
8. Liu, Y.; Liang, B.; Zhang, X.; Hu, N.; Li, K.; Chiavaioli, F.; Gui, X.; Guo, T. Plasmonic Fiber-Optic Photothermal Anemometers With Carbon Nanotube Coatings. *J. Lightwave Technol.* **2019**, *37*, 3373–3380. [CrossRef]
9. Shen, F.; Zhou, K.; Wang, C.; Jiang, H.; Peng, D.; Xia, H.; Xie, K.; Zhang, L. Polarization dependent cladding modes coupling and spectral analyses of excessively tilted fiber grating. *Opt. Express* **2020**, *28*, 1076–1083. [CrossRef] [PubMed]
10. Yan, Z.; Sun, Q.; Wang, C.; Sun, Z.; Mou, C.; Zhou, K.; Liu, D.; Zhang, L. Refractive index and temperature sensitivity characterization of excessively tilted fiber grating. *Opt. Express* **2017**, *25*, 3336–3346. [CrossRef]
11. Zhou, K.; Simpson, G.; Chen, X.; Zhang, L.; Bennion, I. High extinction ratio in-fiber polarizers based on 45° tilted fiber Bragg gratings. *Opt. Lett.* **2005**, *30*, 1285–1287. [CrossRef]
12. Tian, X.; Zhao, X.; Wang, M.; Wang, Z. Suppression of stimulated Brillouin scattering in optical fibers by tilted fiber Bragg gratings. *Opt. Lett.* **2020**, *45*, 4802–4805. [CrossRef]
13. Zhang, Z.; Mou, C.; Yan, Z.; Zhou, K.; Zhang, L.; Turitsyn, S. Sub-100 fs mode-locked erbium-doped fiber laser using a 45°-tilted fiber grating. *Opt. Express* **2013**, *21*, 28297–28303. [CrossRef] [PubMed]
14. Huang, Z.; Huang, Q.; Theodosiou, A.; Cheng, X.; Zou, C.; Dai, L.; Kalli, K.; Mou, C. All-fiber passively mode-locked ultrafast laser based on a femtosecond-laser-inscribed in-fiber Brewster device. *Opt. Lett.* **2019**, *44*, 5177–5180. [CrossRef] [PubMed]
15. Wen, H.; Hsu, Y.; Chen, S.; Chiang, C. The manufacturing process and spectral features of tilted fiber Bragg gratings. *Opt. Laser Technol.* **2021**, *134*, 106615. [CrossRef]
16. Pham, X.; Si, J.; Chen, T.; Qin, F.; Hou, X. Wide Range Refractive Index Measurement Based on Off-Axis Tilted Fiber Bragg Gratings Fabricated Using Femtosecond Laser. *J. Lightwave Technol.* **2019**, *37*, 3027–3034. [CrossRef]
17. Ioannou, A.; Theodosiou, A.; Caucheteur, C.; Kalli, K. Direct writing of plane-by-plane tilted fiber Bragg gratings using a femtosecond laser. *Opt. Lett.* **2017**, *42*, 5198–5201. [CrossRef] [PubMed]
18. Lu, P.; Mihailov, S.J.; Ding, H.; Grobnic, D.; Walker, R.B.; Coulas, D.; Hnatovsky, C.; Naumov, A.Y. Plane-by-Plane Inscription of Grating Structures in Optical Fibers. *J. Lightwave Technol.* **2018**, *36*, 926–931. [CrossRef]
19. Wang, G.; Wang, J.; HUANG, Y.; Wu, Y. Study on Characteristics of Tilted Fiber Grating. In Proceedings of the 2018 International Conference on Modeling, Simulation and Optimization, Xiamen, China, 25–26 November 2018.
20. Guo, T.; Fu, L.; Guan, B.; Albert, J. [INVITED]Tilted fiber grating mechanical and biochemical sensors. *Opt. Laser Technol.* **2016**, *78*, 19–33. [CrossRef]
21. Zhou, K.; Dubov, M.; Mou, C.; Zhang, L.; Mezentsev, V.K.; Bennion, I. Line-by-Line Fiber Bragg Grating Made by Femtosecond Laser. *IEEE Photonics Technol. Lett.* **2010**, *22*, 1190–1192. [CrossRef]
22. Huang, B.; Xu, Z.; Shu, X. Dual interference effects in a line-by-line inscribed fiber Bragg grating. *Opt. Lett.* **2020**, *45*, 2950–2953. [CrossRef] [PubMed]
23. Liu, B.; Yang, K.; Liao, C.; Cai, Z.; Liu, Y.; Sun, B.; Wang, Y. Localized tilted fiber Bragg gratings induced by femtosecond laser line-by-line inscription. *Opt. Lett.* **2021**, *46*, 2204–2207. [CrossRef] [PubMed]

24. Ioannou, A.; Theodosiou, A.; Kalli, K.; Caucheteur, C. Higher-order cladding mode excitation of femtosecond-laser-inscribed tilted FBGs. *Opt. Lett.* **2018**, *43*, 2169–2172. [CrossRef] [PubMed]
25. Ioannou, A.; Theodosiou, A.; Caucheteur, C.; Kalli, K. Femtosecond Laser Inscribed Tilted Gratings for Leaky Mode Excitation in Optical Fibers. *J. Lightwave Technol.* **2020**, *38*, 1921–1928. [CrossRef]

sensors MDPI

Review

# Radiation Effects on Fiber Bragg Gratings: Vulnerability and Hardening Studies

Adriana Morana [1], Emmanuel Marin [1], Laurent Lablonde [2], Thomas Blanchet [3], Thierry Robin [2], Guy Cheymol [4], Guillaume Laffont [3], Aziz Boukenter [1], Youcef Ouerdane [1] and Sylvain Girard [1,*]

[1] UJM, CNRS, IOGS, Laboratoire Hubert Curien, University of Lyon, UMR 5516, 18 rue Prof. B. Lauras, F-42000 Saint-Etienne, France
[2] iXblue Photonics, F-22300 Lannion, France
[3] CEA List, Université Paris-Saclay, F-91120 Palaiseau, France
[4] CEA, Service d'Études Analytiques et de Réactivité des Surfaces, Université Paris-Saclay, F-91191 Gif-sur-Yvette, France
* Correspondence: sylvain.girard@univ-st-etienne.fr

**Abstract:** Fiber Bragg gratings (FBGs) are point optical fiber sensors that allow the monitoring of a diversity of environmental parameters, e.g., temperature or strain. Several research groups have studied radiation effects on the grating response, as they are implemented in harsh environments: high energy physics, space, and nuclear facilities. We report here the advances made to date in studies regarding the vulnerability and hardening of this sensor under radiation. First, we introduce its principle of operation. Second, the different grating inscription techniques are briefly illustrated as well as the differences among the various types. Then, we focus on the radiation effects induced on different FBGs. Radiation induces a shift in their Bragg wavelengths, which is a property serving to measure environmental parameters. This radiation-induced Bragg wavelength shift (RI-BWS) leads to a measurement error, whose amplitude and kinetics depend on many parameters: inscription conditions, fiber type, pre- or post-treatments, and irradiation conditions (nature, dose, dose rate, and temperature). Indeed, the radiation hardness of an FBG is not directly related to that of the fiber where it has been photo-inscribed by a laser. We review the influence of all these parameters and discuss how it is possible to manufacture FBGs with limited RI-BWS, opening the way to their implementation in radiation-rich environments.

**Keywords:** fiber gratings; fiber Bragg gratings; fiber sensors; optical fibers; radiation effects; harsh environments



**Citation:** Morana, A.; Marin, E.; Lablonde, L.; Blanchet, T.; Robin, T.; Cheymol, G.; Laffont, G.; Boukenter, A.; Ouerdane, Y.; Girard, S. Radiation Effects on Fiber Bragg Gratings: Vulnerability and Hardening Studies. *Sensors* **2022**, *22*, 8175. https://doi.org/10.3390/s22218175

Academic Editor: Oleg G. Morozov

Received: 26 September 2022
Accepted: 20 October 2022
Published: 25 October 2022



## 1. Introduction

Silica-based optical fibers (OFs) have attracted a lot of interest from research groups thanks to their properties, such as light weight, small volume, high bandwidth, and resistance to most electromagnetic perturbations, for both telecommunication and sensing applications [1,2]. Among the optical fiber sensors (OFSs) is the fiber Bragg grating (FBG) [3]. It is a punctual sensor, with a length of a few mm to a few cm, photo-inscribed with a laser beam inside the core of a single-mode optical fiber. It is also possible to write FBGs in multimode fibers. However, in this case, the gratings will give rise to several Bragg peaks, as studied for the first time by Wanser et al. in 1994 [4]. In the FBGs, the information is "wavelength encoded"; indeed, each grating causes a dip or peak in the fiber transmission or reflection spectra, respectively, whose peak position is known as the "Bragg wavelength". The Bragg wavelength depends on the selected FBG characteristics and it spectrally shifts when the environmental parameters around the fiber, such as temperature, strain, pressure, and humidity, evolve. This property makes the FBG a good sensor (From the same family of the FBGs, there is another OFS type: long period grating, or briefly, LPG. Recently, a review about the radiation effects on LPGs has been published [5]; consequently, this subject will not be addressed in this article) as, after calibration, the real-time monitoring of the Bragg wavelength allows the determination of the changes in these measurands (see review [6]).

This sensor type is easily multiplexable; indeed, we can easily write FBGs at different places along the same optical fiber, under the condition that each of them could be distinguished by using wavelength division multiplexing (WDM) or time division multiplexing (TDM) [7]. However, WDM is easier to be employed, since several FBGs can be written in series and investigated with a broadband light source. The only conditions are that each grating must have a different Bragg wavelength and that the spectral ranges of variation of the FBGs must not overlap. This can be achieved by tailoring the FBG characteristics at the manufacturing stage. Typically, about ten FBGs can be implemented using this approach in one fiber sample: this number depends on the spectral range of the acquisition system. TDM, instead, can identify a series of several FBGs, characterized initially by identical (or near) Bragg wavelength values, by combining a pulsed tunable laser in the wavelength sensing range of the gratings and a photodiode with a very high acquisition speed. The two techniques WDM and TDM can be combined, opening the way to quasi-distributed measurements with thousands of FBGs along one fiber (see [8] as an example).

Moreover, compared to other OFSs, such as distributed OFS [9], FBGs present a fast response, opening the way to monitor fast dynamics: the acquisition rate of recent interrogation systems can reach 100 MHz [10]. Such high acquisition rate systems are fundamental to developing vibrational (see review [11]) or ultrasonic sensors [12].

Since the information is wavelength encoded, the FBGs have been largely investigated for their applications in harsh environments, combining radiation and sometimes extreme temperatures [13]. Examples of harsh environments are space, nuclear power plants, and nuclear waste storage, reactor dismantling, and high-energy physics facilities [14]. Each of them is characterized by different temperature ranges and irradiation conditions, such as the nature of particles, fluence (dose), and flux (dose rate) (The dose is the quantity of energy deposited inside in the material of interest, here silica. 1 Gy($SiO_2$) = 1 J/kg. The dose rate is the deposition speed of the energy, expressed in Gy/s. In some articles, another older dose unit is used, the rad: 1 Gy = 100 rad). For example, space is characterized by a low dose rate and low doses (lower than $10^{-3}$ Gy/h and 10 kGy, respectively) but large temperature variations (between −200 °C and 300 °C). Fusion-devoted facilities, such as the Laser Megajoule (LMJ, CEA, France) or the National Ignition Facility (NIF, Lawrence Livermore National Laboratory, California), instead, are characterized by low doses (less than 1 kGy) and a very high dose rate (up to MGy/s) but are operating at room temperature (RT), whereas the nuclear reactor core is associated with very high doses (up to GGy) and high temperature (up to 800 °C) (see references in [14]).

Radiation can generate point defects, sometimes called color centers, inside the pure or doped silica matrix of the fiber core and cladding by ionization or displacement damage [15]. At very high total ionizing doses or high neutron fluences, densification is also observed [16]. These two radiation effects will degrade the optical fiber and then the fiber grating performances. For example, each induced defect is characterized by its own absorption bands that degrade the optical fiber transmission. Concerning the grating, because of the radiation, the Bragg peak amplitude can be reduced and its position, the Bragg wavelength, can shift, inducing an error in the measurement parameters. The grating radiation response depends on several parameters, such as the inscription conditions, the fiber, the treatments performed before or after the FBG inscription, and the irradiation conditions (nature, dose, dose rate, and temperature). More than one hundred papers have been published about the FBG response to radiation. A complete review of radiation responses of FBGs was published in 2013 by Gusarov and Hoeffgen [13], whereas in [17] the grating response in radiation-free harsh environments was reviewed. Since then, the research has advanced considerably, pushed by new manufacturing techniques for FBGs and by new needs, in particular at very high temperatures. The aim of this paper is to highlight the different parameters that influence the radiation vulnerability of the different grating types and to explain how radiation affects their performance in real applications involving harsh environments. For this purpose, Table 1 reports the main characteristics of radiation environments with a list of references reporting the results of tests carried out in these environments.

**Table 1.** Main characteristics of radiation environments (extracted from [14]) and list of references where the FBGs were tested in real applications.

| Environment | | Radiation Nature | Dose | Dose-Rate | Temperature | References |
|---|---|---|---|---|---|---|
| Nuclear Reactor Core | | γ-rays<br>neutrons | GGy<br>$10^{20}$ n·cm$^{-2}$ | <$10^{15}$<br>n·cm$^{-2}$·s$^{-1}$ | RT → 800 °C | [18–31] |
| Fusion-devoted facilities | Tokamak (e.g., ITER) | γ-rays<br>14 MeV neutrons | <10 MGy<br><$10^{18}$ n·cm$^{-2}$ | 1 kGy/h<br><$10^{14}$<br>n·cm$^{-2}$·s$^{-1}$ | RT → 400 °C | [32] |
| | LMJ, NIF | X-rays<br>γ-rays<br>14 MeV neutrons | <1 kGy | >MGy/s | RT | |
| High-energy physics facilities | LHC | photons<br>electrons<br>other particles | <100 kGy | <0.1 Gy/h | RT | [33–36] |
| Nuclear Waste Storage | | γ-rays | <10 MGy | <10 Gy/h | RT → 90 °C | [37,38] |
| Medicine | | X-rays<br>protons | $10^{-2}$ Gy → 50 Gy | <1 Gy/s | RT | [39] |
| Space | | X-rays<br>γ-rays<br>protons<br>electrons | <10 kGy | $10^{-5}$ → $10^{-3}$ Gy/h | −200 °C → 300 °C | [40] |

## 2. Operating Principle

An FBG consists of a periodical structuration of the fiber core refractive index, induced by its exposure to laser light (generally on a short fiber length from a few mm to a few cm) [41], as reported in Figure 1. The photo-inscribed grating allows light transmission at all wavelengths except within a small spectral range, for which the light is reflected by the grating. The dip (or peak) in the transmission (or reflection) spectrum, shown in Figure 1, is centered at the wavelength, known as the Bragg wavelength ($\lambda_B$) and defined as:

$$\lambda_B = \frac{2 \times n_{eff} \times \Lambda}{m} \tag{1}$$

where $n_{eff}$ is the effective refractive index of the fundamental mode measured at $\lambda_B$, $\Lambda$ is the grating period, and m is an integer, indicating the grating or diffraction order. $\Delta n^{mod}$ is the refractive index modulation amplitude, defined as the difference between the refractive index of the zones that are illuminated and the ones that are not.

Because of the dependence of the effective refractive index and the grating period on the temperature (respectively, due to the thermo-optic effect [42] and to the thermal expansion [43]) and on the strain applied on the fiber (due to the elasto-optic properties [44]), the FBG can be used as a temperature and/or strain sensor, since its Bragg wavelength shifts with the temperature and strain variations, respectively, $\Delta T$ and $\Delta \varepsilon$:

$$\Delta\lambda_B = C_T \times \Delta T + C_\varepsilon \times \Delta\varepsilon \tag{2}$$

where $C_T$ is the temperature coefficient defined in a temperature range smaller than 100 °C (since the dependence of the Bragg wavelength on the temperature variations has to be described with a polynomial function in larger temperature ranges), and $C_\varepsilon$ is the strain coefficient [3]. For example, for an FBG written in a silica-based optical fiber and having a Bragg peak around 1550 nm, $C_T$ is about 10 pm/°C in the temperature range between RT and 100 °C [3], whereas the axial strain coefficient is around 1.2 pm/με [45]. Concerning the sensitivity to the transverse load (or pressure), it is negligible for a classical FBG.

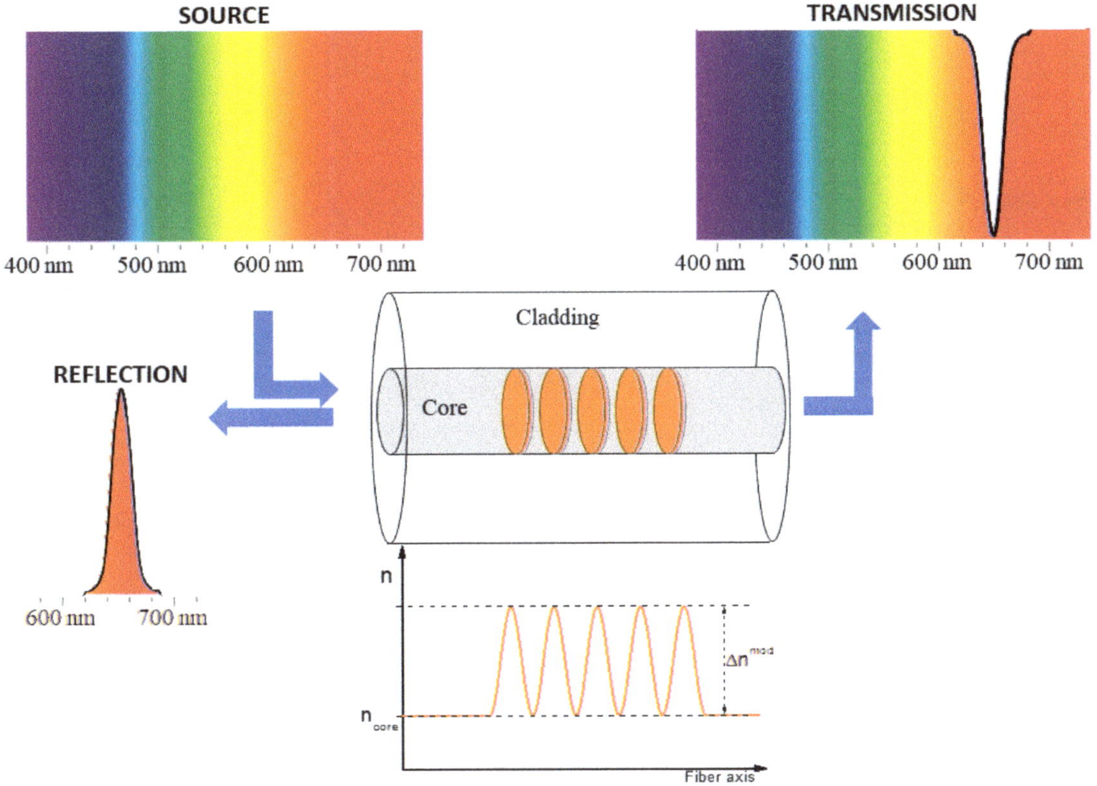


**Figure 1.** Transmission and reflection spectra of a fiber Bragg grating, with the periodical structure of the refractive index in the fiber core.

As introduced in Section 0, gratings written in highly birefringent optical fibers are characterized by two Bragg peaks, which both shift towards the same direction under a temperature or axial strain change but shift in the opposite direction under a transverse mechanical load, making these FBGs very good pressure sensors with a sensitivity of hundreds of pm/(N/mm) [46].

## 3. Inscription Techniques

The periodical refractive index modulation, giving rise to the grating, originates from the silica matrix modifications induced by its exposure to laser light, which can generate defects or structural changes. To create this periodical structuration, several inscription techniques exist.

The first method is known as "point by point" (briefly PbP) [47] and consists in focusing the laser beam through an objective on a point of the fiber core, to change its refractive index locally, then the fiber (or the laser beam) is translated for a certain distance, corresponding to the grating period (Λ), to write the following point. After repeating the process N times, a grating of length $L = N \times \Lambda$ is formed.

The other techniques, by contrast, need an interferometric pattern, such as the one used to manufacture the first FBG by Hill [41] or the ones based on the "phase mask" (PM) [48], the Lloyd's mirror [49], or the Talbot interferometer [50].

The phase mask technique is the most common today. It is based on a PM, which is a one-dimensional periodical structure, photolithographically etched on one of the surfaces of a slat of a material transparent to the laser light, such as silica glass. When the laser beam goes through the PM, the phase is spatially modulated and diffracted: the diffracted orders give rise to an interference pattern, with a period equal to half of the one of the PM (The amplitude of the PM structure is optimized in order to reduce the light transmitted in the zero order (less than 5%) and to divide the whole beam energy between the orders $-1$ and $+1$, or $-2$ and $+2$ [3]). This interference pattern creates the refractive index modulation in the fiber core of the optical fiber placed near the PM if the laser beam energy is high enough to create color centers and/or densification in the lightning area. In this case, the Bragg wavelength depends mainly on the PM period for fibers with the same composition.

The laser employed for the FBG inscription can be a continuous wave (CW) or pulsed laser in the domain of ns or fs. Whereas the wavelength of CW or ns-pulsed lasers has to be in the UV spectral range since one photon must be enough energetic to modify the silica matrix, this condition is not mandatory for fs-pulsed lasers. FBGs can be written with fs-lasers working in the spectral range from UV to IR: if the laser wavelength is in the IR domain, multi-photon processes can take place.

Because of the fiber-coating shielding effects during the grating manufacturing especially with UV lasers, FBGs are often written in bare (uncoated) fibers: the coating is stripped before the grating inscription and reapplied just after. With the development of fs-laser-based techniques, today it is possible to inscribe FBGs also through the various types of coating [51].

Moreover, the inscription set-up can also be incorporated in the optical fiber drawing tower, so that the FBGs are written directly on the bare fiber before its coating deposition. Such FBGs are known as "draw tower gratings", or DTGs [52].

## 4. FBG Classification

The FBG properties, such as their response in harsh environments characterized by extreme temperatures and/or radiations, depend on the fiber characteristics, inscription processes, and pre- or post-inscription treatments. Depending on the manufacturing choices, the gratings are generally classified into different types; each one is characterized by a different origin for the refractive index modulation, which gives rise to a different resistance to high temperatures and also to different radiation responses.

As an example, by combining the PbP or PM technique and an fs-laser, both type I and type II gratings can be written in all fiber types. One FBG type or the other will be inscribed by increasing the laser power; indeed, for a fiber having a Ge-doped core, with an fs-laser at 800 nm and a PM, the thresholds of the pulse peak intensity to write type I or type II FBGs are $2 \times 10^{13}$ W/cm$^2$ and $\sim 5 \times 10^{13}$ W/cm$^2$, respectively [53]. These thresholds depend on the fiber composition, the possible fiber treatments undertaken to improve its photo-sensitivity (such as the $H_2$ loading), the inscription technique, and the laser wavelength.

Table 2 reports the most common grating types defined that have been defined, with the most important inscription conditions, the cause of the refractive index modification, and the maximum temperature each type can withstand, together with some of the most important references (for a more complete classification, see reference [54]).

**Table 2.** Fiber Bragg grating classification with the inscription conditions, their origins, and their temperature operating range.

| Type | Inscription | Origins | Temperature Resistance | References |
|---|---|---|---|---|
| I | UV (continuous or ns-pulsed) laser in photosensitive fibers, such as a Ge-doped one; OR fs-pulsed lasers (generally IR) in all fiber types. | Color centers [1]. In the case of fs-pulsed IR laser, the defects are induced by multi-photon absorption processes. | T < 500 °C | [3,53,55] |
| II | fs-laser, whose power is higher than the damage threshold (~4 × $10^{13}$ W/cm$^2$ for silica-based fibers), through phase mask or point by point | Densification and nano-structuration | T > 800 °C | [53] |
| Regenerated (R) | 1. Seed FBG: type I grating in a photosensitive fiber, $H_2$-loaded before or after the inscription;<br>2. High-temperature treatment (T > 650 °C) that will erase the seed FBG before the appearance of the R-FBG. | Cristobalite, a crystalline polymorphic silica, generated by the high temperature and high pressure due to the hydrogen presence at the core/cladding interface (still debated). | T > 1000 °C | [56–58] |
| III or voids | fs-laser (whose power is higher than $10^{14}$ W/cm$^2$) with the point-by-point technique. | Micro-voids surrounded by a shell of densified silica. | T > 1000 °C | [59,60] |

[1] It is worth noticing that, according some authors, type I FBGs can be due, along with color centers, to structural changes and densification [54]; however, the latter should not be erased by thermal treatments at temperatures lower than 500 °C. So, the subject is still controversial.

## 5. Radiation Effects on Optical Fibers

As highlighted in Table 1, each harsh environment is characterized by different irradiation conditions:

- Nature of radiation: X-rays, γ-rays, protons, electrons, neutrons;
- Total ionizing dose (TID): quantity of energy deposited for the unit of mass of the material; it will be measured in Gy($SiO_2$) in all the manuscripts, except when specified differently;
- Dose rate: quantity of energy deposited per unit of time, measured in Gy($SiO_2$)/s;
- Irradiation temperature.

Radiation breaks bonds by ionization or knock-on processes. The main effect is the generation of point defects from regular or strained Si-O-Si bonds and from precursor centers. Their nature and concentration depend on the composition and also the manufacturing and drawing processes of the fiber; consequently, these parameters govern its radiation sensitivity. A list of the defects induced in silica-based OFs with their description and characteristics is reported in a review [61].

It has been demonstrated that ionization is the predominant effect even in the presence of non-ionizing radiation as neutrons, at least for fluences up to $10^{16}$ neutrons·cm$^{-2}$ (or n·cm$^{-2}$). Indeed, not only are the same defects generated independently of the radiation nature but also their concentration showed similar kinetics and levels at the same TID [62]. Nevertheless, some differences can be observed because of the dose enhancement induced by energetic proton recoils out of the H-containing coatings [63]. In addition to this effect, higher neutron fluences lead to significative structural changes, i.e., densification [16], causing new intrinsic defects or changes in the spectroscopic signatures of already known centers [64].

These phenomena at the microscopic level are at the origin of the degradation of the optical properties of the fibers and show up in three macroscopic effects, described hereafter.

*5.1. Radiation-Induced Attenuation*

The main effect induced by point defects is "radiation-induced attenuation", or briefly RIA, which consists of an increase in the fiber attenuation due to the appearance of their associated absorption bands in the fiber transmission windows. RIA levels and kinetics depend on several parameters [15]:

- Those of the harsh environment: dose, dose rate, temperature, and radiation nature, as already explained, but also the presence of gases that could diffuse inside the optical fiber;
- The characteristics of the optical fiber itself: its core and cladding compositions, the manufacturing process of the preform, and fiber drawing conditions;
- The test conditions: injected signal wavelength and power.

Clearly, fibers with different compositions present different radiation responses, and they are generally divided into three main classes [15], even if the radiation hardness of a fiber is only valid for specific environments and, sometimes, for specific wavelength ranges [65]:

- The "radiation-hardened" OFs, having a pure-silica core (PSC) and fluorine-doped cladding or both core and cladding doped with F, since they show the lowest sensitivity under high-dose (MGy levels), steady-state irradiation among all the fiber types [61];
- The "radiation-tolerant" OFs, such as Ge-doped ones, whose losses are low enough to be used for some applications, such as telecommunications, in harsh environments characterized by low TIDs (<10 kGy), i.e., space [14,66,67];
- The "radiation-sensitive" OFs, which are mainly doped or co-doped with phosphorus or aluminum in their core and/or cladding and present high RIA levels, both in the visible and infrared spectral domains; they could be used for point or distributed radiation detection and dosimetry applications [68,69].

However, it is very important to test the radiation response of a fiber, before using it in a real environment. For example, it has been recently found an ultra-low losses PSC OF (whose losses at 1550 nm are of ~0.15 dB/km before irradiation) that is very radiation sensitive, with RIA levels higher than the ones induced in the P-doped fiber at 2 kGy TID (dose rate of 175 mGy/s) at RT: RIA reached ~2000 dB/km at 1550 nm [70].

*5.2. Radiation-Induced Emission*

The "radiation-induced emission", or RIE, consists of two contributions: Cerenkov radiation [71] and radio-luminescence (RL) [15]. The latter is the emission of light from pre-existing or radiation-induced centers that will be excited by the radiation itself and that will emit a luminescence signal when coming back to the fundamental level. The Cerenkov light and also the RL, at least in most cases, affect the visible spectral range and should not interfere with the FBGs designed to operate in the IR spectral range.

*5.3. Radiation-Induced Compaction*

Concerning the "radiation-induced compaction", or RIC, for the silica-based materials, it was observed that the density ρ increases with the dose D following a power law:

$$\frac{\Delta\rho}{\rho} \propto D^k \quad (3)$$

where k depends on the irradiation nature, being ~2/3 for γ-rays and ~1 for fast neutrons [72], before saturating around 3% at very high doses [16]. For the silica-based OFs, very few data are reported in the literature: linear compaction of ~0.3% (corresponding to a density increase of about 1%) was observed in pure silica samples irradiated with a fast neutron fluence of ~$5 \times 10^{19}$ n/cm$^2$ and a total $\gamma$-dose of about 4 GGy, at a temperature around 290 °C [23].

**6. Radiation Effects on FBGs**

As radiation degrades the fiber transmission through the RIA phenomenon, it can also degrade the FBG performance, as shown in Figure 2, through the two effects highlighted in Figure 3 [13].

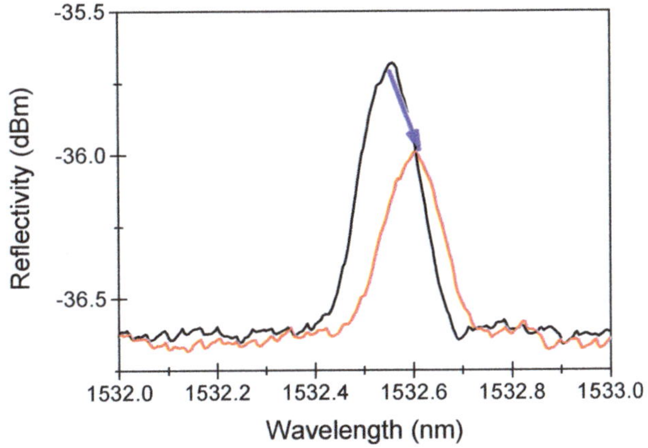


**Figure 2.** An example of the radiation-induced effects on a Bragg peak reflection spectrum, before (black line) and after (red line) X-ray irradiation at 1.5 MGy TID at RT. Data extracted from [73].

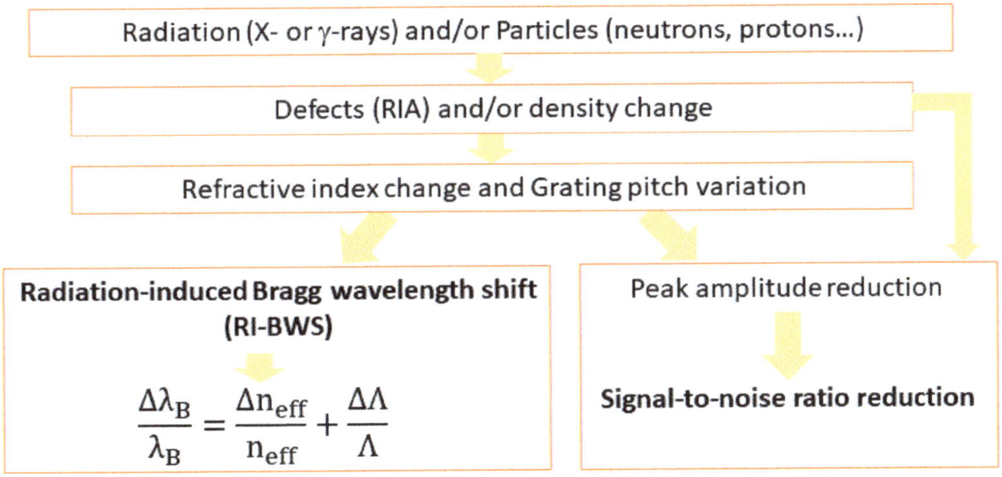


**Figure 3.** Origins of the radiation effects on the FBG peak: radiation-induced Bragg wavelength shift (RI-BWS) and signal-to-noise reduction.

The first one is the "radiation-induced Bragg wavelength shift", hereafter indicated as RI-BWS. As shown in Equation (1), the Bragg wavelength depends on the effective refractive index at $\lambda_B$ and the grating period. The refractive index change is caused by the RIA and the RIC and is known as the "radiation-induced refractive index change" (RIRIC). The grating period modification, instead, can be reduced as a consequence of the compaction. As already mentioned, the latter effect is negligible in most cases; then, RI-BWS can be mainly associated with RIRIC.

The second effect induced on the FBG is a variation of the grating reflectivity, caused by:

- Changes in the grating parameters, such as $n_{eff}$, $\Delta n^{mod}$ and $\Lambda$,
- Degradation of the fiber transmission due to RIA.

Both can lead to a reduction in the signal-to-noise ratio and, in the worst cases, to an increase in the uncertainty associated with the measurement. At very high TID and/or neutron fluence, depending on the initial grating reflectivity, the peak can also be completely erased and the sensor stops working [24]. It is worth noticing that, if the periodical structuration is still present, the Bragg peak can be completely imperceptible within the noise in the transmission spectrum, but it should be detectable in the reflection one, despite its very small reflectivity. An example is reported in [24], where the reflectivity of an FBG (written in a Ge-doped fiber with an fs-laser at 800 nm and treated at 372 °C for 72 h) was reduced from 5% to 0.2% after a neutron fluence of $3 \times 10^{19}$ $n \cdot cm^{-2}$ (neutron energy of ~1 MeV) and a $\gamma$-dose of 1.5 GGy at 250 °C: the peak was still evident in reflection but not in transmission.

Sometimes, instead, an increase in the reflectivity is observed, due to the different kinetics of radiation-induced defects in the laser-modified and unmodified (or weakly modified) areas of the grating (photoinduced refractive index peaks and valleys), as observed in [74]. Finally, when RIA is the cause of the reflectivity decrease because the fiber in which the FBG has been inscribed is characterized by a very high RIA level, the short fiber length containing the grating can be spliced to more radiation-hardened fiber pigtails to transport the signal to reduce the RIA impact [13].

In the next sections, we will focus on the RI-BWS. Since the Bragg peak translates into the value of the external parameter changes that the FBG is sensing, RI-BWS corresponds to an error in the measurement. For example, for an FBG having a thermal sensitivity coefficient $C_T$ of 10 pm/°C and a strain coefficient $C_\varepsilon$ of 1.2 pm/$\mu\varepsilon$, a RI-BWS of about 10 pm will correspond to an error of about 1 °C or 8 $\mu\varepsilon$ in the temperature or strain measurements, respectively. Moreover, it has been demonstrated several times that the thermal sensitivity and strain coefficients are not significantly influenced by radiation [75].

As already mentioned, the grating is a periodical structuration of the refractive index induced by a laser in the fiber core. Since the laser modifies the silica matrix, e.g., generating defects and inducing densification, it is understandable that the radiation effects induced on the bright and dark fringes of the FBG can differ from the ones induced on the fiber itself, which did not undergo any laser treatment. Consequently, *the radiation response of an FBG cannot be directly linked to that of the fiber* and depends on several parameters [76]. Each FBG is particular as it has been written under specific conditions with particular characteristics; therefore, it will show a unique radiation response. For this reason, more than one hundred papers have been published concerning the radiation effects on different FBG types, and a review article has been published in 2013 by Gusarov and Hoeffgen [13]. Since then, new tests on new types of gratings have been performed and in the following, we present the main parameters that influence their radiation response and the associated basic mechanisms. It is worth noticing that, unless otherwise specified, the gratings were bare, meaning without a coating during the irradiation tests.

### 6.1. Optical Fiber Composition

As the radiation response of the OFs strongly depends on the compositions of their core and cladding, the fiber composition also influences the response of the gratings under irradiation. However, it is impossible to determine how, since the grating inscription process itself changes the fiber properties locally. This has been highlighted by Henschel et al., studying the BWS induced by $\gamma$-rays on type I-UV gratings inscribed in fibers with differently doped cores: germanium, phosphorous, and cerium [76]. Figure 4 shows a few of their results (extracted from [76]). It can be observed that the grating written in the radiation-sensitive (in terms of RIA) optical fiber co-doped with Al and P shows the smallest RI-BWS among all the reported gratings.

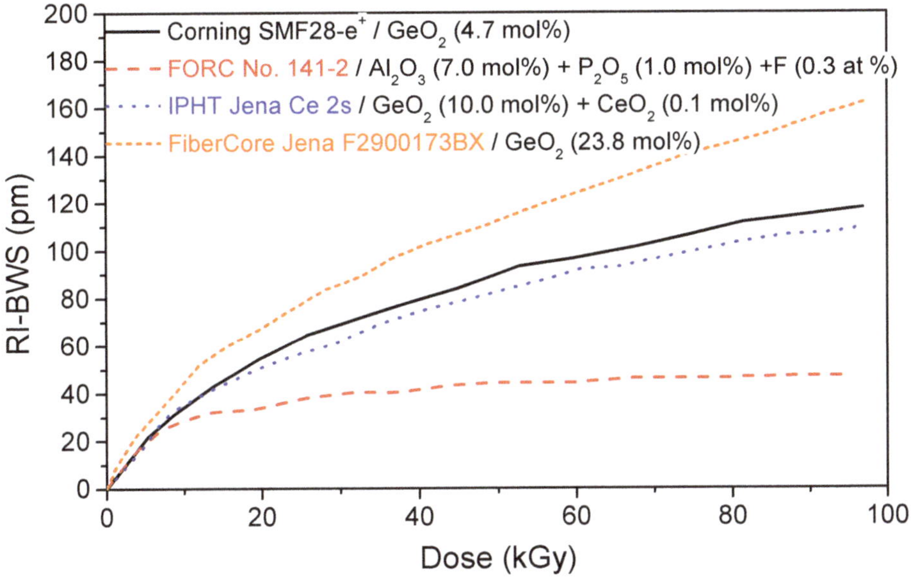


**Figure 4.** BWS induced by γ-rays on type I-UV gratings, written by different manufacturers on uncoated fibers with different core compositions that were recoated after inscription, the dose rate being 0.9 Gy/s up to 100 kGy at RT. Data extracted from [76].

As a consequence, *writing a grating into a radiation-resistant fiber does not ensure a radiation-resistant FBG*. However, as Figure 4 suggests, it seems that the higher the $GeO_2$ concentration in the fiber core, the higher the FBG radiation sensitivity [77,78]. This is highlighted in Figure 5, which compares the radiation responses of two 10 mm long gratings with the same reflectivity, written with a UV laser at 244 nm through a PM in two $H_2$-loaded Ge-doped fibers (at 160 bars and RT, for 10 days): a standard Corning SMF28 (5 wt%) and a photosensitive (>15 wt%). The two gratings were also subjected to the same thermal treatment (at 60 °C for 24 h) to highlight only the Ge-content influence.

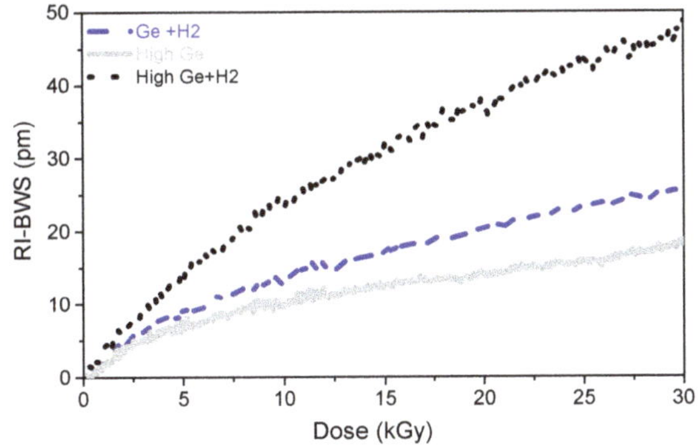


**Figure 5.** BWS induced by X-rays at RT with a dose rate of 5 Gy/s, up to 30 kGy, on three type I gratings written with a CW UV laser (at 244 nm) into two different fibers: a standard fiber (the Corning SMF28, [Ge] = 5 wt%) and a photosensitive one ([Ge] > 15 wt%). Ten-millimeter-long FBGs were written in both fibers after a pre-inscription $H_2$ loading. A 20 mm long FBG was also written in the unloaded photosensitive fiber with higher laser power (170 mW against ~90 mW). Data extracted from [78].

### 6.2. Pre-Inscription $H_2$ Loading

Figure 5 also compares the BWS induced by X-rays on two gratings written with the UV laser in the same fiber, pre-hydrogenated or not before the FBG inscription. In this case, the two gratings were not identical because of the reduced photosensitivity of the unloaded highly Ge-doped fiber compared to the loaded one. Indeed, the FBG written in the $H_2$-loaded fiber (with a laser power of ~90 mW) was only 10 mm long and had a reflectivity of 90% (transmission of $-10$ dB), whereas the grating inscribed in the unloaded fiber with a higher laser power (about 170 mW) was 20 mm long in order to reach a reflectivity of 50% (transmission of $-3$ dB). Conscious of these differences, it can be stated that *the $H_2$ loading used to enhance the fiber sensitivity to UV light increases also the FBG radiation sensitivity*, independently of the fiber composition. With pre-hydrogenation, the RI-BWS saturates at higher levels and at higher doses [75,77]. The probable cause is the radiolytic rupture of the OH-bonds, which are present at higher concentrations in the gratings written in $H_2$-loaded fiber [79].

This agrees with the following observation: the higher the $H_2$ concentration in the fiber core at the time of the inscription, and consequently the OH concentration present in the grating, the more radiation-sensitive is the FBG. Indeed, Henschel et al. highlighted that the RI-BWS increases by increasing the pressure during the fiber $H_2$ loading: since the fibers were loaded for the same duration and at the same temperature (one week at 50 °C), a higher pressure leads to a higher $H_2$ concentration at saturation during the inscription [80].

Moreover, *the grating response does not change significantly, if the loading is performed with $H_2$ or $D_2$*, provided that the same concentration of molecules ($H_2$ or $D_2$) is reached at the moment of the grating inscription, as it was observed for regenerated gratings (no data are available for type I FBGs) [81]. The differences recorded in the RI-BWSs were within 10%.

Concerning the pre-hydrogenation, it must be pointed out that a grating written in a loaded fiber has to undergo thermal treatment to ensure the out-gassing of the remaining $H_2$ molecules. For example, for the gratings, whose results under X-rays are reported in Figure 5, both were subjected to annealing at 60 °C for 24 h to outgas the remaining $H_2$ molecules. This is an important point, since, as we will see in the next paragraph, the post-inscription thermal treatments also influence the FBG radiation response.

These conclusions about the $H_2$-loading effect on the grating radiation sensitivity drawn from the type I gratings written with a UV-laser, for which the hydrogenation is necessary in most cases, can also be extended to other FBG types, as illustrated in Figure 6 for type I and type II FBGs manufactured with an fs-laser [82] under $\gamma$-rays. Indeed, it is worth mentioning that a different behavior will be highlighted under neutrons in Section 0.

### 6.3. Inscription Conditions

Beyond the $H_2$-loading influence, Figure 6 highlights a clear difference in the radiation response of two gratings written with different laser powers. Indeed, among all the parameters, the radiation response of the grating strongly depends on its inscription conditions, which rule the grating type. Figure 7 reports the Bragg wavelength shift induced by X-rays on several FBG types written in the same Ge-doped optical fiber: the Corning SMF28e+ [74].

It should be pointed out that, for the type I FBGs and the seed grating of the regenerated FBG that were written with a UV laser, it was necessary to increase the fiber photosensitivity with pre-hydrogenation before their inscription, because of its low Ge-core concentration. As already highlighted in the previous section, the $H_2$ loading increases the fiber radiation sensitivity but, as shown in Figure 7, even if the two gratings have the same length (10 mm) and are characterized by the same refractive index modulation amplitude ($\Delta n_{mod}$ of about $1.2 \times 10^{-4}$), the inscription conditions of type I gratings (laser wavelength and power) influence the radiation response, reaching different RI-BWS values by a factor of three. In particular, 150 pm and about 50 pm were observed, at 1 MGy TID, for the gratings inscribed with the 10 ns pulsed laser at 248 nm and the CW laser at 244 nm, respectively.

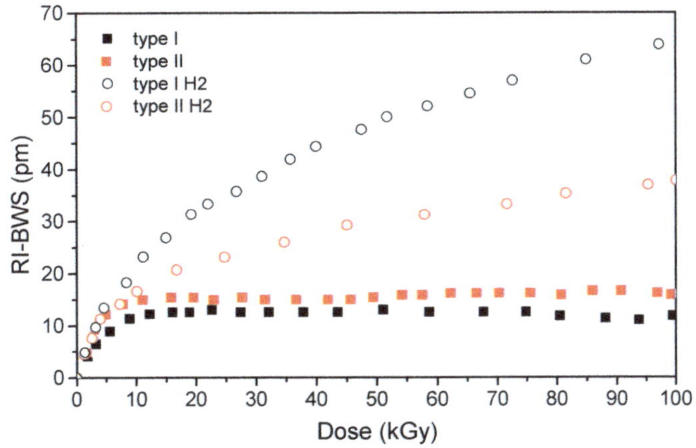


**Figure 6.** BWS induced by γ-rays at RT with a dose rate of 0.94 Gy/s, up to 100 kGy TID (irradiation temperature between 25 °C and 65 °C), on two type I and two type II gratings written with an fs-laser (at 800 nm, a pulse width of 120 fs, and repetition rate of 100 Hz) into the acrylate coated Ge-doped fiber Corning SMF28e. For each grating type, one FBG was inscribed into an $H_2$-loaded fiber and the other in an unloaded one. All the gratings were thermally treated for 4 days at 100 °C. To write type I FBG, the laser power was reduced by increasing the distance between the PM and the fiber. Data extracted from [82].

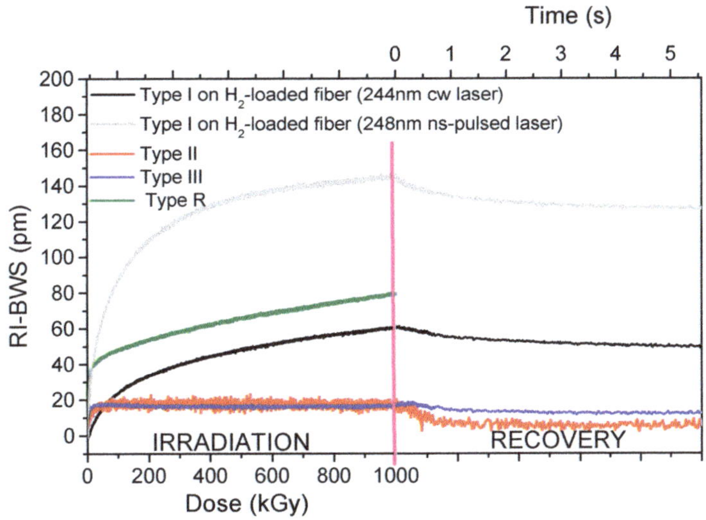


**Figure 7.** RI-BWS recorded up to 1 MGy at RT on different FBG types written in the same Corning SMF28e+ fiber: two type I FBGs, one written with a CW UV laser (at 244 nm, power density of 4.15 W/cm$^2$) and the other with a 10 ns-pulsed UV laser (at 248 nm, pulse energy of 380 mJ/cm$^2$), in a pre-$H_2$-loaded sample; a type II FBG, written through a PM with an fs-IR laser (at 800 nm, a pulse width of 150 fs, and pulse energy of 560 μJ); a type III FBG, written PbP with an fs-IR laser (at 800 nm, a pulse width < 120 fs, and pulse energy of 230 nJ); an R-FBG, manufactured from a type I grating written with a 20 ns-pulsed UV laser (at 248 nm and pulse energy of 9 mJ) and regenerated at 750 °C. The vertical pink line indicates the irradiation end. The dose rate was always 50 Gy/s for all the gratings except for the regenerated one at 10 Gy/s. Data extracted from [74,83].

Moreover, in addition to the higher BWS values, the different types of FBGs present different kinetics for the RI-BWS. The type I UV grating is characterized by a RI-BWS increasing with the dose, without reaching a saturating level at 1 MGy TID. Both types II and III FBGs, instead, present a strong increase in the Bragg peak shift up to ~20 pm at about 50 kGy and then stabilize. Additionally, the type R grating shows a fast and strong RI-BWS increase at the irradiation start (40 pm for a TID of 50 kGy), but, contrary to types II and III, at higher TID the shift seems to increase linearly with the dose. In conclusion, at RT, type II and type III show the highest radiation tolerance, compared to type I or type R.

As for the RIA, the RI-BWS value reduces after irradiation: this phase is known as "recovery". This effect is more or less important depending on many parameters, such as the inscription ones. In the case reported in Figure 7, the RI-BWS is reduced by about 15% for type I FBGs, 5% for type III, and more than 50% for type II. For type R, a recovery of around 7% has been observed (not reported here, since the TID reached in the experiment was 2 MGy) [83]. It is also worth noticing that the dose rate employed for the irradiation of the latter FBG is lower than the one used for the other samples; however, the RI-BWS should be larger at the higher dose rate, as highlighted in Section 0. Consequently, the change of this parameter in the comparison reported in Figure 7 does not invalidate our statements.

### 6.4. Bragg Wavelength

Among the inscription conditions, there is also the grating period, which rules the Bragg wavelength. Figure 8 reports the shift induced on three similar type I gratings, having the same amplitude of the refractive index modulation and submitted to the same treatments before and after inscription, but written at different Bragg wavelengths: 1300 nm, 1430 nm, and 1560 nm. By increasing the latter, the RI-BWS increases (from ~85 pm for 1300 nm to ~125 pm for 1560 nm at 1 MGy); however, once normalized by the initial Bragg wavelength of each grating (see inset of Figure 8), only a slight dependence can be observed, suggesting that the same variation of the refractive index occurs in all the gratings.

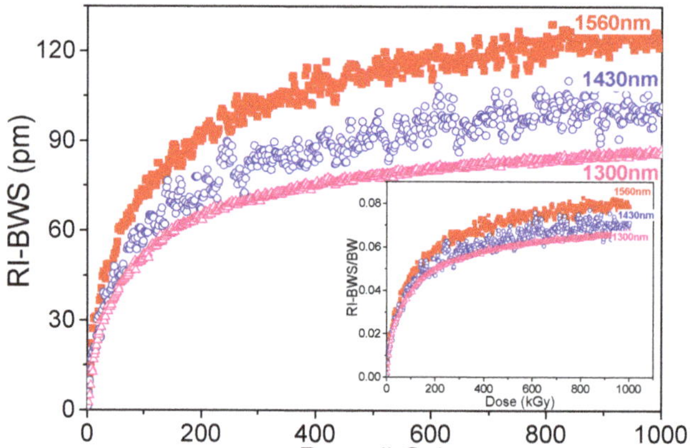


**Figure 8.** BWS induced by X-rays at RT up to 1 MGy (dose rate of 50 Gy/s) on three similar gratings (having the same $\Delta n^{mod}$ of about $2 \times 10^{-4}$) written with a CW UV laser (at 244 nm) in $H_2$-loaded Corning SMF28 at different Bragg wavelengths. In the inset, RI-BWS normalized with respect to the initial Bragg wavelength. Data extracted from [49].

### 6.5. Post-Inscription Thermal Treatment

As for the pre-inscription treatments, e.g., the $H_2$ loading, also the treatments performed on the FBG after its inscription influence its radiation response. The most common is a thermal treatment, used to stabilize the grating. Figure 9 gives some examples of the impact of such thermal treatments on the different FBG types.

The type I gratings in Figure 9a were written with an fs-laser at 800 nm in a Ge-doped fiber [84]. Indeed, since the grating inscription with such a laser does not require a photosensitive fiber, no $H_2$ loading was performed. Consequently, the different effects due to a different thermal treatment performed after inscription and reported in Figure 9a cannot be associated with a phenomenon of hydrogen diffusion. For this FBG type, it is clear that by increasing the temperature of the pre-irradiation thermal treatment, the radiation sensitivity increases. Probably, this annealing recombines part of the defects generated during the grating manufacturing, creating more precursors available during the irradiation run.

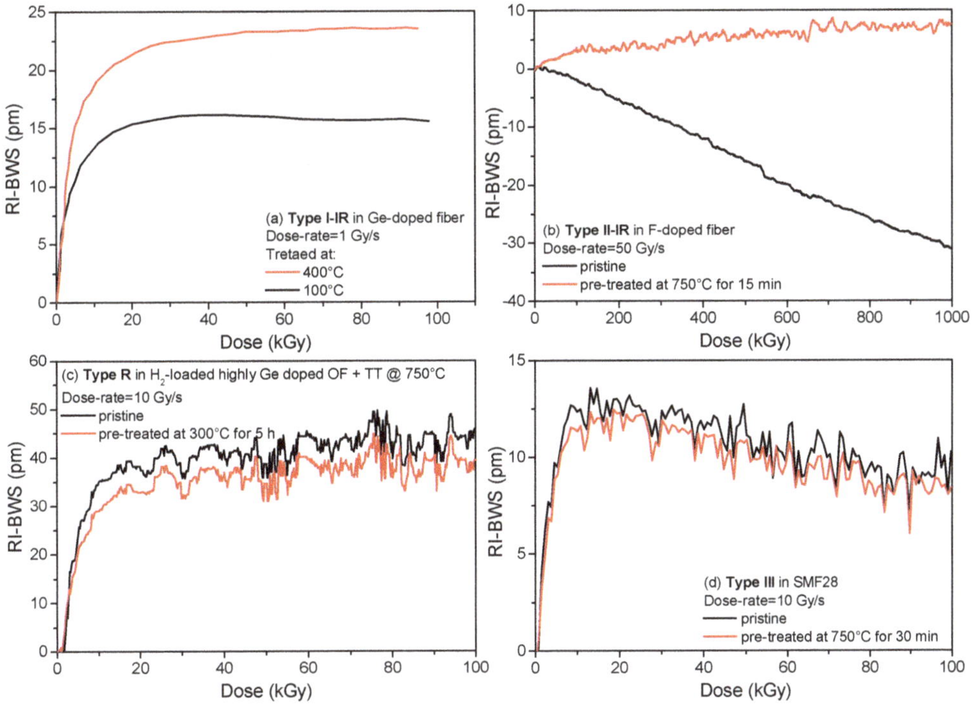


**Figure 9.** Effect of a pre-thermal treatment on the RI-BWS induced on different FBG types: (**a**) a type I FBG written in a Ge-doped fiber (the Corning SMF28e) with an IR fs-laser (at 800 nm, pulse energy of 1200 μJ); (**b**) a type II FBG in an F-doped fiber with an fs-IR laser (at 800 nm, a pulse width of 150 fs, and pulse energy of ~500 μJ); (**c**) type R FBGs written in $H_2$-loaded highly Ge-doped fiber and regenerated at 750 °C; (**d**) a type III in the SMF28 fiber written PbP with an fs-IR laser (at 515 nm, with a pulse width < 290 fs). The gratings reported in (**a**) were investigated under γ-rays, whereas all the others were investigated under X-rays. Data extracted from [84–87].

Type R FBGs undergo high thermal treatment for the regeneration process. Consequently, thermal treatments at temperatures lower than the regeneration one do not change their radiation response (Figure 9c) [85].

The radiation sensitivity of type III gratings is also not influenced by thermal treatment, even if performed at very high temperatures, i.e., 750 °C, as shown in Figure 9d. This agrees with their stability at high temperatures [31].

Concerning the type II FBGs, in Figure 9b the pre-irradiation high-temperature treatment significantly improves its radiation resistance. Indeed, depending on the inscription conditions and on the quality of the set-up alignment, during the type II grating inscription, there can be regions (bright fringes) in which the laser intensity is above the threshold for type II and regions where it will be below this threshold. In these latter zones, only a type I modification will be generated. Consequently, the radiation response of such type II gratings will be influenced by its type I contribution. Thermal treatment at high temperature (~750 °C) can erase this latter contribution, based on point defects, leaving only the type II one, which results in a more radiation-resistant FBG [86].

### 6.6. Pre-Irradiation of the Grating

Contrary to the pre-thermal treatment, whose effect can be positive or negative according to the grating type, pre-irradiation seems to always have a positive effect, and indeed it reduces the radiation sensitivity of all the grating types [77], as shown in Figure 10. The pre-irradiation converts most of the precursors, making them unavailable for further irradiation. However, the amplitude of this effect depends on the grating type, the conditions of the pre-irradiation (mainly the TID), the conditions of the second irradiation, and above all, the conditions (temperature and duration) of the FBG storage between the two irradiation runs.

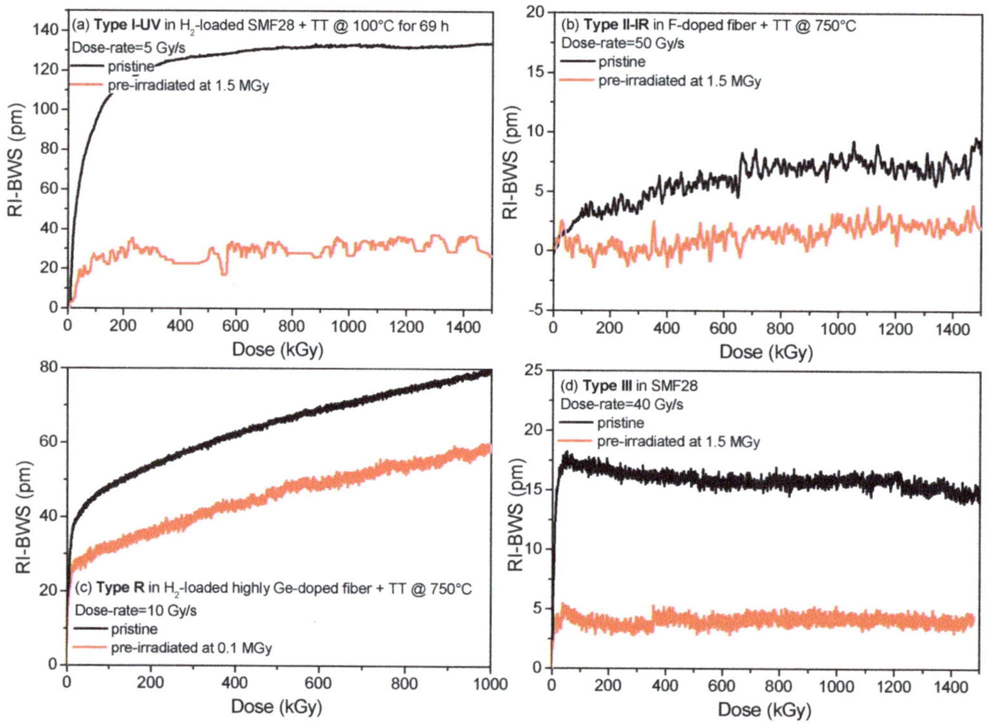


**Figure 10.** Effect of a pre-irradiation on the BWS induced by X-rays on different FBG types: (**a**) a type I FBG written in an $H_2$-loaded SMF28 fiber with a UV CW laser (at 244 nm, power of 120 mW); (**b**) a type II FBG in an F-doped fiber with an fs-IR laser (at 800 nm, a pulse width of 150 fs, and pulse energy of ~500 µJ); (**c**) type R FBGs written in $H_2$-loaded highly Ge-doped fiber and regenerated at 750 °C; (**d**) a type III in the SMF28 fiber written PbP with an fs-IR laser (at 800 nm, a pulse width < 120 fs, and pulse energy of 230 nJ). Data extracted from [83,86,88,89].

### 6.7. Total Ionizing Dose

As shown in all the previous graphs, the RI-BWS depends on the dose. In most cases, the Bragg wavelength red-shifts with increasing dose: Sometimes the shift exhibits a fast increase at the irradiation start and then continues to grow with a smaller slope, and sometimes it quickly reaches a saturation. Such trends are similar to those observed for the RIA as a function of the dose, for which several models, both empirical and semi-empirical, i.e., the power law [90] and the fractal kinetics [91], have been developed to predict their growth kinetics (see reference [15] for an exhaustive list). Such models can also be applied to describe the RI-BWS versus dose curves, as presented in [78].

Although less frequent, cases have been reported in the literature where the Bragg wavelength shifts toward the shorter wavelengths with the increasing dose. The first example was observed by Gusarov et al. on the most classical type I FBGs, written with an ns-pulsed UV laser in an unloaded highly Ge-doped core that was not thermally treated, under γ-rays [92].

As illustrated in Figure 11, at the beginning of the γ-rays exposure, the Bragg peak shifts toward the red as expected, but then, after a TID of about 50 kGy, the direction of the shift is inverted and the peak shifts towards the blue. No detectable shift was observed on the gratings when kept at the same temperature as the irradiation, of about 34 °C, implying an instability of the grating. The authors suggested the presence of two different kinds of defects: one responsible for the refractive index increase and another for the index decrease. It is worth noticing that at the end of the irradiation run, the BWS continues decreasing.

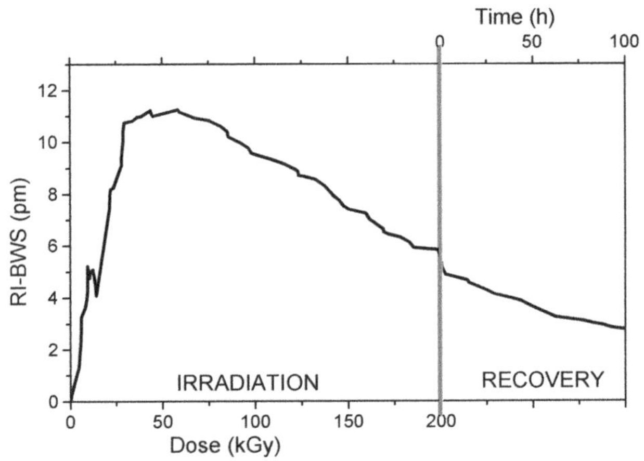


**Figure 11.** γ-rays-induced Bragg wavelength shift on type I FBGs, inscribed with an ns-pulsed UV laser (at 248 nm, a pulse width of 15 ns, and pulse energy of 8 mJ) in an unloaded highly Ge-doped core (~8 mol% of $GeO_2$). The vertical pink line indicates the irradiation end. The accumulated dose reached in the only first irradiation run, here reported, was around 200 kGy, the dose rate being of about 0.28 Gy/s and a temperature of about 34 °C. Data extracted from [92].

Another example of a shift towards the shorter wavelengths has already been shown in Figure 9b on type II FBGs not thermally treated. The origin of this blue shift is very difficult to explain, but the hypothesis of a radiation-induced release of stress should also be considered, since the gratings were not thermally treated in both reported examples.

### 6.8. Dose-Rate

Concerning the dependence of the OF radiation response on the dose rate, it was observed that, in most cases, the higher the dose rate, the higher the RIA [93] at a given TID, since, by increasing the dose rate, the number of defects that could recombine during the irradiation is lowered. It has to be noted that this is not true for all the defects; some, such as the P-related P1 center absorbing around 1.5 µm, seem dose-rate independent, at least for TID up to 500 Gy [69].

A similar rule applies to the FBGs: *the higher the dose rate, the larger the RI-BWS* [94]. Of course, the amplitude of this effect depends on the FBG type, as highlighted in Figure 12.

Type I and type R gratings generally follow this rule; however, the impact of this parameter on the RI-BWS depends on the fiber composition and the treatment the FBG goes through, as the regeneration temperature for the type R gratings. This is evident in Figure 12c for two types of R FBGs: the one written in a B/Ge co-doped fiber shows a higher dependence on the dose rate than the other one written in a Ge-doped fiber [81].

The type II FBGs, instead, are more radiation-resistant at RT and they are characterized by very low RI-BWSs. Consequently, the dose-rate dependence, if present, has not been highlighted yet, as reported in Figure 12b, where the RI-BWS varies between −5 pm and +5 pm, independently of dose and dose rate. The same conclusions can be deduced for the type III FBGs, in Figure 12d, even if a slight difference can be observed between the lowest investigated dose rate (1 Gy/s) and the highest one (40 Gy/s), in agreement with the general rule.

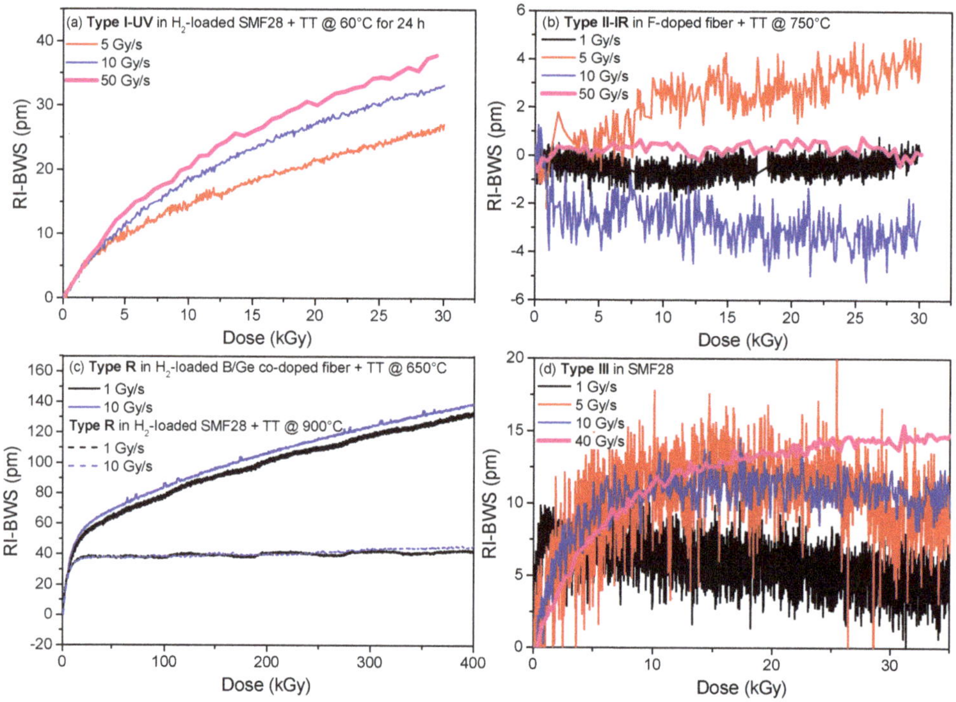


**Figure 12.** BWS induced by X-rays as a function of the dose, for different dose rates and different FBG types: (**a**) a type I FBG written in an $H_2$-loaded SMF28 fiber with a UV CW laser (at 244 nm, power of 120 mW) and treated at 60 °C for 24 h; (**b**) a type II FBG in an F-doped fiber with an fs-IR laser (at 800 nm, a pulse width of 150 fs, and power density of $3 \times 10^{13}$ W/cm$^2$) treated at 750 °C for 15 min; (**c**) type R FBGs in $H_2$-loaded B/Ge codoped fiber (continuous lines) or SMF28 fiber (dashed lines) regenerated at 650 °C or 900 °C, respectively; (**d**) a type III in the SMF28 fiber written PbP with an fs-IR laser (at 800 nm, a pulse width < 120 fs, and pulse energy of 230 nJ). Data extracted from [78,81,89,95].

### *6.9. Irradiation Temperature*

Combining temperature and radiation can lead to different effects [96]; for example, increasing the irradiation temperature could promote:

- The thermal bleaching of the radiation-induced defects, reducing the RIA due to these color centers,
- The conversion from unstable defects to more stable ones, giving rise to an RIA increase or a decrease depending on the investigated spectral range,

- The defect generation rate, increasing the defect concentration and then the associated RIA at a given dose.

Obviously, the impact of temperature depends on time and consequently on dose and dose rate. For example, more time is needed with a lower dose rate than with a higher one,# to reach a certain TID; consequently, a center will have more time to recombine (thermal bleaching) or to be converted into another defect because of the temperature. Then, a lower concentration of this center will be observed at a lower dose rate. This illustrates how difficult the quantification of the impact of the temperature of irradiation is [96]. The grating response as a function of the irradiation temperature strongly depends on the grating type, as highlighted in Figure 13. For type I and type R (Figure 13a,c), the higher the temperature, the smaller the RI-BWS [80,81]. Concerning the type II FBGs, as for the dose rate, they show very low RI-BWSs at all the irradiation temperatures; for example in Figure 13b, the induced shift is between −10 pm and +10 pm at RT and at around 230 °C. The type III FBG, instead, showed a complex behavior. Despite its stability at high temperatures and its good radiation resistance, when the irradiation is performed at RT, it was observed that its radiation sensitivity increases by increasing the irradiation temperature [89], as reported in Figure 13d for a type III FBG written in the Corning SMF28e+. In this case, whereas at RT, the Bragg wavelength shifts towards the red and quickly saturates at 15 pm, after only 20 kGy TID (dose rate being of 40 Gy/s); at a high temperature of 250 °C, it blue-shifts down to −20 pm, without showing a saturating behavior, at least up to 1 MGy. Moreover, the gratings irradiated at high temperature lose their high temperature stability [89]. However, recently, a new result has been published, showing that the radiation resistance of type III FBGs improves by increasing the irradiation temperature, when the gratings are written with an optimized inscription set-up [87].

As the high temperature changes the radiation response of the gratings, a temperature lower than the RT value could also influence the induced BWS. Indeed, by changing the defect generation and recombination rate, at low temperatures, new centers could appear that were unstable at RT. Consequently, the RIA generally increases by decreasing the irradiation temperature, but obviously it depends on several parameters, among others, the fiber composition. For example, whereas the RIA at 1550 nm strongly increases at low temperatures for Ge-doped fibers (by a factor of 20 when the temperature decreases from RT to −80 °C at 10 kGy TID) [97], this effect is less dramatic for the F-doped fibers (with a factor of four, from RT to −80 °C at 100 kGy TID) [98]. Instead, the 1.55 µm RIA is stable within 15% from −80 °C and +120 °C for the P-doped OFs [99].

Figure 14 compares the BWSs induced on different FBGs when irradiated at RT and at a lower one, −120 °C. As for the high temperatures, the behavior changes with the grating types. The type I-UV here reported, an FBG written in an $H_2$-loaded SMF28e+ with a UV CW laser (at 244 nm), shows a higher BWS [100], in agreement with the higher IR-RIA observed on the same Ge-doped fiber [97]. However, it should be pointed out that in [100] for another type I FBG written with an ns-pulsed UV laser in the same fiber, the BWS induced by X-rays at RT and at −120 °C showed similar kinetics and levels, leading to the conclusion that the response strongly depends on the inscription conditions. For type II and type III FBGs written in radiation-resistant fibers, instead, the BWS is very small at both the investigated temperatures, as illustrated in Figure 14b,c, showing that both FBG types are very radiation-resistant in this temperature range [100]. No data are available for the R-FBG behavior at temperatures below RT.

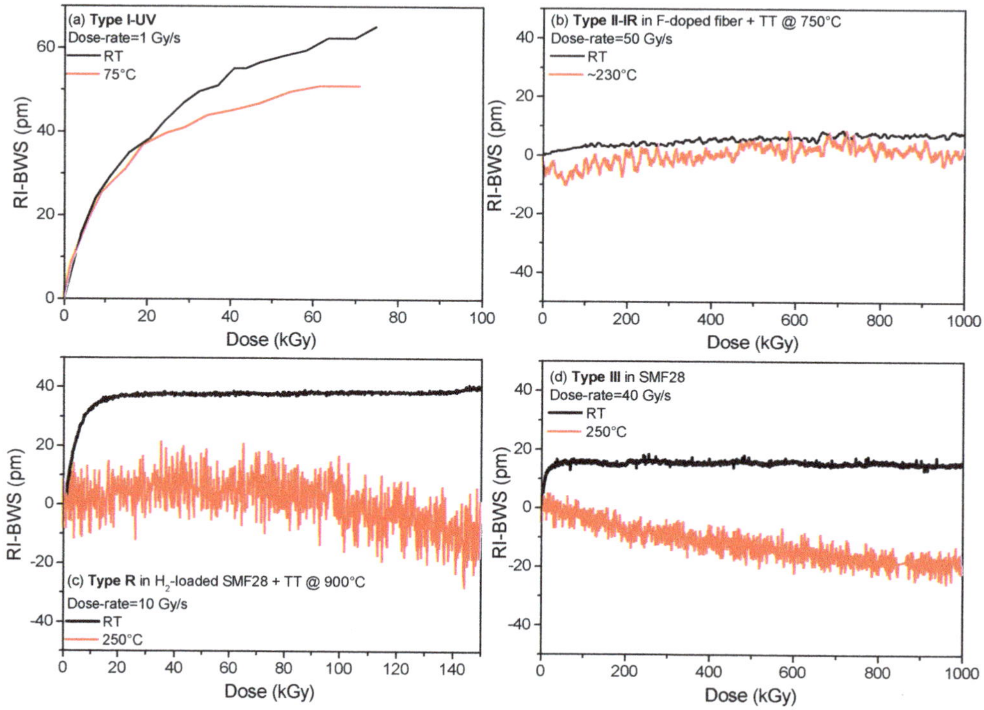


**Figure 13.** RI- BWS as a function of the dose, at two different temperatures, for different FBG types: (**a**) a type I FBG written with a UV laser (at 248 nm) in an $H_2$-loaded Ge-doped fiber (SMF28e) treated at 240 °C for 3 min and at 100 °C for 3 days, with a dose rate of about 1 Gy/s; (**b**) a type II FBG in an F-doped fiber with an fs-IR laser (centered at 800 nm, with a pulse width of 50 fs) treated at 750 °C for 15 min, with a dose rate of 50 Gy/s; (**c**) type R FBGs in SMF28 fiber regenerated at 900 °C, with a dose rate of 10 Gy/s; (**d**) a type III in the SMF28 fiber written PbP with an fs-IR laser (at 800 nm, a pulse width < 120 fs, and pulse energy of 230 nJ), with a dose rate of 40 Gy/s. The gratings reported in (**a**) were investigated under γ-rays, whereas all the others were obtained under X-rays. Data extracted from [80,81,86,89].

### 6.10. Nature of Radiation

When it comes to radiation, X-rays and γ-rays come to mind first. However, as reported in Table 1, harsh environments can be characterized also by the presence of other particles, such as protons, electrons, and neutrons. The first two are common in space, whereas neutrons are found in nuclear reactor cores or fusion-devoted facilities. Test facilities offering these types of beams are less accessible and only a few studies have been published about the FBG response under protons [39,101–105], electrons [104], or neutrons [19,22–24,27,28,106].

Even if particles can induce displacement damages, the response under radiation of a fiber depends on the defects created during the irradiation, and their main generation processes are associated with ionizing, at least for low to moderate fluences. Consequently, the RIA is mainly independent of the radiation nature [38] for most of the targeted environments (except for, e.g., nuclear core instrumentation). For example, the P1 is a center induced in the P-doped silica, absorbing around 1.5 µm: the growth kinetics of its absorption band is independent of the nature of irradiation, i.e., γ- or X-rays, protons, and neutrons, which makes it, together with other properties, a very good dosimeter of total ionizing dose in mixed environments [61,69,107].

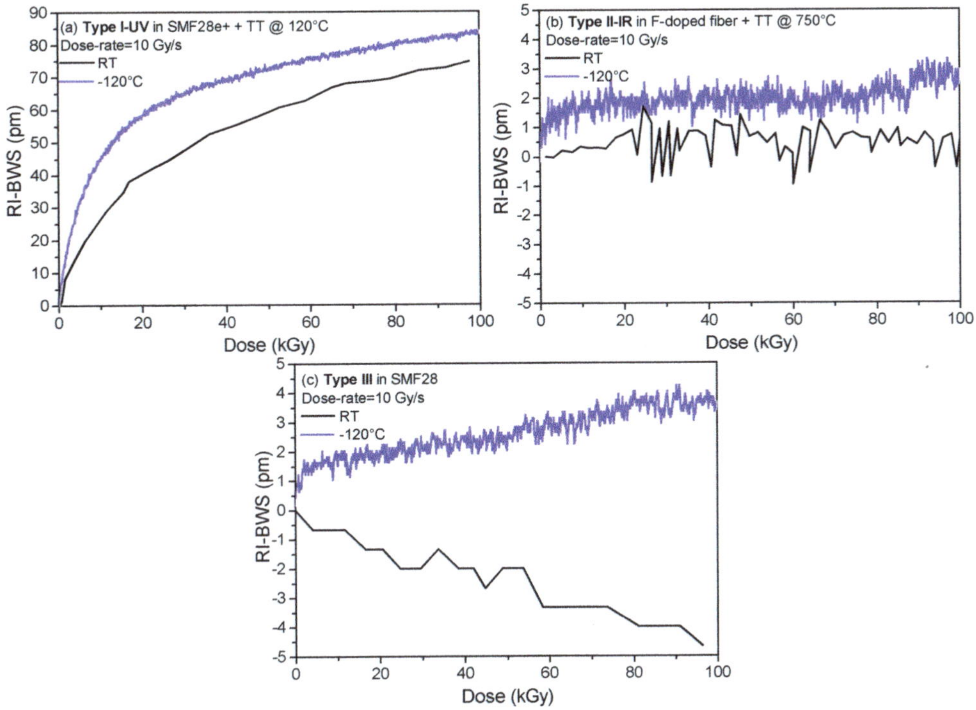


**Figure 14.** BWS induced by X-rays as a function of the dose at two different temperatures, for different FBG types: (**a**) a type I FBG written with a UV laser (at 244 nm) in an $H_2$-loaded Ge-doped fiber (SMF28e) treated at 120 °C for 8 h, with a dose rate of about 1 Gy/s; (**b**) a type II FBG in an F-doped fiber with an fs-IR laser (at 800 nm, with a pulse width of 50 fs) treated at 750 °C for 15 min, with a dose rate of 10 Gy/s; (**c**) a type III in the same F-doped OF written PbP with an fs-IR laser (centered at 800 nm, with a pulse width 120 fs), with a dose rate of 10 Gy/s. Data extracted from [100].

Even if tests under $\gamma$- or X-rays can help us predict the RI-BWS, a comparison performed on the response of type I FBGs written with a UV laser in photosensitive fibers under protons or electrons and X-rays reported in [104] lets us conclude that the physics inside the gratings is more complex. It is worth noticing that the irradiations were carried out at the same dose rate and temperature up to the same TID, in order not to observe strange combined effects. For example, under protons of 63 MeV, at least up to 10 kGy (fluence of about $10^{10}$ p·cm$^{-2}$), the BWS induced on gratings written in Ge-doped or B/Ge co-doped fibers is slightly smaller than the one induced by X-rays under the same conditions. On the contrary, the response induced under electrons is strongly dependent on the fiber composition. Furthermore, as already mentioned, even if the radiation response of a P-doped fiber in the IR does not depend on the radiation nature, it is not the case for a grating written in a P/Ce co-doped fiber, where the X-rays induce a larger shift than electrons or protons.

Concerning the neutrons, whereas several campaigns have been carried out at low flux, up to a fluence of about $10^{17}$ n·cm$^{-2}$ [19,20,22,31], the research targets higher fluences for applications in nuclear reactor cores. The shift induced by a neutron fluence of $10^{19}$ n·cm$^{-2}$ (neutron energy of 1 MeV) at a temperature of about 250 °C with an accumulated gamma dose of 0.5 GGy was investigated on different gratings after irradiation [23,24]. Contrary to what was reported before, the gratings written in an $H_2$-loaded Ge-doped fiber with an fs-laser were more radiation-resistant than the ones in unloaded fibers (all these FBGs were thermally treated at ~370 °C for 72 h). Indeed, the first ones showed a lower reflectivity reduction and also smaller BWS: this varies from −10 to −100 pm for the loaded FBGs, whereas it is about −130 pm for the unloaded ones. A type II FBG written in an unloaded Ge-doped fiber with an fs-laser and pre-treated at 750 °C for 2 h reached a BWS of −22 pm and −299 pm at the neutron fluence of $10^{19}$ n·cm$^{-2}$ and $3 \times 10^{19}$ n·cm$^{-2}$, respectively. Moreover, it is worth noticing that writing such a grating in a pure silica core fiber pre-irradiated at $5 \times 10^{19}$ n·cm$^{-2}$ and subjected to the same thermal treatment does not improve its radiation resistance, since the induced shift at the highest investigated fluence is still of about −200 pm.

Recently, FBGs were written in random air-line (RAL) fibers with an fs-laser (at 800 nm, with a pulse width of 80 fs) and tested under neutron fluence up to ~$3 \times 10^{20}$ n·cm$^{-2}$ (neutron energy of ~1 MeV), a flux of $6 \times 10^{13}$ n·cm$^{-2}$·s$^{-1}$, and at a temperature of about 600 °C [26]. The Bragg wavelength showed a blue shift and a reflectivity reduction, both depending linearly on the time (or neutron fluence): their slope was about 0.096 nm/day and 0.125 dB/day, respectively. The BWS induced at the highest fluence was −4.47 nm.

Such a large shift was also observed on a particular regenerated grating written in a Ge/F-codoped fiber. In [30], this is referred to as a chemical composition grating, or briefly, CCG, since the refractive index periodical structure is associated with the fluorine migration out of the bright fringes: the hydrogen present as OH groups (induced during the seed grating inscription in the $H_2$-loaded fiber) will react with the fluorine, creating volatile HF molecules during the regeneration. Under neutron fluence up to ~$10^{19}$ n·cm$^{-2}$ (neutron energy of 1 MeV) and flux of $7 \times 10^{12}$ n·cm$^{-2}$·s$^{-1}$, with an associated γ-dose of ~2 GGy at 150 °C, this CCG showed a red shift with a saturating behavior up to ~14 nm, very different from the previous results that reported a blue shift.

### 6.11. Coating and Embedding

Most of the results here reported have been obtained on uncoated FBGs, in order to study only the radiation influence on the glass structure of the grating. However, for real applications, the gratings cannot be bare, since they will be more fragile. However, *the coating can influence the grating radiation sensitivity*. Gusarov et al. carried out a systematic study on the γ-rays' effects on identical type I DTGs, written with a UV laser in a highly Ge-doped fiber during fiber drawing [108]. Their results are illustrated in Figure 15 and show that the BWS induced on the recoated FBGs is larger than in the bare sample. Indeed, whereas the shift in the latter is due to radiation effects on the glass, the difference between the coated and the bare FBGs has been attributed to the radiation effect on the coating: radiation makes the coating shrink or swell, inducing a change in the stress on the fiber and consequently a shift in the Bragg peak. From Figure 15, it is clear that the larger effects are observed for the ormocer and then the polyimide. The same authors also show that the shift induced on an acrylate-coated grating was between the ones recorded on the bare sample and the polyimide-coated one [108].

The same conclusions can be drawn from the study under proton irradiation by Curras et al. [103]. The authors explain the difference in the measured RI-BWSs with a two-stage process. Initially, the bonds of the coating polymeric chains are broken by radiation and induce a gas release, which will be trapped within the polymeric matrix, inducing the coating swelling. However, the gas release changes the polymeric structure, promoting transversal covalence bonds between the linear chains and consequently increasing the coating rigidity and inducing the coating shrinking.

However, in other studies under γ-rays [80], under X-rays [88], or under protons [105], the RI-BWS is smaller on coated gratings than on uncoated ones. In all cases, the coating does not shield the radiation (this could be the case for low-energy particles); moreover, it was demonstrated that the phase of UV curing (necessary for the bare FBG recoating with acrylate) does not influence its radiation response [88,109]. One of the remaining hypotheses is that the radiation continues the coating polymerization. Indeed, in [88], by performing at least three UV cycles on the recoated FBG, the authors show that such a recoated FBG will be characterized by the same RI-BWS as the bare grating.

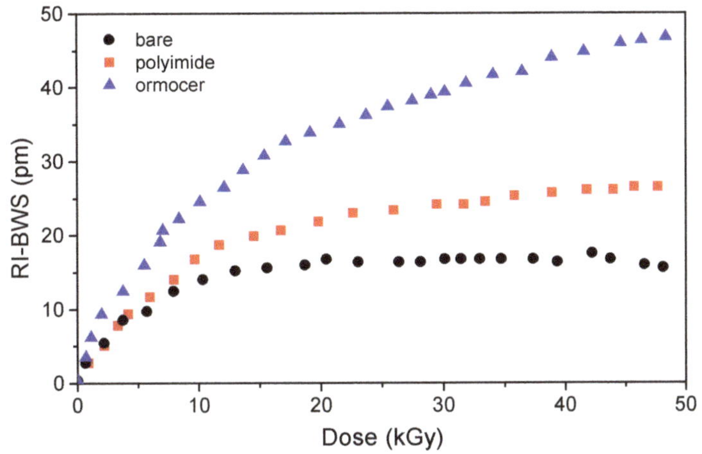


**Figure 15.** γ-rays induced BWS as a function of the dose, for identical type I FBGs written with a UV laser (at 248 nm, a pulse width of 15 ns, and power energy of 400 mJ/cm$^2$) in a highly Ge-doped fiber and recoated after inscription with polyimide (which was thermally cured) or ormocer, with a dose rate of ~0.11 Gy/s, at 35 °C. A bare grating was also reported for the sake of comparison. Data extracted from [108].

As with the coating, the embedding or packaging of the grating can also change its radiation sensitivity; an example of this is reported in [105]. Moreover, Lebel-Cormier et al. recently reported a radiation dosimeter based on polymer-embedded FBGs [110]: they showed that, by gluing the FBG inside a square prism in different polymer materials and reducing the error in the peak detection to 0.03 pm, the induced shift depends linearly on the dose, up to 20 Gy, with a slope of about ~0.06 pm/Gy, which depends on the material choice and not on the fiber composition.

## 7. Radiation Effects on Exotic FBGs

### *7.1. π-Phase Shifted Grating*

A π-phase shifted grating is a grating with a phase step within the FBG length, which gives rise to a very narrow Lorentzian-shape band-pass peak (generally with a width of −3 dB of a few pm) in the band-stop spectral region of the FBG [111]. It is a very narrow filter, with a width of only a few pm; consequently, it can be used as a high-resolution sensor or spectral filter, to realize a distributed-feedback (or DFB) laser, an optical function for RF filtering.

During the inscription, a small birefringence can be induced on the fiber, inducing two peaks instead of only one. For this reason, these gratings are often written in polarization-maintaining fibers, which allow propagation of only one polarization and therefore excite only one peak. The radiation response of a π-phase shifted grating written in an $H_2$-loaded Ge-doped core PANDA fiber with a UV laser and then subjected to thermal treatment at 120 °C for 8 h was studied under X-rays at RT up to 1 kGy TID, with a dose rate of ~30 mGy/s [112]. Both the main Bragg peak and the two peaks characterizing the π-phase shifted grating shift towards the red of the same quantity. For example, in this case, the BWS observed at the highest investigated TID was ~5 pm.

It is worth noticing that the band-pass peak within the large Bragg notch depends on the phase-step size [111]; consequently, if this parameter does not change, the band-pass peak has to shift together with the main one. Nevertheless, it could also depend on the writing process used to incorporate the phase step in the grating. Moreover, at very high doses or neutron fluence, when a compaction phenomenon takes place [16], the phase-step size could change, leading to a different shift from the one induced on the main peak.

### 7.2. Fiber Random Gratings

A fiber random grating, or FRG, is a grating whose refractive index modulation is not periodical but random [113]. An FRG is inscribed in the fiber core by the point-by-point technique, varying the distance between two consecutive spots randomly between 0 and a few µm, over a length of several cm, generally longer than an FBG [113]. Depending on the laser power density, type I or type II can be written.

Contrary to FBGs, the FRG gives rise to a very large reflection band, over 200 nm, caused by the superposition of the interference patterns of all the Mach–Zehnder interferometers constituted by each pair of two consecutive spots. A change in temperature or axial strain (but also in the surrounding refractive index) modifies the interferometer length and effective refractive indices of the core and cladding modes, causing a phase shift and, consequently, a spectral shift (*SS*) in the reflection spectrum [113], as shown in Figure 16. The spectral shift can be easily calculated by comparing two reflection spectra, acquired before and after a perturbation is applied, i.e., by performing a cross-correlation between the two.

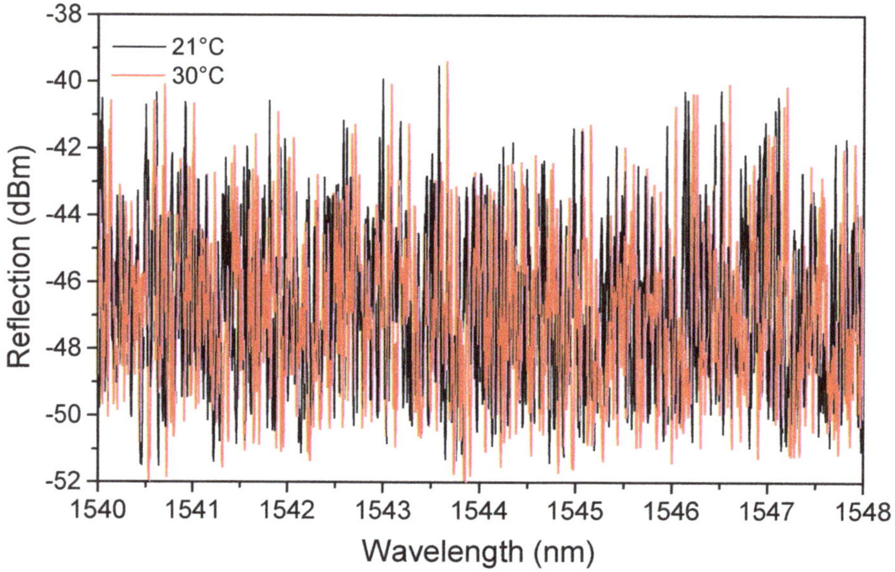


**Figure 16.** A part of reflection spectra of a type II 10 cm long FRG, inscribed in a Ge-doped fiber (the Corning SMF28e) with an fs-laser (at 800 nm), acquired at two temperatures, in order to highlight the changes induced by a temperature variation of about 10 °C. Data extracted from [114].

As for an FBG, the spectral shift depends linearly on temperature (in the 100 °C range) and strain:

$$SS = C_T \times \Delta T + C_\varepsilon \times \Delta\varepsilon, \tag{4}$$

with values of the same order of magnitude as those of a classic FBG for the temperature and strain coefficients, $C_T$ and $C_\varepsilon$, respectively [113]. However, the FRG can be employed as a multiparameter sensor, allowing the simultaneous measurement of the temperature and axial strain, through the "wavelength-division cross-correlation method": the reflection spectrum has to be divided into N large subregions (N depending on the number of parameters to be measured, generally two or three), characterized by temperature and strain coefficients differing by up to 10%. The larger the subregions, the better the sensing resolution; however, the larger the spectral distance between the subregions, the larger the difference between the coefficients, and then the better the discrimination capability of the sensor [113]. This will allow writing a system of N equations, such as the one in

Equation (4), with N unknown variables, one for each sensing parameter, ΔT, and Δε (more information about the analysis of simultaneous parameters is available in [115]).

Type I and type II FRGs, written in a Ge-doped fiber, were tested under X-rays up to about 150 kGy, with a dose rate of 1 Gy/s, at RT [114]. Just as for an FBG, radiation induces an additional spectral shift, which depends on the fiber composition and inscription parameters. However, this shift is exactly the same for all the subregions. Therefore, for example with three subregions, it would be possible to discriminate two parameters even under irradiation, since one equation of the system will be sufficient to isolate the radiation-induced shift.

## 8. Radiation Effects on FBGs in Exotic Fibers

### *8.1. FBGs Inscribed in Highly Birefringent Photonic Crystal Fibers*

Photonics crystal fibers, or briefly PCFs, are OFs characterized by a periodic transverse microstructure made up of air and glass (see review [116]). This structure allows the manufacture of highly birefringent fibers, and a grating written in it presents two Bragg peaks when investigated with an unpolarized light [117]. Indeed, the two orthogonally polarized modes propagate with two different phase velocities and consequently are characterized by different effective refractive indices, which depend on the mean effective index ($n_{eff}$) and on the phase modal birefringence $B$. Therefore, the spectral distance between the two peaks is defined as [117]:

$$\Delta\lambda_B = \lambda_{B2} - \lambda_{B1} = 2 \times B \times \Lambda. \quad (5)$$

Whereas the two peaks shift towards the same direction, keeping $\Delta\lambda_B$ constant, when the grating is subjected to a temperature or a longitudinal strain, under pressure or a transverse mechanical load, they move in opposite directions, reducing their spectral distance, as illustrated in Figure 17.

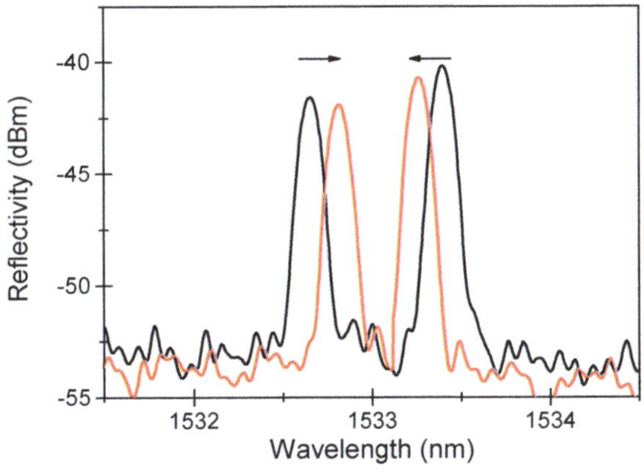


**Figure 17.** Reflection spectra of an FBG written in a highly birefringent PCF, unstressed (black curve), and when a transverse load is applied (red curve). The arrows highlight the direction of the shift in the peaks.

This characteristic makes this FBG a good pressure and transverse strain sensor for structural health monitoring [46]. It is worth noticing that such gratings are characterized by classical values for temperature and strain coefficients, whereas the transverse sensitivity varies between 150 pm/(N/mm) and 550 pm/(N/mm), depending on the fiber orientation during the compression test [46] (for more information, see review [118]).

The radiation response of such a grating was investigated under X-rays [73]. The FBG was written in a highly birefringent PCF known as 'butterfly', having a core in Ge-doped silica to facilitate the grating inscription, with an fs laser at 267 nm, and underwent a 16 h thermal treatment at 80 °C. The grating presented two peaks with a spectral distance of ~0.74 nm, which corresponds to a phase modal birefringence B of ~$7 \times 10^{-4}$. This FBG was irradiated at RT up to 1.5 MGy TID, with a dose rate of 24 Gy/s. As observed for type I gratings, both Bragg peaks shift towards longer wavelengths during the irradiation, as shown in Figure 18, quickly reaching the saturation value at a TID of about 15 kGy. The spectral distance, instead, remains unaffected by the radiation, within 10 pm; such a small change corresponds to a variation of the phase modal birefringence B of ~$10^{-5}$. Moreover, if this grating is used as a transverse mechanical load sensor, the radiation-induced error in the load measurement is less than 0.1 N/mm. In conclusion, *this transverse mechanical load sensor is intrinsically radiation-resistant up to MGy dose levels.*

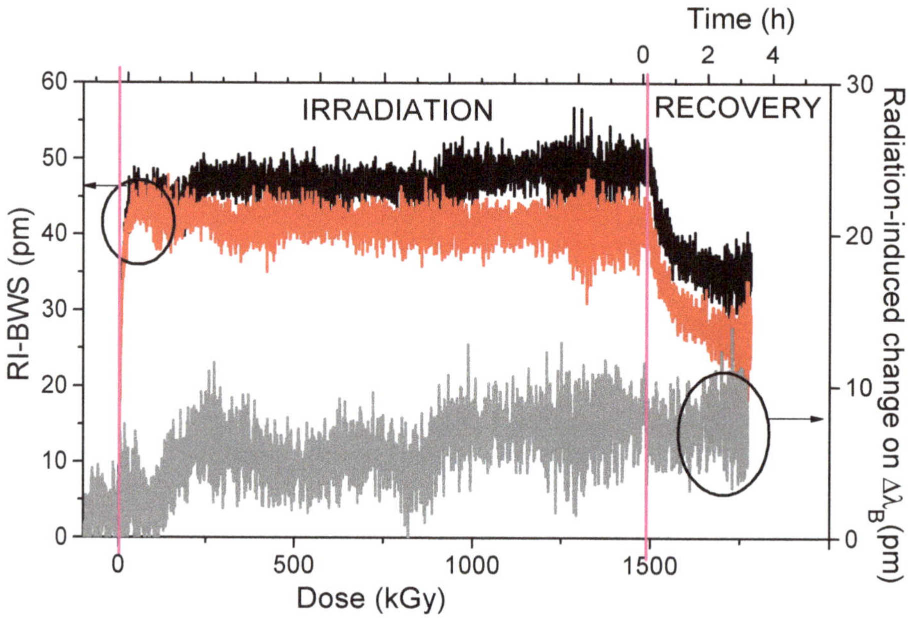


**Figure 18.** On the left axis, BWS induced by X-rays on the two peaks (peak 1—black points; peak 2—red points) of a grating written with an fs laser (at 267 nm and a pulse width of 120 fs) inside a butterfly PCF with a Ge-doped core, at RT up to 1.5 MGy TID, with a dose rate of 24 Gy/s. On the right axis, changes are induced by the radiation on the spectral distance between the two Bragg peaks. The two pink vertical lines indicate the start and stop of the irradiation run. Data extracted from [73].

### *8.2. FBGs Inscribed in Multicore Fibers*

Multicore fibers, or MCFs, consist of fibers where multiple separate cores have been incorporated in their claddings. They were proposed and manufactured, and their optical properties were characterized, for the first time in 1979 [119]: seven preforms were inserted in a jacket tube, that was collapsed to create an unique preform before the fiber drawing process. Their main aim was to increase the fiber transmission capability for telecommunication. In this case, each core should be an individual waveguide, so the crosstalk was a phenomenon to avoid. The crosstalk depends on the distance between the fiber cores and the transmission distance and it consists in coupling a signal launched into one core with the neighboring ones. Nowadays, the crosstalk can be suppressed or enhanced (see review [120]).

MCFs can have homogeneous or heterogeneous cores: in the first case, the refractive index profile of each core is the same, whereas in the second it is not and then the cores can present different types of doping.

Such fibers have also attracted more and more interest for sensing applications, both as distributed sensors (see review [121]) and as point sensors based on FBGs (see review [122]), i.e., to discriminate strain and temperature or for shape sensing. Indeed, during fiber bending, whereas the central core (as the one in a classical SMF) is on a strain neutral axis and then insensitive, the cores on the outer and inner sides will be, respectively, stretched and compressed, so they will undergo a strain change. Consequently, monitoring the strain variation in at least three cores, for example with three FBGs, will give rise to a three-axis shape sensor [123].

Writing FBGs in these types of fiber is more complicated because of their geometry, as with the PCFs. With a UV laser and the phase mask techniques, gratings will be inscribed in all cores at the same time (if they are sufficiently photo-sensitive); instead, to write an FBG in only one core, an fs-laser should be used with the point-by-point technique [122].

Under radiation, the Bragg peaks of identical gratings written in all the cores of a homogeneous MCF have to shift by the same amount; consequently, *the shape sensors based on FBGs inscribed on homogenous MCFs are intrinsically radiation-resistant.* This was confirmed by Barrera et al. investigating the response under γ-rays of two shape sensors based on Bragg gratings written with an UV laser in two 7 Ge-doped cores MCFs, one $H_2$-loaded, in order to increase the radiation sensitivityof each FBG, and one unloaded [124].

### 8.3. FBGs Inscribed in Polymer Optical Fibers

Polymer Optical Fibers (POFs), also known as plastic fibers, have several advantages compared to silica fibers, such as lower cost and higher flexibility, despite their higher optical losses [125]. Consequently, the Bragg gratings (or the 'quasi-single mode' Bragg gratings, since most of these fibers are multimode [126]) written in such fibers, briefly indicated as POFBGs, are interesting for several applications, thanks to their higher strain, temperature and humidity sensitivity (see review [127]). Even if more studies have been dedicated to the radiation-effects on the POF properties, also the radiation response of the FBGs inscribed in these fibers has been investigated by some researchers.

As a first example, we report here a study about the response of a FBG written with an fs-laser in the CYTOP POF, which presents lower intrinsic losses at 1550 nm than the other types of POFs [38]. In this work, Broadways et al. observed, despite a peak shape change with the appearing of two sub-peaks, a small amplitude reduction of only 3 dB and a blue shift under γ-rays up to ~40 kGy TID, as illustrated in Figure 19.

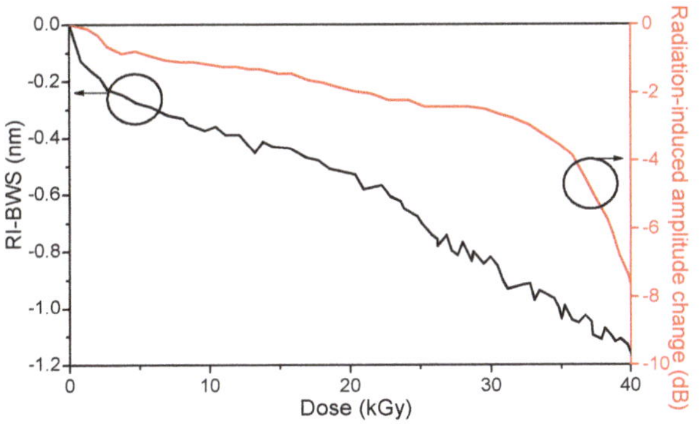


**Figure 19.** RI-BWS on the left (black curve) and peak amplitude reduction on the right (red curve) as a function of the dose, for an FBG written with an fs-laser in a CYTOP, under γ-rays. Data extracted from [38].

The grating sensitivity was −26.2 pm/kGy, with a maximum BWS of about −1 nm at the highest investigated dose. Such a shift corresponds to a 70% increase in relative humidity or to a variation of 700 με or 55 °C if the grating is used as a humidity, strain, or temperature sensor, respectively. However, the authors suggest its use as a possible dosimeter, whose resolution was about 40 Gy with their acquisition system. Figure 19 highlights a clear change in the peak amplitude of about 33 kGy: this was explained by a degradation of the splice between the CYTOP sample and the silica-based transport fibers used to connect the FBG to the acquisition system. Indeed, splicing still remains one of the most difficult tasks when working with POFs.

It is worth noticing that sometimes a red shift for the POFBGs is observed. Indeed, in a work dealing with the response of an FBG written with a CW UV laser (at 325 nm) in a PMMA fiber under neutron fluence with energy from 2 MeV to 10 MeV [128], the Bragg peak shifts towards the red and reaches 7 pm at the highest investigated TID of about 720 Gy.

## 9. Radiation-Resistant FBGs for Applications in Harsh Environments

All the studies published until now, and here summarized, on the radiation response of the FBGs, allow us to conclude which grating type will be the best choice for a given application in a given radiation-rich environment. It is very important to keep in mind that writing a grating in a radiation-resistant fiber does not assure its radiation resistance. However, using radiation-resistant or tolerant fibers will reduce the peak reflectivity decreases associated with the RIA phenomena.

As reported in Table 1, each environment is characterized by different doses, dose rates, and temperatures, parameters that will influence the various FBGs differently.

The type I gratings, especially the ones written with a UV laser, are the easier ones to manufacture but are also the ones with larger RI-BWS. By decreasing dose and dose rate, this shift reduces, and this might suggest them as an acceptable option for low doses, e.g., for space applications. However, the very low temperatures that can be reached in such environments will worsen the sensor response, inducing bigger errors in the measurements. Even in this case, they are not an adequate choice. A solution to improve their performance could be pre-irradiation, but, in this case, their response will depend on the conditions of such a treatment and of the storage between pre-irradiation and real use: too many parameters to control. In conclusion, type I FBGs are not suitable for radiation-rich environments.

The regenerated gratings have been considered for nuclear reactor applications, which means high dose, dose rate, and temperature, from RT to 800 °C, with a high neutron fluence to complicate the situation. Even if very small shifts have been observed at high temperatures, it is not always the case at temperatures as low as RT: in all the graphs here reported, a RI-BWS of at least 40 pm has been observed on different R-FBGs, which corresponds to an error in the temperature measurement of about 4 °C. Almost no tests have been performed under a high neutron fluence. The only one reported by Fernandez et al. showed a shift of about 10 nm at a neutron fluence of ~$10^{19}$ $n{\cdot}cm^{-2}$ (neutron energy of ~1 MeV) with an associated $\gamma$-dose of ~2 GGy at 150 °C [30]. Consequently, until new tests are carried out, the radiation resistance of such gratings is not guaranteed.

Two grating types are left.

Type II gratings, written with an fs-laser, showed good radiation resistance, but this can be strongly dependent on the inscription conditions and set-up, especially when they are inscribed through a phase mask. With a PM, indeed, all the bright fringes are not illuminated with the same laser intensity, supposing that the laser beam is Gaussian. Therefore, as explained in Section 0, both type I and type II components could be generated. In order to obtain only a type II FBG, after its inscription in a radiation-resistant fiber with the optimized power conditions necessary to obtain a type II [53], thermal treatment should be performed to erase the type I component and improve the radiation resistance of the final FBG [86]. The good performance of such a grating was tested up to 3 MGy TID under X-rays or $\gamma$-rays, with temperatures from −100 °C to +350 °C, under protons for space applications, and even in a nuclear reactor core with a fast neutron fluence of $3 \times 10^{19}$ $n/cm^2$ and a total $\gamma$-dose of about 4 GGy, at a temperature of about 290 °C (in this case, a RI-BWS of 40 pm was measured after irradiation and corresponds to a temperature error of only ~4 °C [23]).

Finally, the type III gratings seem promising but still more investigations are needed. RI-BWSs of the order of magnitude of 10 pm have been observed on Ge-doped fiber (slightly lower on an F-doped fiber) at 1 MGy TID. Their response does not seem to change drastically with the dose rate or at low temperatures. However, what still remains to understand is their behavior at high temperatures. Indeed, a blue shift, almost linear with the dose, was observed during irradiation at a high temperature, about 200 °C, [89] whereas in another study, the grating showed a good radiation resistance [87]. The two investigated type III gratings were manufactured by different laboratories, and until now it is not clear how they differed. So, we cannot conclude yet which is a good process to obtain radiation-resistant type III FBGs.

## 10. Conclusions

Fiber Bragg gratings are point sensors suitable for monitoring temperature and/or strain. Various types exist with different inscription processes and different thermal and radiation resistances. Under radiation, indeed, the performance of such sensors can be degraded and an error in measurements can be induced. The magnitude of this error will depend on the conditions of the radiation-rich environment and on the grating itself. The purpose of this article was not to list all the published works about gratings under radiation—they are too many–but to provide a guide on how the radiation response of different grating types depends on the parameters of the environment, the FBG manufacturing, or the treatments performed before or after the inscription. It is important to highlight first that writing a grating into a radiation-resistant fiber does not ensure a radiation-resistant FBG, and second that the radiation resistance of an FBG strongly depends on its application. For example, a grating whose radiation response could be acceptable for a space application, as a type I, could not resist the environment of the nuclear reactor core. Consequently, it is not possible to determine a classification of FBGs based on their radiation response. However, among all the FBG types, the type II and the type III gratings seem to be the best option for applications having to operate in a wide temperature range and at high radiation doses. Furthermore, it is worth mentioning that, if the fiber is not radiation-resistant, the signal will not be transmitted from/to the instrumentation, because of the RIA. A solution, in this case, can be to splice a small piece of this fiber containing the FBG with a radiation-hardened one.

To conclude, despite the fact that radiation alters the performance of FBGs, thanks to the various hardening studies led by the scientific community, radiation-hardened FBG-based sensors appear today as a very promising solution for monitoring environmental parameters in the most challenging radiation environments. Thanks to a better understanding of the basic mechanisms related to FBG photoinscription and radiation effects, we could imagine that even more radiation-tolerant FBGs could be designed in the future, also able to survive combined high temperature and high radiation dose constraints.

**Author Contributions:** Conceptualization: A.M., E.M., S.G., A.B. and Y.O.; writing—original draft preparation: A.M., E.M. and S.G.; writing—review and editing: A.M., E.M., L.L., T.B., T.R., G.C., G.L., A.B., Y.O. and S.G.; data curation: A.M. and T.B.; resources: A.M., E.M., T.B., G.C., G.L. and S.G. All authors have read and agreed to the published version of the manuscript.

**Funding:** This research received no external funding.

**Conflicts of Interest:** The authors declare no conflict of interest.

## References

1. Keiser. *Optical Fiber Communications*; McGraw-Hill Education (India) Pvt Limited: Karnataka, India, 2008; ISBN 978-0-07-064810-4.
2. Rajan, G. *Optical Fiber Sensors: Advanced Techniques and Applications*; CRC Press: Boca Raton, FL, USA, 2017; ISBN 978-1-4822-2829-8.
3. Hill, K.O.; Meltz, G. Fiber Bragg Grating Technology Fundamentals and Overview. *J. Light. Technol.* **1997**, *15*, 1263–1276. [CrossRef]

4. Wanser, K.H.; Voss, K.F.; Kersey, A.D. Novel Fiber Devices and Sensors Based on Multimode Fiber Bragg Gratings. In Proceedings of the Tenth International Conference on Optical Fibre Sensors, Glasgow, Scotland, 11–13 October 1994; SPIE: Bellingham, WA, USA, 14 September 1994; Volume 2360, pp. 265–268.
5. Esposito, F.; Srivastava, A.; Campopiano, S.; Iadicicco, A. Radiation Effects on Long Period Fiber Gratings: A Review. *Sensors* **2020**, *20*, 2729. [CrossRef] [PubMed]
6. Sahota, J.K.; Gupta, N.; Dhawan, D. Fiber Bragg Grating Sensors for Monitoring of Physical Parameters: A Comprehensive Review. *Opt. Eng.* **2020**, *59*, 060901. [CrossRef]
7. Morey, W.W.; Dunphy, J.R.; Meltz, G. Multiplexing Fiber Bragg Grating Sensors. In Proceedings of the Distributed and Multiplexed Fiber Optic Sensors, Boston, MA, USA, 8–9 September 1993; SPIE: Bellingham, WA, USA, 1 January 1992; Volume 1586, pp. 216–224.
8. Sun, Q.; Ai, F.; Liu, D.; Cheng, J.; Luo, H.; Peng, K.; Luo, Y.; Yan, Z.; Shum, P.P. M-OTDR Sensing System Based on 3D Encoded Microstructures. *Sci. Rep.* **2017**, *7*, 41137. [CrossRef] [PubMed]
9. Hartog, A.H. An Introduction to Distributed Optical Fibre Sensors. Available online: https://www.taylorfrancis.com/books/mono/10.1201/9781315119014/introduction-distributed-optical-fibre-sensors-arthur-hartog (accessed on 9 September 2022).
10. Barbarin, Y.; Lefrançois, A.; Rougier, B.; Sinatti, F.; Lassalle, O.; Osmont, A.; Luc, J. Shocks Sensing by Fiber Bragg Gratings and a 100 MHz Dynamic Dispersive Interrogator. In Proceedings of the 2018 IEEE Research and Applications of Photonics In Defense Conference (RAPID), Miramar Beach, FL, USA, 22–24 August 2018; pp. 1–3.
11. Li, T.; Guo, J.; Tan, Y.; Zhou, Z. Recent Advances and Tendency in Fiber Bragg Grating-Based Vibration Sensor: A Review. *IEEE Sens. J.* **2020**, *20*, 12074–12087. [CrossRef]
12. Minardo, A.; Cusano, A.; Bernini, R.; Zeni, L.; Giordano, M. Response of Fiber Bragg Gratings to Longitudinal Ultrasonic Waves. *IEEE Trans. Ultrason. Ferroelectr. Freq. Control* **2005**, *52*, 304–312. [CrossRef] [PubMed]
13. Gusarov, A.; Hoeffgen, S.K. Radiation Effects on Fiber Gratings. *IEEE Trans. Nucl. Sci.* **2013**, *60*, 2037–2053. [CrossRef]
14. Girard, S.; Morana, A.; Ladaci, A.; Robin, T.; Mescia, L.; Bonnefois, J.-J.; Mekki, J.; Paveau, A.; Cadier, B.; Marin, E.; et al. Recent Advances in Radiation Hardened Fiber-Based Technologies for Space Applications. *J. Opt.* **2018**, *20*, 49. [CrossRef]
15. Girard, S.; Kuhnhenn, J.; Gusarov, A.; Brichard, B.; Van Uffelen, M.; Ouerdane, Y.; Boukenter, A.; Marcandella, C. Radiation Effects on Silica-Based Optical Fibers: Recent Advances and Future Challenges. *IEEE Trans. Nucl. Sci.* **2013**, *60*, 2015–2036. [CrossRef]
16. Primak, W. Fast-Neutron-Induced Changes in Quartz and Vitreous Silica. *Phys. Rev.* **1958**, *110*, 1240–1254. [CrossRef]
17. Mihailov, S.J. Fiber Bragg Grating Sensors for Harsh Environments. *Sensors* **2012**, *12*, 1898–1918. [CrossRef]
18. Fernandez Fernandez, A.; Berghmans, F.; Brichard, B.; Decréton, M.; Gusarov, A.; Deparis, O.; Mégret, P.; Blondel, M.; Delchambre, A. Multiplexed Fibre Bragg Grating Sensors for In-Core Thermometry in Nuclear Reactors. In Proceedings of the Environmental and Industrial Sensing, Boston, MA, USA, 9–11 September 2000.
19. Fernandez, A.F.; Gusarov, A.I.; Bodart, S.; Lammens, K.; Berghmans, F.; Decreton, M.; Megret, P.; Blondel, M.; Delchambre, A. Temperature Monitoring of Nuclear Reactor Cores with Multiplexed Fiber Bragg Grating Sensors. *Opt. Eng.* **2002**, *41*, 1246–1254.
20. Gusarov, A.; Fernandez, A.F.; Vasiliev, S.; Medvedkov, O.; Blondel, M.; Berghmans, F. Effect of Gamma–Neutron Nuclear Reactor Radiation on the Properties of Bragg Gratings Written in Photosensitive Ge-Doped Optical Fiber. *Nucl. Instrum. Methods Phys. Res. Sect. B Beam Interact. Mater. At.* **2002**, *187*, 79–86. [CrossRef]
21. Fernandez, A.F.; Gusarov, A.; Brichard, B.; Decréton, M.; Berghmans, F.; Mégret, P.; Delchambre, A. Long-Term Radiation Effects on Fibre Bragg Grating Temperature Sensors in a Low Flux Nuclear Reactor. *Meas. Sci. Technol.* **2004**, *15*, 1506–1511. [CrossRef]
22. Gusarov, A. Long-term exposure of fiber Bragg gratings in the BR1 low-flux nuclear reactor. *IEEE Trans. Nucl. Sci.* **2010**, *57*, 2044–2048. [CrossRef]
23. Remy, L.; Cheymol, G.; Gusarov, A.; Morana, A.; Marin, E.; Girard, S. Compaction in Optical Fibres and Fibre Bragg Gratings Under Nuclear Reactor High Neutron and Gamma Fluence. *IEEE Trans. Nucl. Sci.* **2016**, *63*, 2317–2322. [CrossRef]
24. Cheymol, G.; Remy, L.; Gusarov, A.; Kinet, D.; Mégret, P.; Laffont, G.; Blanchet, T.; Morana, A.; Marin, E.; Girard, S. Study of Fiber Bragg Grating Samples Exposed to High Fast Neutron Fluences. *IEEE Trans. Nucl. Sci.* **2018**, *65*, 2494–2501. [CrossRef]
25. Laffont, G.; Cotillard, R.; Roussel, N.; Desmarchelier, R.; Rougeault, S. Temperature Resistant Fiber Bragg Gratings for On-Line and Structural Health Monitoring of the Next-Generation of Nuclear Reactors. *Sensors* **2018**, *18*, 1791. [CrossRef]
26. Zaghloul, M.A.S.; Wang, M.; Huang, S.; Hnatovsky, C.; Grobnic, D.; Mihailov, S.; Li, M.-J.; Carpenter, D.; Hu, L.-W.; Daw, J.; et al. Radiation Resistant Fiber Bragg Grating in Random Air-Line Fibers for Sensing Applications in Nuclear Reactor Cores. *Opt. Express* **2018**, *26*, 11775. [CrossRef]
27. Fielder, R.S.; Klemer, D.; Stinson-Bagby, K.L. High Neutron Fluence Survivability Testing of Advanced Fiber Bragg Grating Sensors. *AIP Conf. Proc.* **2004**, *699*, 650–657. [CrossRef]
28. Fujita, K.; Kimura, A.; Nakazawa, M.; Takahashi, H. Bragg Peak Shifts of Fiber Bragg Gratings in Radiation Environment. In *Fiber Optic Sensor Technology II*; SPIE: Bellingham, WA, USA, 2001; pp. 184–191. [CrossRef]
29. Wu, Z.; Zaghloul, M.A.S.; Carpenter, D.; Li, M.-J.; Daw, J.; Mao, Z.-H.; Hnatovsky, C.; Mihailov, S.J.; Chen, K.P. Mitigation of Radiation-Induced Fiber Bragg Grating (FBG) Sensor Drifts in Intense Radiation Environments Based on Long-Short-Term Memory (LSTM) Network. *IEEE Access* **2021**, *9*, 148296–148301. [CrossRef]
30. Fernandez, A.F.; Brichard, B.; Berghmans, F.; Rabii, H.E.; Fokine, M.; Popov, M. Chemical Composition Fiber Gratings in a High Mixed Gamma Neutron Radiation Field. *IEEE Trans. Nucl. Sci.* **2006**, *53*, 1607–1613. [CrossRef]

31. Nehr, S.; Cotillard, R.; Laffont, G.; Villard, J.-F.; Destouches, C.; Carpenter, D.; Daw, J.; Calderoni, P. On-Line Monitoring of Multiplexed Femtosecond Fiber Bragg Gratings Exposed to High Temperature and High Neutron Fluence. In Proceedings of the 26th International Conference on Optical Fiber Sensors, Lausanne, Switzerland, 24–28 September 2018; Optica Publishing Group: Washington, DC, USA, 2018; p. WF56.
32. Corre, Y.; Laffont, G.; Pocheau, C.; Cotillard, R.; Gaspar, J.; Roussel, N.; Firdaouss, M.; Gardarein, J.-L.; Guilhem, D.; Missirlian, M. Integration of Fiber Bragg Grating Temperature Sensors in Plasma Facing Components of the WEST Tokamak. *Rev. Sci. Instrum.* **2018**, *89*, 063508. [CrossRef] [PubMed]
33. Berruti, G.; Consales, M.; Giordano, M.; Sansone, L.; Petagna, P.; Buontempo, S.; Breglio, G.; Cusano, A. Radiation Hard Humidity Sensors for High Energy Physics Applications Using Polyimide-Coated Fiber Bragg Gratings Sensors. *Sens. Actuators B Chem.* **2013**, *177*, 94–102. [CrossRef]
34. Makovec, A.; Berruti, G.; Consales, M.; Giordano, M.; Petagna, P.; Buontempo, S.; Breglio, G.; Szillasi, Z.; Beni, N.; Cusano, A. Radiation Hard Polyimide-Coated FBG Optical Sensors for Relative Humidity Monitoring in the CMS Experiment at CERN. *J. Instrum.* **2014**, *9*, C03040. [CrossRef]
35. Fienga, F.; Marrazzo, V.R.; Spedding, S.B.; Szillasi, Z.; Beni, N.; Irace, A.; Zeuner, W.; Ball, A.; Vaccaro, V.G.; Salvant, B.; et al. Fiber Bragg Grating Sensors as Innovative Monitoring Tool for Beam Induced RF Heating on LHC Beam Pipe. *J. Light. Technol.* **2021**, *39*, 4145–4150. [CrossRef]
36. A Comparative Study of Radiation-Tolerant Fiber Optic Sensors for Relative Humidity Monitoring in High-Radiation Environments at CERN | IEEE Journals & Magazine | IEEE Xplore. Available online: https://ieeexplore.ieee.org/abstract/document/6909002 (accessed on 12 June 2022).
37. Kinet, D.; Chah, K.; Gusarov, A.; Faustov, A.; Areias, L.; Troullinos, I.; Van Marcke, P.; Craeye, B.; Coppens, E.; Raymaekers, D.; et al. Proof of Concept for Temperature and Strain Measurements With Fiber Bragg Gratings Embedded in Supercontainers Designed for Nuclear Waste Storage. *IEEE Trans. Nucl. Sci.* **2016**, *63*, 1955–1962. [CrossRef]
38. Broadway, C.; Kinet, D.; Theodosiou, A.; Kalli, K.; Gusarov, A.; Caucheteur, C.; Mégret, P. CYTOP Fibre Bragg Grating Sensors for Harsh Radiation Environments. *Sensors* **2019**, *19*, 2853. [CrossRef]
39. Bakaic, M.; Hanna, M.; Hnatovsky, C.; Grobnic, D.; Mihailov, S.; Zeisler, S.S.; Hoehr, C. Fiber-Optic Bragg Gratings for Temperature and Pressure Measurements in Isotope Production Targets for Nuclear Medicine. *Appl. Sci.* **2020**, *10*, 4610. [CrossRef]
40. McKenzie, I.; Ibrahim, S.; Haddad, E.; Abad, S.; Hurni, A.; Cheng, L.K. Fiber Optic Sensing in Spacecraft Engineering: An Historical Perspective From the European Space Agency. *Front. Phys.* **2021**, *9*, 719441. [CrossRef]
41. Hill, K.O.; Fujii, Y.; Johnson, D.C.; Kawasaki, B.S. Photosensitivity in Optical Fiber Waveguides: Application to Reflection Filter Fabrication. *Appl. Phys. Lett.* **1978**, *32*, 647–649. [CrossRef]
42. Leviton, D.B.; Frey, B.J. Temperature-Dependent Absolute Refractive Index Measurements of Synthetic Fused Silica. In Proceedings of the Optomechanical Technologies for Astronomy, Orlando, FL, USA, 24–31 May 2006; SPIE: Bellingham, WA, USA, 2006; Volume 6273, pp. 800–810.
43. Brückner, R. Properties and Structure of Vitreous Silica. I. *J. Non-Cryst. Solids* **1970**, *5*, 123–175. [CrossRef]
44. Hocker, G.B. Fiber-Optic Sensing of Pressure and Temperature. *Appl. Opt.* **1979**, *18*, 1445–1448. [CrossRef] [PubMed]
45. Black, R.J.; Zare, D.; Oblea, L.; Park, Y.-L.; Moslehi, B.; Neslen, C. On the Gage Factor for Optical Fiber Grating Strain Gages. In Proceedings of the Society for the Advancement of Materials and Process Engineering, Long Beach, CA, USA, 5–9 May 2008; pp. 18–22.
46. Sonnenfeld, C.; Sulejmani, S.; Geernaert, T.; Eve, S.; Lammens, N.; Luyckx, G.; Voet, E.; Degrieck, J.; Urbanczyk, W.; Mergo, P.; et al. Microstructured Optical Fiber Sensors Embedded in a Laminate Composite for Smart Material Applications. *Sensors* **2011**, *11*, 2566–2579. [CrossRef] [PubMed]
47. Malo, B.; Hill, K.O.; Bilodeau, F.; Johnson, D.C.; Albert, J. Point-by-Point Fabrication of Micro-Bragg Gratings in Photosensitive Fibre Using Single Excimer Pulse Refractive Index Modification Techniques. *Electron. Lett.* **1993**, *29*, 1668–1669. [CrossRef]
48. Hill, K.O.; Malo, B.; Bilodeau, F.; Johnson, D.C.; Albert, J. Bragg Gratings Fabricated in Monomode Photosensitive Optical Fiber by UV Exposure through a Phase Mask. *Appl. Phys. Lett.* **1993**, *62*, 1035–1037. [CrossRef]
49. Li, X.; Gao, W.; Shimizu, Y.; Ito, S. A Two-Axis Lloyd's Mirror Interferometer for Fabrication of Two-Dimensional Diffraction Gratings. *CIRP Ann.* **2014**, *63*, 461–464. [CrossRef]
50. Gribaev, A.I.; Pavlishin, I.V.; Stam, A.M.; Idrisov, R.F.; Varzhel, S.V.; Konnov, K.A. Laboratory Setup for Fiber Bragg Gratings Inscription Based on Talbot Interferometer. *Opt. Quantum Electron.* **2016**, *48*, 540. [CrossRef]
51. Martinez, A.; Khrushchev, I.Y.; Bennion, I. Direct Inscription of Bragg Gratings in Coated Fibers by an Infrared Femtosecond Laser. *Opt. Lett.* **2006**, *31*, 1603–1605. [CrossRef]
52. Lindner, E.; Mörbitz, J.; Chojetzki, C.; Becker, M.; Brückner, S.; Schuster, K.; Rothhardt, M.; Bartelt, H. Draw Tower Fiber Bragg Gratings and Their Use in Sensing Technology. In Proceedings of the Fiber Optic Sensors and Applications VIII, Bellingham, WA, USA, 2 June 2011; SPIE: Bellingham, WA, USA, 2011; Volume 8028, pp. 69–75.
53. Smelser, C.W.; Mihailov, S.J.; Grobnic, D. Formation of Type I-IR and Type II-IR Gratings with an Ultrafast IR Laser and a Phase Mask. *Opt. Express* **2005**, *13*, 5377–5386. [CrossRef]
54. Canning, J. Fibre Gratings and Devices for Sensors and Lasers. *Laser Photonics Rev.* **2008**, *2*, 275–289. [CrossRef]
55. Askins, C.G.; Tsai, T.-E.; Williams, G.M.; Putnam, M.A.; Bashkansky, M.; Friebele, E.J. Fiber Bragg Reflectors Prepared by a Single Excimer Pulse. *Opt. Lett.* **1992**, *17*, 833–835. [CrossRef] [PubMed]

56. Lai, M.-H.; Gunawardena, D.S.; Lim, K.-S.; Yang, H.-Z.; Ahmad, H. Observation of Grating Regeneration by Direct $CO_2$ Laser Annealing. *Opt. Express* **2015**, *23*, 452. [CrossRef]
57. Bueno, A.; Kinet, D.; Mégret, P.; Caucheteur, C. Fast Thermal Regeneration of Fiber Bragg Gratings. *Opt. Lett.* **2013**, *38*, 4178. [CrossRef]
58. Canning, J.; Stevenson, M.; Bandyopadhyay, S.; Cook, K. Extreme Silica Optical Fibre Gratings. *Sensors* **2008**, *8*, 6448–6452. [CrossRef]
59. Martinez, A.; Khrushchev, I.Y.; Bennion, I. Thermal Properties of Fibre Bragg Gratings Inscribed Point-by-Point by Infrared Femtosecond Laser. *Electron. Lett.* **2005**, *41*, 176–178. [CrossRef]
60. Martinez, A.; Dubov, M.; Khrushchev, I.; Bennion, I. Photoinduced Modifications in Fiber Gratings Inscribed Directly by Infrared Femtosecond Irradiation. *IEEE Photonics Technol. Lett.* **2006**, *18*, 2266–2268. [CrossRef]
61. Girard, S.; Alessi, A.; Richard, N.; Martin-Samos, L.; De Michele, V.; Giacomazzi, L.; Agnello, S.; Francesca, D.D.; Morana, A.; Winkler, B.; et al. Overview of Radiation Induced Point Defects in Silica-Based Optical Fibers. *Rev. Phys.* **2019**, *4*, 100032. [CrossRef]
62. Girard, S.; Baggio, J.; Bisutti, J. 14-MeV Neutron g-Ray, and Pulsed X-Ray Radiation-Induced Effects on Multimode Silica-Based Optical Fibers. *IEEE Trans. Nucl. Sci.* **2006**, *53*, 3750–3757. [CrossRef]
63. Henschel, H.; Kohn, O.; Lennartz, W.; Metzger, S.; Schmidt, H.U.; Rosenkranz, J.; Glessner, B.; Siebert, B.R.L. Comparison between Fast Neutron and Gamma Irradiation of Optical Fibres. In Proceedings of the Radiation and Its Effects on Components and Systems, Cannes, France, 15–19 September 1997; IEEE: Manhattan, NY, USA, 1997; pp. 430–438.
64. Morana, A.; Girard, S.; Cannas, M.; Marin, E.; Marcandella, C.; Paillet, P.; Périsse, J.; Macé, J.-R.; Boscaino, R.; Nacir, B.; et al. Influence of Neutron and Gamma-Ray Irradiations on Rad-Hard Optical Fiber. *Opt. Mater. Express* **2015**, *5*, 898–911. [CrossRef]
65. Girard, S.; Keurinck, J.; Ouerdane, Y.; Meunier, J.-P.; Boukenter, A. Gamma-Rays and Pulsed X-Ray Radiation Responses of Germanosilicate Single-Mode Optical Fibers: Influence of Cladding Codopants. *J. Light. Technol.* **2004**, *22*, 1915. [CrossRef]
66. Space Radiation Effects in High Performance Fiber Optic Data Links for Satellite Data Management | IEEE Journals & Magazine | IEEE Xplore. Available online: https://ieeexplore.ieee.org/abstract/document/490907 (accessed on 10 September 2022).
67. Barnes, C.E.; Dorsky, L.I.; Johnston, A.R.; Bergman, L.A.; Stassinopoulos, E.G. *Overview of Fiber Optics in the Natural Space Environment*; Greenwell, R.A., Paul, D.K., Eds.; SPIE: San Jose, CA, USA, 1991; pp. 9–16.
68. Henschel, H.; Köhn, O.; Körfer, M.; Stegmann, T.; Wittenburg, K.; Wulf, F. Optical Fiber Dosimetry at the Tesla Test Facility (TTF). *AIP Conf. Proc.* **2000**, *546*, 647–649. [CrossRef]
69. Qualification and Calibration of Single-Mode Phosphosilicate Optical Fiber for Dosimetry at CERN | IEEE Journals & Magazine | IEEE Xplore. Available online: https://ieeexplore.ieee.org/document/8709781 (accessed on 11 June 2022).
70. Morana, A.; Campanella, C.; Vidalot, J.; De Michele, V.; Marin, E.; Reghioua, I.; Boukenter, A.; Ouerdane, Y.; Paillet, P.; Girard, S. Extreme Radiation Sensitivity of Ultra-Low Loss Pure-Silica-Core Optical Fibers at Low Dose Levels and Infrared Wavelengths. *Sensors* **2020**, *20*, 7254. [CrossRef]
71. Bolotovskii, B.M. Vavilov—Cherenkov Radiation: Its Discovery and Application. *Phys.-Uspekhi* **2009**, *52*, 1099. [CrossRef]
72. Primak, W.; Kampwirth, R. The Radiation Compaction of Vitreous Silica. *J. Appl. Phys.* **1968**, *39*, 5651–5658. [CrossRef]
73. Morana, A.; Baghdasaryan, T.; Girard, S.; Marin, E.; Geernaert, T.; Thienpont, H.; Berghmans, F.; Boukenter, A.; Ouerdane, Y. Radiation-Induced Effects on Fiber Bragg Gratings Inscribed in Highly Birefringent Photonic Crystal Fiber. *IEEE Trans. Nucl. Sci.* **2019**, *66*, 120–124. [CrossRef]
74. Morana, A.; Girard, S.; Marin, E.; Lancry, M.; Marcandella, C.; Paillet, P.; Lablonde, L.; Robin, T.; Williams, R.J.; Withford, M.J.; et al. Influence of Photo-Inscription Conditions on the Radiation-Response of Fiber Bragg Gratings. *Opt. Express* **2015**, *23*, 8659–8669. [CrossRef] [PubMed]
75. Gusarov, A.; Starodubov, D.; Berghmans, F.; Deparis, O.; Defosse, Y.; Fernandez, A.F.; Decréton, M.; Mégret, P.; Blondel, M. Comparative Study of the MGy Dose Level $\gamma$-Radiation Effect on FBGs Written in Different Fibres. In Proceedings of the Ofs-13: 13th International Conference on Optical Fiber Sensors & Workshop on Device and System Technology Toward Future Optical Fiber Communication and Sensing, Kyongju, Korea, 12–16 April 1999; p. 3746.
76. Henschel, H.; Hoeffgen, S.K.; Krebber, K.; Kuhnhenn, J.; Weinand, U. Influence of Fiber Composition and Grating Fabrication on the Radiation Sensitivity of Fiber Bragg Gratings. *IEEE Trans. Nucl. Sci.* **2008**, *55*, 2235–2242. [CrossRef]
77. Lin, S.; Song, N.; Jin, J.; Wang, X.; Yang, G. Effect of Grating Fabrication on Radiation Sensitivity of Fiber Bragg Gratings in Gamma Radiation Field. *IEEE Trans. Nucl. Sci.* **2011**, *58*, 2111–2117. [CrossRef]
78. Morana, A.; Marin, E.; Girard, S.; Marcandella, C.; Paillet, P.; Boukenter, A.; Ouerdane, Y. Dose Rate Effect Comparison on the Radiation Response of Type I Fiber Bragg Gratings Written With UV Cw Laser. *IEEE Trans. Nucl. Sci.* **2016**, *63*, 2046–2050. [CrossRef]
79. Gusarov, A.I.; Berghmans, F.; Fernandez Fernandez, A.; Deparis, O.; Defosse, Y.; Starodubov, D.S.; Decréton, M.; Mégret, P.; Blondel, M. Behavior of Fibre Bragg Gratings Under High Total Dose Gamma Radiation. In Proceedings of the 1999 Fifth European Conference on Radiation and Its Effects on Components and Systems, RADECS, Fontevraud, France, 13–17 September 1999; Volume 5, p. 461.
80. Henschel, H.; Hoeffgen, S.K.; Kuhnhenn, J.; Weinand, U. Influence of Manufacturing Parameters and Temperature on the Radiation Sensitivity of Fiber Bragg Gratings. *IEEE Trans. Nucl. Sci.* **2010**, *57*, 2029–2034. [CrossRef]

81. Blanchet, T.; Morana, A.; Marin, E.; Ouerdane, Y.; Boukenter, A.; Girard, S. Regeneration of Fiber Bragg Gratings and Their Responses Under X-Rays. *IEEE Trans. Nucl. Sci.* **2021**, *68*, 1681–1687. [CrossRef]
82. Grobnic, D.; Henschel, H.; Hoeffgen, S.K.; Kuhnhenn, J.; Mihailov, S.J.; Weinand, U. Radiation Sensitivity of Bragg Gratings Written with Femtosecond IR Lasers. In *Fiber Optic Sensors and Applications VI*; Udd, E., Du, H.H., Wang, A., Eds.; SPIE: Bellingham, WA, USA, 2009.
83. Blanchet, T.; Desmarchelier, R.; Morana, A.; Boukenter, A.; Ouerdane, Y.; Marin, E.; Laffont, G.; Girard, S. Radiation and High Temperature Effects on Regenerated Fiber Bragg Grating. *J. Light. Technol.* **2019**, *37*, 4763–4769. [CrossRef]
84. Henschel, H.; Grobnic, D.; Hoeffgen, S.K.; Kuhnhenn, J.; Mihailov, S.J.; Weinand, U. Development of Highly Radiation Resistant Fiber Bragg Gratings. *IEEE Trans. Nucl. Sci.* **2011**, *58*, 2103–2110. [CrossRef]
85. Blanchet, T.; Desmarchelier, R.; Morana, A.; Laffont, G.; Marin, E.; Boukenter, A.; Ouerdane, Y.; Girard, S. Regenerated Fiber Bragg Gratings under High Temperature and Radiations. In Proceedings of the 26th Optical Fiber Sensors, Lausanne, Switzerland, 24–28 September 2018; Optical Society of America: Washington, DC, USA, 2018; p. WF65.
86. Morana, A.; Girard, S.; Marin, E.; Marcandella, C.; Paillet, P.; Périsse, J.; Macé, J.-R.; Boukenter, A.; Cannas, M.; Ouerdane, Y. Radiation Tolerant Fiber Bragg Gratings for High Temperature Monitoring at MGy Dose Levels. *Opt. Lett.* **2014**, *39*, 5313–5316. [CrossRef] [PubMed]
87. Blanchet, T.; Cotillard, R.; Morana, A.; Desmarchelier, R.; Marin, E.; Ouerdane, Y.; Boukenter, A.; Fourmentel, D.; Bréaud, S.; Gussarov, A.; et al. Effect of Radiation and Temperature on High Temperature Resistant Fiber Bragg Gratings. In Proceedings of the Optical Sensing and Detection VII, Strasbourg, France, 3–7 April 2022; SPIE: Bellingham, WA, USA, 2022; Volume 12139, pp. 305–309.
88. Blanchet, T.; Morana, A.; Laffont, G.; Cotillard, R.; Marin, E.; Boukenter, A.; Ouerdane, Y.; Girard, S. Radiation Effects on Type I Fiber Bragg Gratings: Influence of Recoating and Irradiation Conditions. *J. Light. Technol.* **2018**, *36*, 998–1004. [CrossRef]
89. Morana, A.; Girard, S.; Marin, E.; Lancry, M.; Grelin, J.; Marcandella, C.; Paillet, P.; Boukenter, A.; Ouerdane, Y. Dependence of the Voids-Fiber Bragg Grating Radiation Response on Temperature, Dose and Dose-Rate. *IEEE Trans. Nucl. Sci.* **2017**, *65*, 1619–1623. [CrossRef]
90. Griscom, D.L.; Gingerich, M.E.; Friebele, E.J. Model for the Dose, Dose-Rate and Temperature Dependence of Radiation-Induced Loss in Optical Fibers. *IEEE Trans. Nucl. Sci.* **1994**, *41*, 523–527. [CrossRef]
91. Griscom, D.L. Fractal Kinetics of Radiation-Induced Point-Defect Formation and Decay in Amorphous Insulators: Application to Color Centers in Silica-Based Optical Fibers. *Phys. Rev. B* **2001**, *64*, 174201. [CrossRef]
92. Gusarov, A.; Kinet, D.; Caucheteur, C.; Wuilpart, M.; Mégret, P. Gamma Radiation Induced Short-Wavelength Shift of the Bragg Peak in Type I Fiber Gratings. *IEEE Trans. Nucl. Sci.* **2010**, *57*, 3775–3778. [CrossRef]
93. Van Uffelen, M. Modélisation de systèmes d'acquisition et de transmission à fibres optiques destinés à fonctionner en environnement nucléaire. Ph.D. Thesis, Paris XI Orsay, Paris, France, 2001.
94. Fernandez Fernandez, A.; Brichard, B.; Berghmans, F.; Décreton, M. Dose-Rate Dependencies in Gamma-Irradiated Fiber Bragg Grating Filters. *IEEE Trans. Nucl. Sci.* **2002**, *49*, 2874–2878. [CrossRef]
95. Morana, A.; Girard, S.; Marin, E.; Périsse, J.; Genot, J.S.; Kuhnhenn, J.; Grelin, J.; Hutter, L.; Mélin, G.; Lablonde, L.; et al. Radiation-Hardened Fiber Bragg Grating Based Sensors for Harsh Environments. *IEEE Trans. Nucl. Sci.* **2017**, *64*, 68–73. [CrossRef]
96. Girard, S.; Marcandella, C.; Morana, A.; Perisse, J.; Di Francesca, D.; Paillet, P.; Mace, J.-R.; Boukenter, A.; Leon, M.; Gaillardin, M.; et al. Combined High Dose and Temperature Radiation Effects on Multimode Silica-Based Optical Fibers. *IEEE Trans. Nucl. Sci.* **2013**, *60*, 4305–4313. [CrossRef]
97. Morana, A.; Campanella, C.; Aubrey, M.; Marin, E.; Boukenter, A.; Ouerdane, Y.; Girard, S. Temperature Dependence of Low-Dose Radiation-Induced Attenuation of Germanium-Doped Optical Fiber at Infrared Wavelengths. *IEEE Trans. Nucl. Sci.* **2022**, *69*, 512–517. [CrossRef]
98. Morana, A.; Roche, M.; Campanella, C.; Mélin, G.; Robin, T.; Marin, E.; Boukenter, A.; Ouerdane, Y.; Girard, S. Temperature Dependence of Radiation-Induced Attenuation of a Fluorine-Doped Single-Mode Optical Fiber at InfraRed Wavelengths. *IEEE Trans. Nucl. Sci.* **2022**, *69*, 1515–1520. [CrossRef]
99. Morana, A.; Campanella, C.; Marin, E.; Mélin, G.; Robin, T.; Li Vecchi, G.; Di Francesca, D.; Boukenter, A.; Ouerdane, Y.; Mady, F.; et al. Operating Temperature Range of Phosphorous-Doped Optical Fiber Dosimeters Exploiting Infrared Radiation-Induced Attenuation. *IEEE Trans. Nucl. Sci.* **2021**, *68*, 906–912. [CrossRef]
100. Morana, A.; Girard, S.; Marin, E.; Lablonde, L.; Robin, T.; Lancry, M.; Boukenter, A.; Ouerdane, Y. Radiation-Response of Fiber Bragg Gratings at Low Temperatures. *IEEE Trans. Nucl. Sci.* **2020**, *67*, 1637–1642. [CrossRef]
101. Morana, A.; Girard, S.; Marin, E.; Trinczek, M.; Hoehr, C.; Blackmore, E.; Périsse, J.; Paillet, P.; Marcandella, C.; Duhamel, O.; et al. Radiation-Hardened Fiber Bragg Gratings for Space Missions. In Proceedings of the Australian Conference on Optical Fibre Technology, Sydeny, Australia, 5–8 September 2016; Optical Society of America: Washington, DC, USA, 2016; p. JT4A-25.
102. Taylor, E.W.; Hulick, K.E.; Battiato, J.M.; Sanchez, A.D.; Winter, J.E.; Pirich, A.R. Response of Germanium-Doped Fiber Bragg Gratings in Radiation Environments. In Proceedings of the Enabling Photonic Technologies for Aerospace Applications; International Society for Optics and Photonics, Orlando, FL, USA, 22–23 April 1999; Volume 3714, pp. 106–114.

103. Curras, E.; Virto, A.L.; Moya, D.; Vila, I.; Carrion, J.G.; Frovel, M.; Garcia-Lopez, J.; Jimenez, M.C.; Morilla, Y.; Palomo, F.R. Influence of the Fiber Coating Type on the Strain Response of Proton-Irradiated Fiber Bragg Gratings. *IEEE Trans. Nucl. Sci.* **2012**, *59*, 937–942. [CrossRef]
104. Blanchet, T.; Morana, A.; Allanche, T.; Sabatier, C.; Reghioua, I.; Marin, E.; Boukenter, A.; Ouerdane, Y.; Paillet, P.; Gaillardin, M.; et al. X-Ray, Proton, and Electron Radiation Effects on Type I Fiber Bragg Gratings. *IEEE Trans. Nucl. Sci.* **2018**, *65*, 1632–1638. [CrossRef]
105. Frövel, M.; Moya, D.; Vila, I.; Currás, E.; Virto, A.L.; Carrión, J.G.; Pintado, J.M.; Garcia-López, J.; Jiménez, M.C.; Morilla, Y.; et al. Proton Radiation Sensitivity of Fiber Bragg Gratings for Particle Colliders. *Struct. Health Monit.* **2013**, *2013*, 361–368.
106. Mitigation of Radiation-Induced Fiber Bragg Grating (FBG) Sensor Drifts in Intense Radiation Environments Based on Long-Short-Term Memory (LSTM) Network | IEEE Journals & Magazine | IEEE Xplore. Available online: https://ieeexplore.ieee.org/abstract/document/9598896 (accessed on 30 August 2022).
107. Girard, S.; Morana, A.; Hoehr, C.; Trinczek, M.; Vidalot, J.; Paillet, P.; Bélanger-Champagne, C.; Mekki, J.; Balcon, N.; Li Vecchi, G.; et al. Atmospheric Neutron Monitoring through Optical Fiber-Based Sensing. *Sensors* **2020**, *20*, 4510. [CrossRef] [PubMed]
108. Gusarov, A.; Chojetzki, C.; Mckenzie, I.; Thienpont, H.; Berghmans, F. Effect of the Fiber Coating on the Radiation Sensitivity of Type I FBGs. *IEEE Photonics Technol. Lett.* **2008**, *20*, 1802–1804. [CrossRef]
109. Gusarov, A.; Vasiliev, S.; Medvedkov, O.; Mckenzie, I.; Berghmans, F. Stabilization of Fiber Bragg Gratings against Gamma Radiation. In Proceedings of the 2007 9th European Conference on Radiation and Its Effects on Components and Systems, Deauville, France, 10–14 September 2007; pp. 1–8.
110. Lebel-Cormier, M.-A.; Boilard, T.; Bernier, M.; Beaulieu, L. Medical Range Radiation Dosimeter Based on Polymer-Embedded Fiber Bragg Gratings. *Sensors* **2021**, *21*, 8139. [CrossRef]
111. Kashyap, R. Fiber Bragg Gratings. In *Optics and Photonics*; Academic Press: Cambridge, MA, USA, 1999; ISBN 0-12-400560-8.
112. Morana, A.; Marin, E.; Girard, S.; Lablonde, L.; Pinsard, E.; Robin, T.; Boukenter, A.; Ouerdane, Y. Radiation Response of Distributed Feedback Bragg Gratings for Space Applications. *IEEE Trans. Nucl. Sci.* **2020**, *67*, 284–288. [CrossRef]
113. Xu, Y.; Lu, P.; Gao, S.; Xiang, D.; Lu, P.; Mihailov, S.; Bao, X. Optical Fiber Random Grating-Based Multiparameter Sensor. *Opt. Lett.* **2015**, *40*, 5514. [CrossRef] [PubMed]
114. Blanchet, T.; Morana, A.; Marin, E.; Ouerdane, Y.; Boukenter, A.; Hnatovsky, C.; Lu, P.; Mihailov, S.; Girard, S. Radiation Responses of Fiber Random Gratings. In Proceedings of the 2020 20th European Conference on Radiation and Its Effects on Components and Systems (RADECS), Online, 19–20 October 2020; pp. 1–4.
115. Jin, W.; Michie, W.C.; Thursby, G.; Konstantaki, M.; Culshaw, B. Simultaneous Measurement of Strain and Temperature: Error Analysis. *Opt. Eng.* **1997**, *36*, 598–609. [CrossRef]
116. Russell, P.S.J. Photonic-Crystal Fibers. *J. Light. Technol.* **2006**, *24*, 4729–4749. [CrossRef]
117. Geernaert, T.; Nasilowski, T.; Chah, K.; Szpulak, M.; Olszewski, J.; Statkiewicz, G.; Wojcik, J.; Poturaj, K.; Urbanczyk, W.; Becker, M.; et al. Fiber Bragg Gratings in Germanium-Doped Highly Birefringent Microstructured Optical Fibers. *IEEE Photonics Technol. Lett.* **2008**, *20*, 554–556. [CrossRef]
118. Berghmans, F.; Geernaert, T.; Sonnenfeld, C.; Sulejmani, S.; Luyckx, G.; Lammens, N.; Degrieck, J.; Chah, K.; Thienpont, H. Microstructured Optical Fiber Bragg Grating Sensors for Structural Health Monitoring Applications. In Proceedings of the EWSHM—7th European Workshop on Structural Health Monitoring, Nantes, France, 8–11 July 2014; p. 11.
119. Inao, S.; Sato, T.; Sentsui, S.; Kuroha, T.; Nishimura, Y. Multicore Optical Fiber. In Proceedings of the Optical Fiber Communication, Washington, DC, USA, 6 March 1979; Optica Publishing Group: Washington, DC, USA, 1979; p. WB1.
120. Saitoh, K.; Matsuo, S. Multicore Fiber Technology. *J. Light. Technol.* **2016**, *34*, 55–66. [CrossRef]
121. Zhao, Z.; Tang, M.; Lu, C. Distributed Multicore Fiber Sensors. *Opto-Electron. Adv.* **2020**, *3*, 190024-17. [CrossRef]
122. Zhao, Z.; Dang, Y.; Tang, M. Advances in Multicore Fiber Grating Sensors. *Photonics* **2022**, *9*, 381. [CrossRef]
123. Moore, J.P.; Rogge, M.D. Shape Sensing Using Multi-Core Fiber Optic Cable and Parametric Curve Solutions. *Opt. Express* **2012**, *20*, 2967–2973. [CrossRef] [PubMed]
124. Barrera, D.; Madrigal, J.; Delepine-Lesoille, S.; Sales, S. Multicore Optical Fiber Shape Sensors Suitable for Use under Gamma Radiation. *Opt. Express* **2019**, *27*, 29026–29033. [CrossRef] [PubMed]
125. SpringerLink. POF Handbook. Available online: https://link.springer.com/book/10.1007/978-3-540-76629-2 (accessed on 1 September 2022).
126. Theodosiou, A.; Lacraz, A.; Stassis, A.; Koutsides, C.; Komodromos, M.; Kalli, K. Plane-by-Plane Femtosecond Laser Inscription Method for Single-Peak Bragg Gratings in Multimode CYTOP Polymer Optical Fiber. *J. Light. Technol.* **2017**, *35*, 5404–5410. [CrossRef]
127. Broadway, C.; Min, R.; Leal-Junior, A.G.; Marques, C.; Caucheteur, C. Toward Commercial Polymer Fiber Bragg Grating Sensors: Review and Applications. *J. Light. Technol.* **2019**, *37*, 2605–2615. [CrossRef]
128. Hamdalla, T.A.; Nafee, S.S. Bragg Wavelength Shift for Irradiated Polymer Fiber Bragg Grating. *Opt. Laser Technol.* **2015**, *74*, 167–172. [CrossRef]

sensors

MDPI

Article

# Short-Term Creep Effect on Strain Transfer from Fiber-Reinforced Polymer Strips to Fiber Bragg Grating-Optical Fiber Sensors

**Hai Van Tran and Soo-Yeon Seo ***

Department of Architectural Engineering, Korea National University of Transportation, Chungju 27389, Republic of Korea
* Correspondence: syseo@ut.ac.kr

**Abstract:** In this study, the short-term creep effect (STCE) on strain transfer from fiber-reinforced polymer (FRP) strips to fiber Bragg grating-optical fiber (FBG-OF) sensors was investigated. Thirty OF sensors attached to FRP strips were investigated through three primary test parameters: bond length (40, 60, 80, 100, 120, and 150 mm); adhesive type (epoxy resin, CN adhesive, and epoxy resin combined with CN adhesive); and bonding method (embedded and external bonding methods). The strain transfer ability of the OF sensors was evaluated based on the strain ratio of the OF sensor to the FRP strip under different sustained stresses of 20, 40, 50, and 60% of the FRP ultimate tensile strength ($f_u$). From the test results, it was found that the debonding phenomenon occurred at the interface between the FBG-OF sensor and the adhesive and was clearly observed after applying a load for three days. It was also found that the CN adhesive showed better strain transfer compared to the other adhesive types. Regarding the OF sensors bonded by epoxy resin, in order to maintain strain transfer ability under a high level of sustained stress ($0.6f_u$), minimum bond lengths of 100 and 120 mm were required for the embedded and external bonding methods, respectively.

**Keywords:** fiber-reinforced polymer strip; optical fiber sensor; fiber Bragg grating; strain transfer; short-term creep effect



**Citation:** Tran, H.V.; Seo, S.-Y. Short-Term Creep Effect on Strain Transfer from Fiber-Reinforced Polymer Strips to Fiber Bragg Grating-Optical Fiber Sensors. *Sensors* **2023**, *23*, 1628. https://doi.org/10.3390/s23031628

Academic Editor: Mario Iodice

Received: 26 December 2022
Revised: 13 January 2023
Accepted: 31 January 2023
Published: 2 February 2023



## 1. Introduction

At present, fiber-reinforced polymer (FRP) composite is well known as an effective retrofitting material for reinforced concrete members due to its advantages, such as lightweight, good corrosion resistance, compatibility with concrete, and high strength-to-weight ratio [1]. Among installing techniques, the most common method, known as the externally bonded (EB) method, has been applied by attaching FRP sheets or plates to concrete surfaces with epoxy resin. In addition, a novel technique is called the near surface-mounted (NSM) method, in which FRP bars or strips are embedded into grooves on the concrete cover and filled with epoxy resin afterward [2,3]. Compared to the EB method, the NSM method with a full bond length along the FRP composite can more effectively improve the strength capacity of a reinforced concrete (RC) member when it is used for flexural strengthening. Moreover, the deformation capacity can be increased by designing to have a partially debonded region in the NSM method [4,5].

Normally, in retrofitting with FRP composite, epoxy resin is used for bonding it to concrete surfaces because epoxy resin has good stress transfer ability and the capacity for concentrated stress redistribution [6]. However, the strength of epoxy adhesive can be decreased because of construction mistakes and harsh environmental conditions, especially temperature. Therefore, a series of studies has been performed on smart composites in which FRP materials were combined with sensors. By retrofitting these smart composites, the performance of the FRP material and the health condition of the structural elements were monitored by means of strain variation of FRP materials, detection of new cracks, and expansion of cracks [7,8].

Recently, most smart composites have been fabricated with OF sensors based on fiber Bragg grating (FBG) technology [9]. Bragg grating, discovered in 1978 by Ken Hill [10], is based on the optical principle of "total internal reflection" in order to confine and transmit light in the core. Later, the industrial photo-inscription technique for fiber Bragg grating sensors was demonstrated in 1989 by Meltz et al. [11]. FBG sensors have been used mainly in the aeronautical industry due to their small sizes with fast response and immunity to electromagnetic interference. Sensing using OF sensors has attracted the interest of many researchers in the development of structural health monitoring (SHM) systems due to their useful inherent advantages, such as high durability, stability in long-term measurement, and the possibility of making multiple sensors along with a single optical fiber [12,13]. A review article dedicated to the research and development activities of FBG sensors for structural health monitoring was published by Majumder et al. [14].

Thus far, many studies have been performed to evaluate OF sensors for structural health monitoring and assessment of reinforced concrete members. Crack-opening displacements in concrete can be detected and measured using a fiber-optic laser speckle-intensity sensor that was developed by Zhang and Ansari [15,16]. Gu et al. [17] adopted OF sensors for concrete structures in order to establish crack detection methodologies based on monitored performance. Zhao et al. [18] investigated the debonding mechanism of FRP systems with concrete surfaces by OF sensors embedded at their interface. Furthermore, OF sensors have also been combined with FRP materials to fabricate a smart composite for retrofitting RC members. Siwowski et al. [19] applied distributed fiber optic sensors (DFOS) in FRP composites for bridge monitoring. Wood et al. [20] adopted the DFOS system substituted for electrical strain gauges to measure the strain distribution in FRP sheets used to retrofit shear wall structures and indicated that the DFOS system can be used to measure two-dimensional spatial strain with high precision. Wang et al. [21] validated the sensing capacity of OF sensors embedded in FRP bars using the tensile, bond, and beam flexural tests.

In order to reduce the fabrication period and simplify the fabrication procedure, Seo et al. [22,23] performed studies to suggest a minimum bond length between OF sensors and FRP materials through the tensile test of FBG-FRP composite using an analytical approach. However, in these studies, the creep effect on the epoxy resin used as an adhesive between the FBG sensor and the FRP material, which may affect strain transfer, was not considered. According to Tam et al. [24], the strength of epoxy resin can be significantly decreased under highly sustained stresses within a short period of time. So far, few studies have investigated the creep effect on strain transfer from an FRP strip to an OF sensor, especially OF sensors bonded with limited bond length. Therefore, more studies related to the creep effect on the shear transfer of FBG-FRP composite are needed.

The primary objective of this study was to experimentally assess the short-term creep effect of different stress levels on the shear transfer of FBG-FRP composites. The FBG-FRP specimens were fabricated and investigated through three different parameters, namely bond length, bonding method, and adhesive type. In addition, strain values from the FBG sensors were validated with those from electrical strain gauge (ESG) sensors attached at the middle position of the bond length.

## 2. Experimental Program

### 2.1. Materials

In this study, carbon fiber-reinforced polymer strips (SK-CPS-0512) supplied by the SK Group (Seoul, Republic of Korea) were utilized to manufacture FBG-FRP specimens. Three tensile specimens of FRP strips with the same cross-section dimensions as the FRP strips used for the FBG-FRP composite specimens (a thickness of 1.2 mm and a width of 15 mm) were fabricated and tested in tension in accordance with ASTM D3039 [25]. Figure 1 illustrates the tensile test setup of the FRP strip. The stress–strain relationship and mechanical properties are presented in Figure 2 and Table 1, respectively.

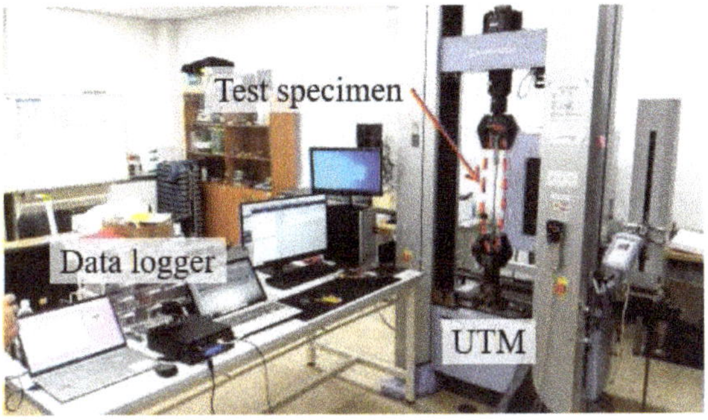


**Figure 1.** Tensile test set-up for the FRP strip.

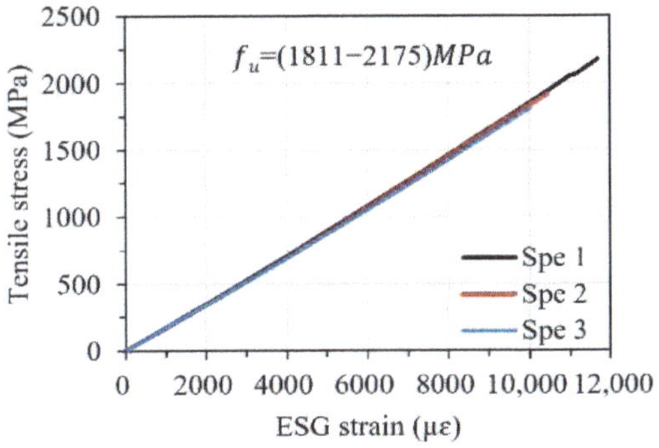


**Figure 2.** Stress–strain relationship of the FRP strip.

**Table 1.** Dimension and mechanical properties of the FRP strip.

| Type | Thickness (mm) | Width (mm) | Tensile Strength (MPa) | Elastic Module (MPa) |
|---|---|---|---|---|
| SK-CPS-0512 (CFRP strip) | 1.2 | 15 | 1813 | 183,000 |

The same type of epoxy resin (SK-CPA10, SK Chemicals Co., Gyeonggi-do, Republic of Korea) used in previous studies by Seo et al. [22,23] was applied as an adhesive. The epoxy resin was mixed with a hardener in a 2:1 ratio in accordance with the manufacturer's guidelines. The mechanical properties of the epoxy resin provided by the manufacturer are presented in Table 2. The epoxy resin was employed for bonding FBG sensors to the surface of the FRP strips and the two overlapped FRP strips where OF sensors were embedded between the two FRP strips. In addition, cyanoacrylate (CN) adhesive (Tokyo Measuring Instruments Laboratory Co., Tokyo, Japan) was used to bond the FBG sensors to the FRP strips. The CN adhesive is considered an effective adhesive to bond strain gauges with a short curing time of 20–60 s. The attachment of the FBG sensors to the surface of the FRP strips can be accomplished in a short period of time. According to the OF sensor manufacturer, CN adhesive can be applied to bond OF sensors on metallic and plastic surfaces. In a study by Motwani et al. [26], CN adhesive even showed greater effectiveness than epoxy resin when bonding OF sensors to carbon fiber-reinforced polyphenylene sulphide.

To fabricate the FBG-FRP specimens, a fiber Bragg grating-optical fiber sensor (Corning SMF-28, FBG Inc., Seoul, Republic of Korea) with a 10 mm sensing part and diameter of 9 μm was utilized (see Figure 3). The glass core sensing part was covered with glass cladding with a diameter of 125 μm and then coated with primary and secondary coating layers with diameters of 187.5 and 250 μm, respectively (see Figure 4). The primary and secondary coating layers were made of polymethyl methacrylate (PMMA) material. Finally, the entire length of the FBG sensor was covered with a jacket layer to avoid any damage due to the brittle characteristics of the glass core part. The coating surface was directly bonded to the epoxy resin after removing the jacket layer.

**Table 2.** Mechanical properties of epoxy resin.

| Type | Compressive Strength (MPa) | Shear Bond Strength (MPa) | Bond Strength to Concrete (MPa) |
|---|---|---|---|
| SK-CPA10 | 90 | 10 | 1.5 |

Data were provided by the manufacturer.

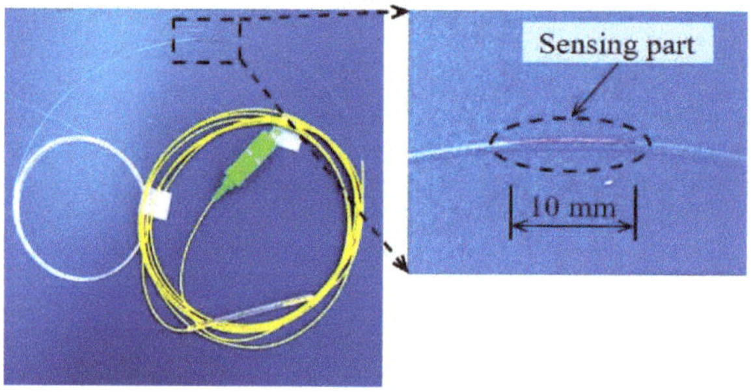


**Figure 3.** A picture of the FBG-optical fiber sensor.

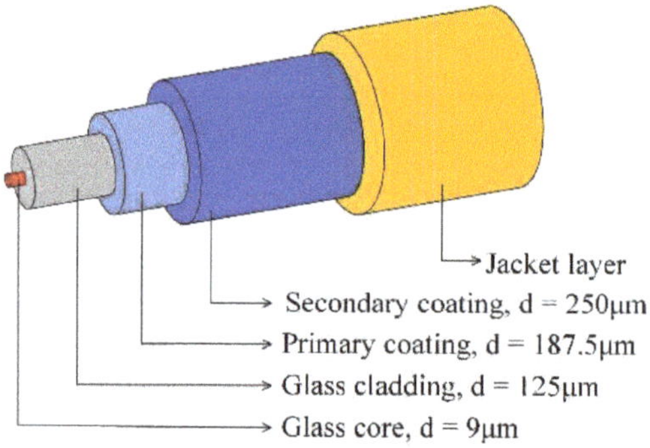


**Figure 4.** Configuration of the FBG-optical fiber sensor.

*2.2. Test Parameters*

To investigate the short-term creep effect on strain transfer from the FRP strips to the OF sensors, 30 OF sensors in total bonded to FRP strips were investigated through three parameters (i.e., bond length, bonding method, and adhesive type). In detail, bond lengths of 40, 60, 80, 100, 120, and 150 mm, two bonding methods of embedded and external bonding, and three types of adhesive (epoxy resin, CN adhesive, and epoxy resin combined with CN adhesive) were studied. Table 3 shows the test parameters of the OF sensors bonded to the FRP strip of each specimen. The "X-Y-Z" format was adopted as abbreviated names to classify the 30 OF sensors: "X" denotes the adhesive type, "Y" denotes the bonding method, and "Z" denotes the bond length. For example, E-EM-80 indicates that the OF sensor was bonded with epoxy resin adhesive following the embedded method with a bond length of 80 mm. It should be noted that the OF sensors of the E-EBM-80b and EC-EBM-120 cases were damaged during fabrication; therefore, no data were collected from these cases.

**Table 3.** List of parameters of FBG-optical fiber sensors bonded on FRP strips.

| Specimen Name | Name of FBG-OF Sensor | Bond Methods | Adhesive Types | Bond Length (mm) |
|---|---|---|---|---|
| E specimen | E-EBM-80a | External bond | Epoxy resin | 80 |
| | E-EBM-80b [(1)] | | | 80 |
| | E-EBM-100a | | | 100 |
| | E-EBM-100b | | | 100 |
| | E-EBM-150a | | | 150 |
| | E-EBM-150b | | | 150 |
| | E-EM-80 | Embedded bond | | 80 |
| | E-EM-100 | | | 100 |
| | E-EM-150 | | | 150 |
| EC specimen | EC-EBM-80a | External bond | Epoxy resin and CN adhesive | 80 |
| | EC-EBM-80b | | | 80 |
| | EC-EBM-100a | | | 100 |
| | EC-EBM-100b | | | 100 |
| | EC-EBM-150a | | | 150 |
| | EC-EBM-150b | | | 150 |
| | EC-EM-80 | Embedded bond | | 80 |
| | EC-EM-100 | | | 100 |
| | EC-EM-150 | | | 150 |
| C specimen | C-EBM-40a | External bond | CN adhesive | 40 |
| | C-EBM-40b | | | 40 |
| | C-EBM-60a | | | 60 |
| | C-EBM-60b | | | 60 |
| | C-EBM-80a | | | 80 |
| | C-EBM-80b | | | 80 |
| | C-EM-40 | Embedded bond | | 40 |
| | C-EM-60 | | | 60 |
| | C-EM-80 | | | 80 |
| A specimen | E-EBM-120 | External bond | Epoxy resin | 120 |
| | EC-EBM-120 [(1)] | | Epoxy resin and CN adhesive | 120 |
| | C-EBM-100 | | CN adhesive | 100 |

[(1)] The FBG-OF sensor was broken during fabrication.

### 2.3. Fabrication of FBG-FRP Specimens

Figure 5 shows the geometrical details of the E specimen. The E specimen was made by combing two FRP strips with widths of 15 mm, thicknesses of 1.2 mm, and lengths of 1800 mm. Between the two FRP strips bonded with epoxy resin, OF sensors were attached at three positions with different bond lengths using the embedded bonding method. In addition, OF sensors were attached to both surfaces using the external bonding method with different bond lengths. The EC and C specimens had similar geometrical details to the E specimens; details of the OF sensor locations can be seen in Figures 6 and 7, which illustrate the FBG-FRP segments for the EC and C specimens, respectively. To obtain sufficient data to determine the minimum bonding length between the OF sensor and the FRP strip, an additional specimen, A, was prepared. The A specimen was fabricated with the same overall geometrical dimensions as the other specimens except for the adhesive and bond length for externally bonded OF sensors. In detail, three types of adhesive were considered for the externally bonded OF sensors: epoxy resin only, epoxy resin and CN adhesive combined, and CN adhesive only. The bond lengths of the OF sensors were 120 mm for the first and second cases and 100 mm for the third case.

Figure 8 shows the fabricating process of the FBG-FRP specimens in the laboratory. The surfaces of each FRP strip were roughened using emery paper to increase the adhesion between substrate and adhesive. Next, they were cleaned with acetone solution before attaching the OF sensors. The OF sensors were fixed using adhesive tape after applying a small amount of pressure, as shown in Figure 8a. This small pressure corresponding to 5% of the strain capacity of the OF sensor was applied in order to keep the OF sensor straight during the fabrication process. The magnitude of the applied prestress can be controlled through the shift wavelength variation of the OF sensor, which was observed on the HYPERION SI-155 optical sensing instrument (see Figure 8b). Then, the FRP strip attached with the OF sensors was placed in a mold and filled with epoxy resin to join it with the other FRP strip (see Figure 8c). Regarding the embedded OF sensors, the adhesive tapes were kept when filling with epoxy resin to achieve the desired bond length. The specimen was demolded after curing for 24 h in order to attach other OF sensors to the surface of the FRP strips.

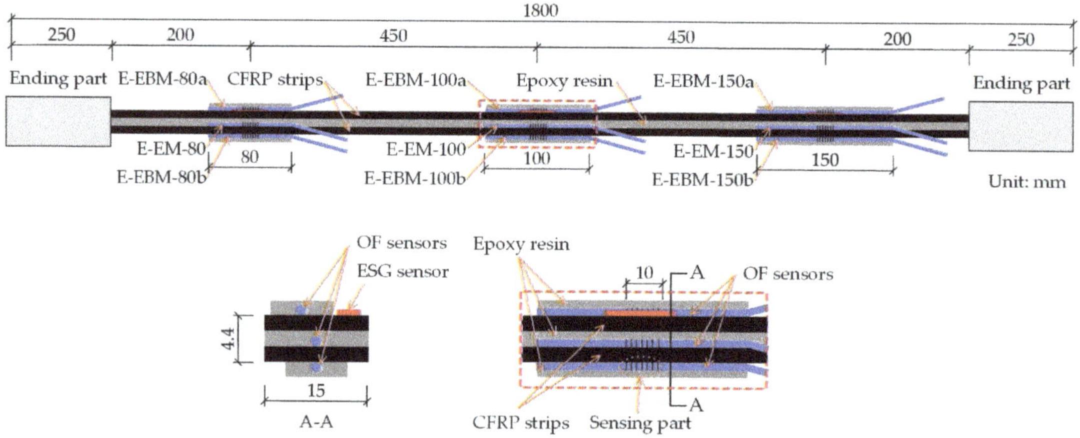


**Figure 5.** Geometrical details of the E specimen.

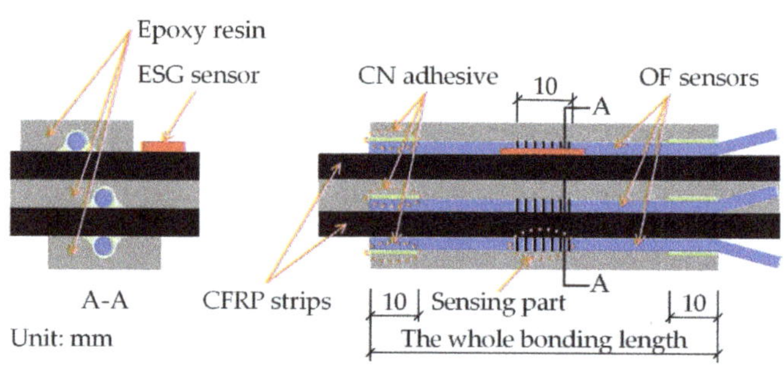


**Figure 6.** Illustration of the FBG-FRP segment of the EC specimen.

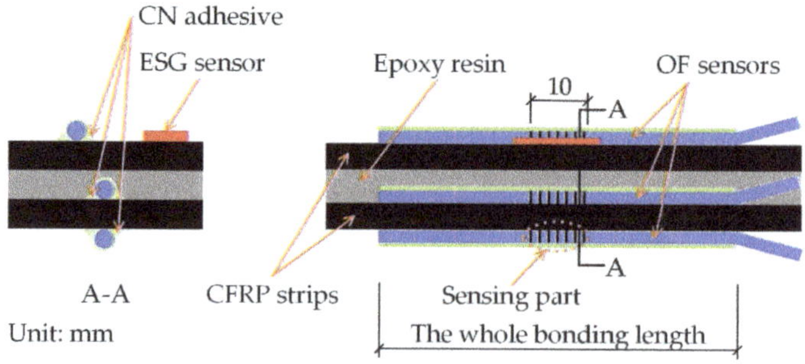


**Figure 7.** Illustration of the FBG-FRP segment of the C specimen.

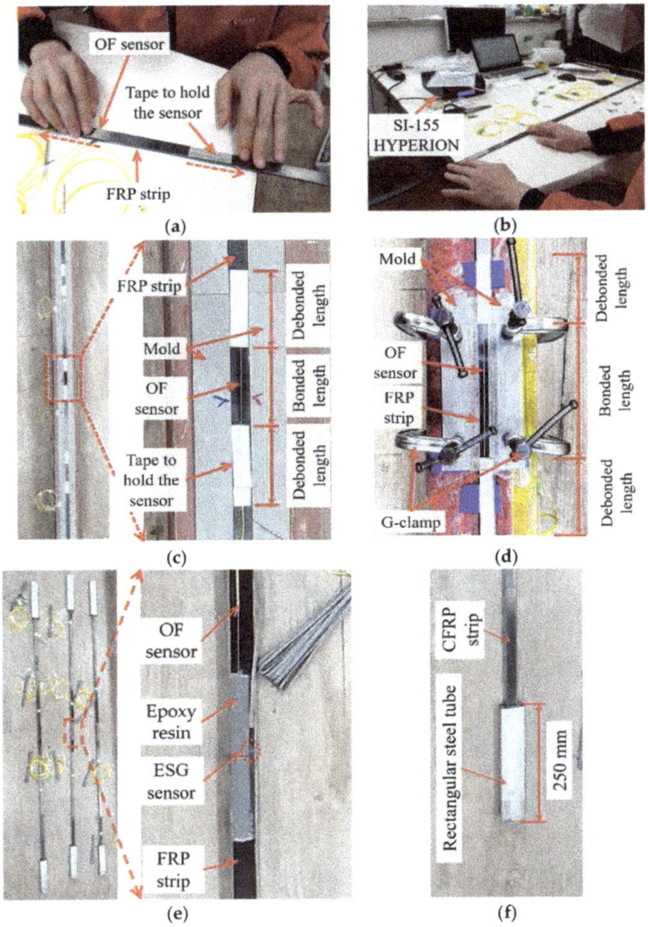


**Figure 8.** Fabrication process of FBG-FRP composite: (**a**) Applying a small prestress on the FBG-OF sensor; (**b**) Controlling the small prestress; (**c**) Embedded bonding method; (**d**) External bonding method; (**e**) FBG-FRP test specimens; (**f**) Ending part.

After embedding three OF sensors between two FRP strips, six OF sensors were externally bonded to the front and back surfaces of the FRP strips (see Figure 8d). The externally bonded OF sensors were fabricated by following the same procedure as the embedded OF sensors. Then, to evaluate strain transfer from the FRP strip to the OF sensors, ESG sensors were attached at the middle position of the bond length (see Figure 8e). To ensure the complete transfer of tensile stresses, the ending parts of the test specimens were bonded into rectangular steel tubes using epoxy, as shown in Figure 8f. The length of the FRP strip embedded inside the rectangular steel tube was calculated in order to prevent debonding, which can happen before fracture failure of the FRP strip. Finally, the FBG-FRP specimens were cured for 7 days under temperature conditions of 23 ± 3 °C in accordance with ASTM 3039 [25] before performing the experiment. Regarding the EC specimen, after using adhesive tape to maintain the small pressure, CN adhesive was used to bond the OF sensors to the FRP strip with a bond length of 10 mm at each ending part. The remaining part of the entire bond length was bonded using epoxy resin (see Figure 6).

*2.4. Short-Term Creep Test*

The setup schema of the short-term tensile test is shown in Figure 9a. One ending part of the test specimen was fixed by the steel frame, while the other was subjected to sustained tensile loads using oil jacks (TECPOS TDC-2015) with a capacity of 200 kN. The applied load was controlled using a load cell (CAS LS-20B) with a capacity of 200 kN. Meanwhile, the SI-155 HYPERION optical sensing instrument and the TDS-530 data logger were used to record the data from the FBG and ESG sensors, respectively (see Figure 9b).

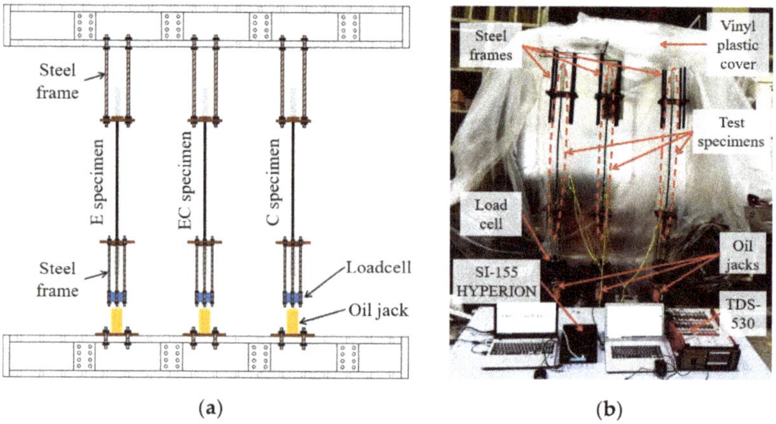


**Figure 9.** The short-term tensile test setup: (**a**) Setup schema; (**b**) Picture of the test setup.

According to ACI 440 [27], the total applied stress for design purposes is limited to approximately 55% of the ultimate tensile strength in CFRP in order to prevent creep rupture failure. At present, a limited number of experimental results have been reported regarding the minimum bond lengths needed to maintain the strain transfer ability from FRP strips to OF sensors corresponding to differently sustained stresses. Therefore, test specimens were subjected to four different stress levels of 20, 40, 50, and 60%, corresponding to the ultimate tensile strength ($f_u$) of the FRP strip. In the progression of stress, a stress level of $0.2f_u$ was maintained for the first 4 weeks, and then the next increments were maintained for 1 week each. Figure 10 illustrates the loading history applied to the test specimens during the short-term creep test. The applied load was checked and controlled every day to maintain the target load before recording the data. The vinyl plastic cover was used to create a small and enclosed space around the testing area to control temperature and humidity (see Figure 9b). During the test, the temperature and humidity conditions were maintained within the ranges of 23 ± 3 °C and 50 ± 10%, respectively, in accordance with ASTM 3039 [25].

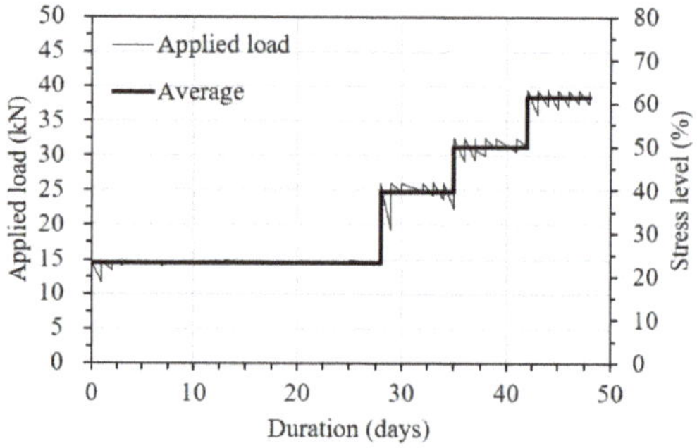


**Figure 10.** Loading history for the short-term creep test.

## 3. Test Results

### 3.1. Creep Effect on the FRP Strip

Figure 11 shows the stress–strain curves of the FRP strips under the short-term creep test. The dotted lines in the figures are for comparison, and the stress–strain curves of the FRP strips obtained from a simple tensile test are shown in Figure 2. The solid line shows the relationship between stress and strain of the FRP strip subjected to loads with a duration of 28 days for $0.2f_u$ and 7 days for each subsequent increments. The stresses were calculated by dividing the applied load on the FRP strip by the cross-sectional area of the FRP strip. It should be noted that the displayed strains are the average values of the three strain gauges attached to each FRP strip.

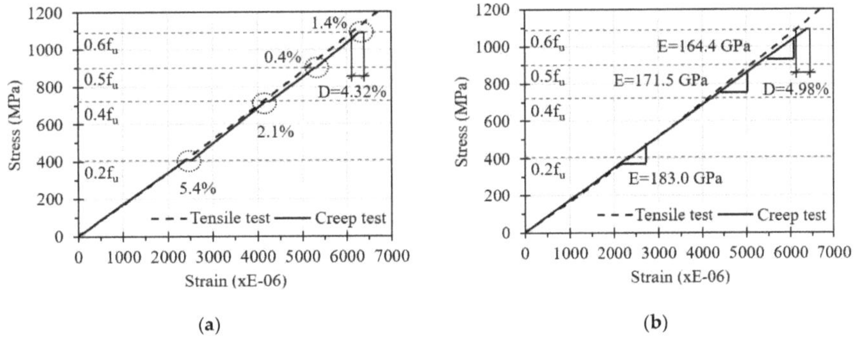


**Figure 11.** Creep effect on the FRP strip: (**a**) The E specimen; (**b**) The C specimen.

From the graphs, it can be seen that the FRP strip was affected by sustained stress. In detail, the strain on the FRP strip in the E specimen increased by 5.4% due to the creep effect after applying a sustained stress of $0.2f_u$ for 28 days (see Figure 11a). In addition, strain increases due to creep effects were observed after loading for 7 days each at $0.4f_u$ and $0.6f_u$. The additional creep strain generated from the sustained stress at $0.4f_u$ became plastic deformation. Compared with the stress–strain curve of the simple tensile test, the accumulated additional plastic strain due to creep at $0.6f_u$ was 4.32% of the total strain. This implies that the FRP strip may show additional creep strain, even under low sustained stress, and this can be a plastic strain. Similar results were observed on the FRP strip in the EC specimen.

The FRP strip in the C specimen did not show a clear increase in strain under sustained low stresses (see Figure 11b). However, the elastic modulus of the FRP strip decreased gradually under high sustained stress levels. In detail, the elastic modulus of 183 GPa was maintained after applying sustained stress up to $0.4f_u$. However, it decreased to 171.5 and 164.4 GPa after applying stresses of $0.5f_u$ and $0.6f_u$ for 7 days, respectively. As shown in Figures 5–7, the locations where the ESG sensors were attached were right next to the adhesives, such as epoxy or CN, to install the OF. Therefore, there is a possibility that the adhesives may have contributed to the difference in creep effect observed on these FRP strips. When the CN adhesive was applied, the creep strain of the FRP strip due to a sustained load at a low stress of $0.2f_u$ was not remarkable. At a high stress of $0.6f_u$, however, the additionally increased plastic strain was 4.98%, which is compatible with that of the E specimen with epoxy resin as the adhesive. From this, it is recommended to consider an additional strain increase of at least 5% due to creep effect when the FRP strip is exposed to sustained tensile stress conditions.

Table 4 summarizes the creep compliances of the FRP strip, calculated according to:

$$J_c(t) = \frac{\varepsilon(t)}{\sigma_0} \tag{1}$$

where $J_c(t)$ and $\varepsilon(t)$ are the creep compliances and strain at time (*t*), respectively, and $\sigma_0$ is the applied stress. The applied stresses were calculated by dividing the applied load by the cross-sectional area of the FRP strip. Figure 12a presents the average creep compliances of the FRP strip in the E specimen under different stress levels. The average creep compliance showed a gradual increase over time at $0.2f_u$. After increasing the stress to $0.4f_u$, $0.5f_u$, and $0.6f_u$, it was found that changes in the average creep compliances over time were not significant. It can be seen that a maximum difference of 5.8% was observed between the average creep compliances at $0.2f_u$ after applying the load for 7 days compared to the others. This can be explained by the significantly higher creep strain in the early stage of duration. Meanwhile, no differences between average creep compliances under different stress levels were observed in the C specimen, as shown in Figure 12b. This can be attributed to the strain on the FRP strip due to the creep effect not being displayed clearly in the C specimen (see Figure 11b).

**Table 4.** Creep compliance of the FRP strip.

| Applied Stress | Specimen Name | Creep Compliances ($10^{-6} \times MPa^{-1}$) | | | | | | | | COV (%) | | $J_{c,end}/J_{c,start}$ (%) |
|---|---|---|---|---|---|---|---|---|---|---|---|---|
| | | ESG-1 | | ESG-2 | | ESG-3 | | Average | | | | |
| | | Start [(1)] | End [(2)] | Start | End | Start | End | Start | End | Start | End | |
| $0.2f_u$ | E specimen | 5.85 | 6.48 | 5.70 | 5.84 | 6.08 | 6.26 | 5.88 | 6.19 | 3.27 | 5.26 | 105.27 |
| | EC specimen | 5.73 | 6.22 | 5.88 | 6.05 | 5.82 | 6.07 | 5.81 | 6.11 | 1.29 | 1.50 | 105.16 |
| | C specimen | 5.82 | 6.33 | 5.65 | 5.81 | 5.73 | 5.91 | 5.73 | 6.02 | 1.46 | 4.59 | 105.06 |
| $0.4f_u$ | E specimen | 5.87 | 5.97 | 5.68 | 5.77 | 5.84 | 6.01 | 5.80 | 5.92 | 1.77 | 2.18 | 102.07 |
| | EC specimen | 5.88 | 6.08 | 5.92 | 6.02 | 5.74 | 5.89 | 5.84 | 6.00 | 1.65 | 1.58 | 102.74 |
| | C specimen | 5.79 | 5.95 | 5.77 | 5.80 | 5.72 | 5.94 | 5.76 | 5.90 | 0.67 | 1.38 | 102.43 |
| $0.5f_u$ | E specimen | 5.91 | 5.96 | 5.72 | 5.75 | 5.97 | 5.97 | 5.87 | 5.89 | 2.27 | 2.17 | 100.34 |
| | EC specimen | 5.96 | 5.99 | 5.93 | 5.95 | 5.84 | 5.91 | 5.91 | 5.95 | 1.05 | 0.71 | 100.68 |
| | C specimen | 5.79 | 5.86 | 5.78 | 5.91 | 5.71 | 5.82 | 5.76 | 5.86 | 0.76 | 0.77 | 101.74 |
| $0.6f_u$ | E specimen | 5.86 | 5.87 | 5.59 | 5.74 | 5.85 | 5.94 | 5.77 | 5.85 | 2.66 | 1.74 | 101.39 |
| | EC specimen | 5.91 | 5.94 | 5.77 | 5.92 | 5.83 | 5.95 | 5.84 | 5.93 | 1.23 | 0.29 | 101.54 |
| | C specimen | 5.83 | 6.09 | 5.84 | 5.91 | 5.87 | 5.90 | 5.85 | 5.97 | 0.41 | 1.80 | 102.05 |

(1) Instantaneous creep compliance after applying load; (2) Creep compliance value before increasing load.

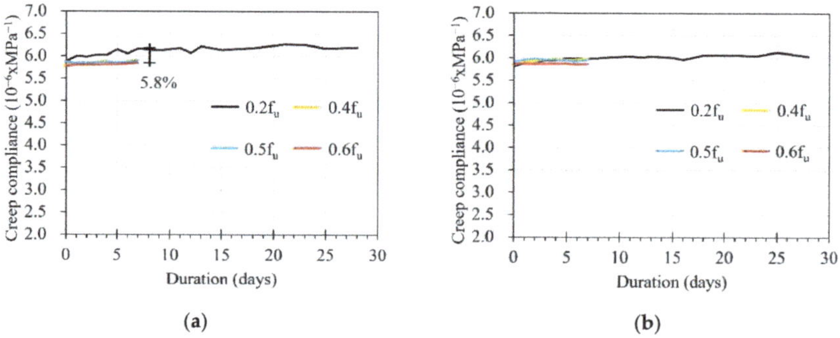


**Figure 12.** Creep compliance of the FRP strip: (**a**) The E specimen; (**b**) The C specimen.

*3.2. Short-Term Creep Effect on the Sensing Capacity of the OF Sensor*

The strain was obtained by converting the reflected wavelength of the OF sensors measured by the SI-155 HYPERION optical sensing instrument, using Equation (2).

$$\Delta\varepsilon = \frac{1}{1-P_e}\left[\frac{\Delta\lambda_B}{\lambda_B} - (\alpha+\xi)\Delta T\right] \quad (2)$$

where $P_e = 0.22$ is the photoelastic modulus,$\lambda_B$ is the reflected wave length, $\alpha = 0.55 \times 10^{-6}$ is the expansion coefficient corresponding to temperature, and $\xi = 8.6 \times 10^{-6}$ is the coefficient considering the change of reflection corresponding to temperature [28]. The transferred strain percentages (TSPs) from the FRP strip to the FBG-OF sensors are summarized in Table 5. The TSP values at the start and end points of each duration were calculated at the moments after applying the load and before increasing to a higher load.

**Table 5.** Transferred strain percentages from the FRP strip to the FBG-optical fiber sensors.

| Specimen Name | Name of FBG-OF Sensor | Transferred Strain Percentage at Stress Level of (%) | | | | | | | |
|---|---|---|---|---|---|---|---|---|---|
| | | $0.2f_u$ | | $0.4f_u$ | | $0.5f_u$ | | $0.6f_u$ | |
| | | Start (1) | End (2) | Start | End | Start | End | Start | End |
| E specimen | E-EBM-80a | 99 | 86 | 83 | 55 | 61 | 44 | 51 | 37 |
| | E-EBM-80b (3) | - | - | - | - | - | - | - | - |
| | E-EBM-100a | 99 | 100 | 97 | 98 | 98 | 98 | 99 | 98 |
| | E-EBM-100b | 98 | 99 | 96 | 96 | 96 | 94 | 95 | 80 |
| | E-EBM-150a | 99 | 99 | 98 | 97 | 96 | 97 | 96 | 97 |
| | E-EBM-150b | 98 | 98 | 97 | 97 | 97 | 97 | 97 | 98 |
| | E-EM-80 | 100 | 94 | 98 | 95 | 96 | 95 | 96 | 3 |
| | E-EM-100 | 99 | 99 | 98 | 97 | 98 | 97 | 99 | 98 |
| | E-EM-150 | 98 | 99 | 98 | 98 | 97 | 97 | 97 | 97 |
| EC specimen | EC-EBM-80a | 99 | 96 | 92 | 47 | 55 | 37 | 42 | 26 |
| | EC-EBM-80b | 98 | 95 | 90 | 87 | 89 | 86 | 86 | 78 |
| | EC-EBM-100a | 98 | 96 | 94 | 91 | 92 | 86 | 86 | 85 |
| | EC-EBM-100b | 99 | 99 | 94 | 94 | 94 | 94 | 94 | 93 |
| | EC-EBM-150a | 99 | 100 | 100 | 99 | 99 | 99 | 97 | 98 |
| | EC-EBM-150b | 100 | 100 | 99 | 99 | 98 | 97 | 97 | 97 |
| | EC-EM-80 | 100 | 99 | 95 | 90 | 91 | 84 | 84 | 76 |
| | EC-EM-100 | 99 | 100 | 95 | 95 | 95 | 90 | 90 | 88 |
| | EC-EM-150 | 100 | 100 | 99 | 98 | 98 | 96 | 98 | 0 |
| C specimen | C-EBM-40a | 97 | 69 | 69 | 58 | 60 | 53 | 57 | 45 |
| | C-EBM-40b | 97 | 79 | 80 | 78 | 81 | 74 | 62 | 51 |
| | C-EBM-60a | 99 | 96 | 96 | 92 | 93 | 92 | 91 | 90 |
| | C-EBM-60b | 99 | 93 | 94 | 91 | 91 | 91 | 90 | 90 |
| | C-EBM-80a | 98 | 96 | 96 | 93 | 93 | 93 | 92 | 55 |
| | C-EBM-80b | 99 | 96 | 97 | 93 | 94 | 94 | 94 | 94 |
| | C-EM-40 | 97 | 60 | 60 | 56 | 61 | 44 | 50 | 36 |
| | C-EM-60 | 100 | 99 | 100 | 96 | 97 | 93 | 92 | 91 |
| | C-EM-80 | 100 | 98 | 99 | 96 | 96 | 96 | 96 | 96 |
| A specimen | E-EBM-120 | 98 | 99 | 98 | 98 | 97 | 97 | 98 | 98 |
| | EC-EBM-120 (3) | - | - | - | - | - | - | - | - |
| | C-EBM-100 | 100 | 98 | 99 | 97 | 98 | 96 | 98 | 97 |

(1) Instantaneous TSP after applying load; (2) TSP value before increasing load; (3) FBG-OF sensor was broken during preparation process.

### 3.2.1. Influence of Bonding Methods

Figures 13–15 show the variation in TSPs from the FRP strip to the OF sensors under different stress levels. It can be seen that the bonding methods, namely the embedded method (EM) and the externally bonded method (EBM), affected the strain transfer from the FRP strip to the OF sensors. E-EM-80 showed an overall constant TSP after a small decrement until debonding occurred, whereas E-EBM-80a showed a gradual decreasing pattern over time (see Figure 13a). In detail, the instantaneous TSP values of E-EBM-80a and E-EM-80 were 99% and 100% after applying $0.2f_u$ and then slightly decreased to 86% and 94%, respectively. However, the TSP of E-EBM-80a decreased more rapidly at a stress level of $0.4f_u$, and this decreasing pattern continued corresponding to increasing time. Conversely, E-EM-80 maintained a stable TSP of approximately 94% after the stress level was increased to $0.4f_u$. Even after the stress level was increased to $0.6f_u$, E-EM-80 did not show any decrement in TSP, but the TSP momentarily became "0" on the fourth day. From Figure 13b,c, it can be seen that when the bond length of the OF sensor by the embedded method was 100 mm or more, the strain could be transferred to the OF sensor very effectively for a loading period of 50 days in total. On the contrary, in the case of the external bonding method, it can be seen that the transmission rate of the strain started to decrease at a stress level of $0.4f_u$ at a bond length of 100 mm, and in the case of 150 mm, it was transmitted very effectively throughout the entire period.

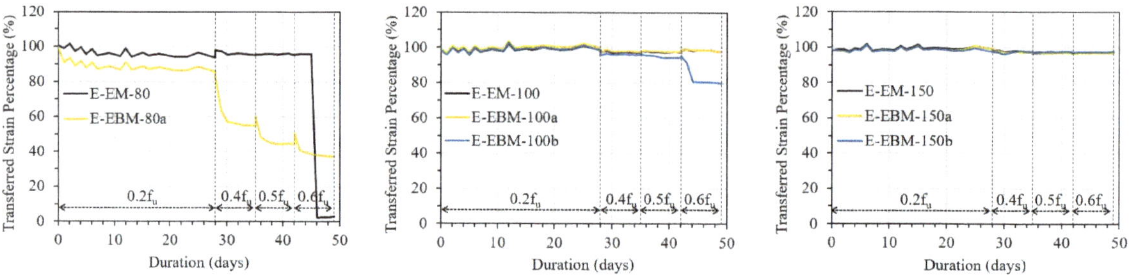


(a) (b) (c)

**Figure 13.** Short-term creep effect on strain transfer of the E specimen with different bonding methods: (**a**) L = 80 mm; (**b**) L = 100 mm; and (**c**) L = 150 mm.

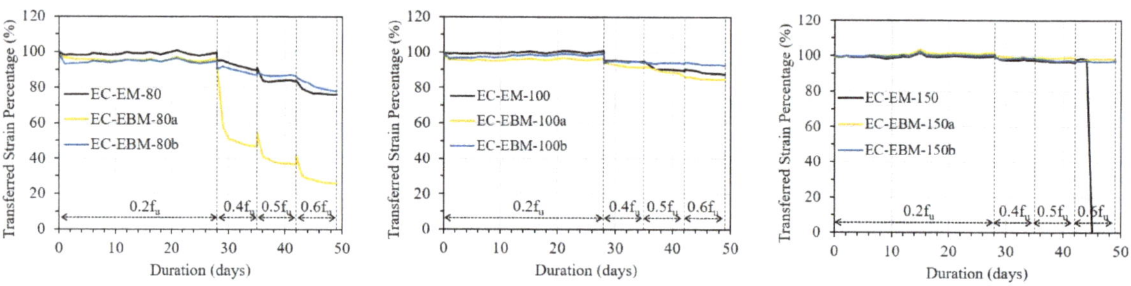


(a) (b) (c)

**Figure 14.** Short-term creep effect on strain transfer in the EC specimen with different bonding methods: (**a**) L = 80 mm; (**b**) L = 100 mm; and (**c**) L = 150 mm.

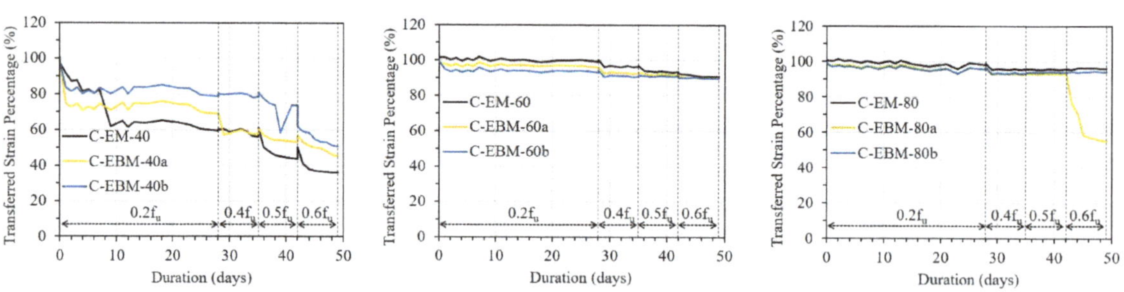


(a) (b) (c)

**Figure 15.** Short-term creep effect on strain transfer of the C specimen with different bonding methods: (**a**) L = 40 mm; (**b**) L = 60 mm; and (**c**) L = 80 mm.

In Figure 14, the TSPs of the EM and EBM in the EC specimen are compared. There was a small drop in TSP in the case of the EBM with a bond length of 80 mm 1 day after applying a load of $0.2f_u$, and it was maintained for 28 days before the load level was increased. However, when the load increased to $0.4f_u$, the TSP of one OF sensor abruptly dropped, and the other showed a gradually decreasing pattern. In the case of the EM with a bond length of 80 mm, a small drop in the early days did not occur, but a gradually decreasing pattern was observed when the load was increased to $0.4f_u$. In the case when the bond length was 100 mm, as shown in Figure 14b, there was no significant drop in the TSP of the two bonding methods. However, a small drop and decreasing pattern of TSP occurred at $0.4f_u$ in both bonding methods. This reduction in TSP did not occur in cases where the bond length was 150 mm for both bonding methods (Figure 14c) before $0.6f_u$. However, a rapid drop in TSP occurred in the EM due to bond failure after the stress was increased to $0.6f_u$.

Figure 15 shows the TSP of the OF sensors in the C specimen. In the case of a bond length of 40 mm, regardless of the bonding method, the TSPs of the OF sensors decreased gradually immediately after $0.2f_u$ stress was applied, reaching less than 80% on the 28th day. For the longer bond lengths, the EM showed a slightly better TSP than the EBM during the creep test. In particular, in the case of the embedded method, more than 95% of the TSP can be obtained at $0.6f_u$, even with a bond length of 80 mm (see Table 5).

Regarding the OF sensors bonded with epoxy resin or epoxy resin combined with CN adhesive, the EM showed better efficiency compared to the EBM in the case of the OF sensors having a bond length of 80 mm. According to the theoretical approaches of Ansari et al. and Seo et al. [22,23,29], the shear transfer coefficient from the FRP strips to the embedded OF sensors can be increased due to shear transfer from the FRP strips on both sides compared to the OF sensors bonded externally on one side. However, when the bond length was 150 mm, the TSP of the OF sensor in both the EM and EBM showed more than 97%, and no difference was observed between them.

The TSP can be considered the sensing capability of the OF sensor, which shows the accuracy level for the strain transfer capability from the FRP strip to the OF sensor. From the studies of Seo et al. [22,23], the strain can be transferred sufficiently from the FRP strip to the OF sensor with a bond length of 40 mm if not exposed to sustained loading conditions. When evaluating the sensing capacity of OF sensors bonded to an FRP strip with a bond length not less than 40 mm, the test result indicated a deviation range of 5%. This deviation range is compatible with that of the Omega strain gauges series (Omega Engineering, Inc., Seoul, Republic of Korea). Moreover, when using epoxy or CN adhesive, as in this study, if the bond length is increased in consideration of the creep effect, the sensing capacity of the OF can be improved within the deviation range of 3%.

### 3.2.2. Influence of Bond Lengths

Figures 16–18 show errors in the strain transfer coefficient (ESTC) of sensors with different bond lengths on strain transfer from the FRP strip to the OF sensor under sustained stresses. The effect of bond length was assessed through the ESTE values calculated by dividing the decrement of the OF sensor strain value by the corresponding ESG strain value.

Figure 16 presents the error ratio in the ESTC of the E specimen corresponding to bond length. In the case of the EM, E-EM-80 showed a critically sharp increment in ESTC under a stress level of $0.6f_u$; meanwhile, E-EM-100 and E-EM-150 did not show any noticeable changes (Figure 16a). Regarding the case of EBM, the ESTCs of E-EBM-80a and E-EBM-100b were significantly increased from the stress levels of $0.4f_u$ and $0.6f_u$, respectively (Figure 16b). This implies that minimum bond lengths of 100 and 120 mm are recommended for the EM and EBM, respectively, in order to sufficiently transfer the strain from the FRP strip to the OF sensor.

From Figure 17 presenting the ESTC of the EC specimen, in the case of the EM, it can be observed that EC-EM-80 and EC-EM-100 showed high ESTC values of 24 and 12%, respectively, under the stress level of $0.6f_u$. Meanwhile, EC-EM-150 lost its strain transfer ability (see Figure 17a). Similar results can be observed in the case of the EBM, in which the

OF sensors with bond lengths of 80 and 100 mm showed high ESTC values of over 70 and 20%, respectively, under a stress level of $0.6f_u$ (see Figure 17b).

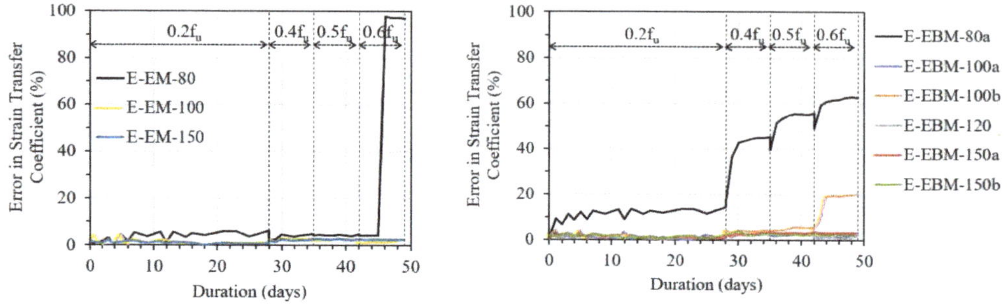


(a) (b)

**Figure 16.** The error in strain transfer coefficient of the E specimen with different bond lengths: (**a**) Embedded method and (**b**) Externally bonded method.

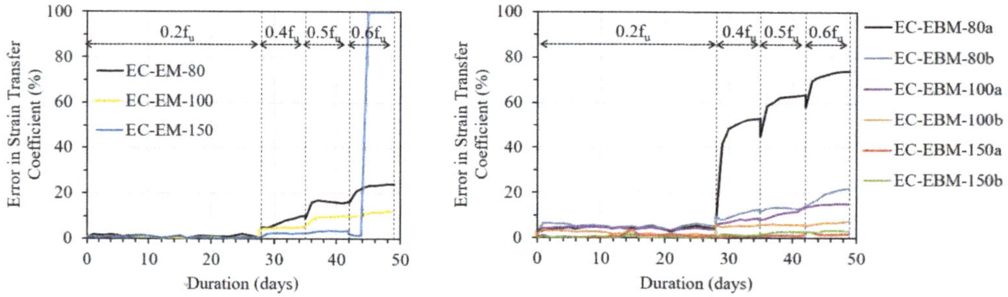


(a) (b)

**Figure 17.** The error in strain transfer coefficient of the EC specimen with different bond lengths: (**a**) Embedded method and (**b**) Externally bonded method.

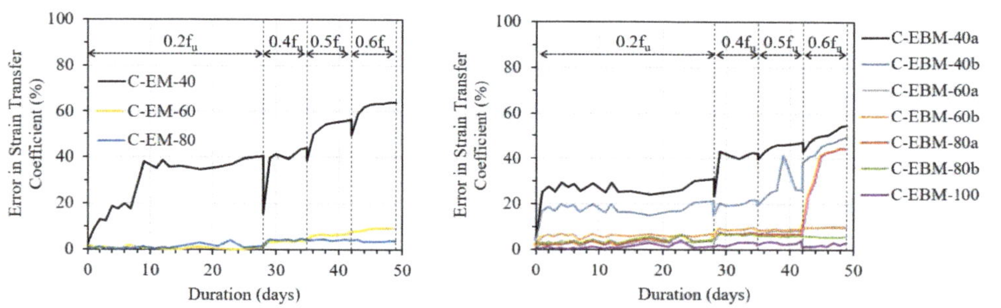


(a) (b)

**Figure 18.** The error in strain transfer coefficient of the C specimen with different bond lengths: (**a**) Embedded method and (**b**) Externally bonded method.

According to Motwani et al. [26], CN adhesive with low viscosity (~20 cP at room temperature) may show different stiffness compared to epoxy resin with high viscosity (~25,000 cP at room temperature). Moreover, the poor water resistance of CN adhesive may cause a change in the chemical bond, leading to a reduction in bonding strength [26]. Conversely, epoxy resin has good water resistance. Because of the incompatibility between the epoxy resin and the CN adhesive, the strain transfer ability of the OF sensor can be decreased. Therefore, EC-EM-150 lost its strain transfer ability, even with a longer bond length compared with EC-EM-80 and EC-EM-100. From the result, the combination of epoxy resin and CN adhesive is not recommended for attaching OF sensors to FRP strips.

Figure 18 shows the influence of bond lengths on the ESTC of the C specimen. For the OF sensors bonded by the EM, the E-EM-40 showed a serious ESTC of 40%, even under a stress level of $0.2f_u$. With longer bond lengths, however, the ESTC values of C-EM-60 and C-EM-80 were maintained at 10 and 5%, respectively, under a stress level of $0.6f_u$ (see Figure 18a). Regarding the OF sensors bonded by the EBM, a minimum bond length of 100 mm was necessary to adequately transfer the strain from the FRP strip to the OF sensor (see Figure 18b).

### 3.2.3. Influence of Different Types of Adhesives

Figures 19 and 20 present the effect of adhesive type on the TSP from the FRP strip to the OF sensor with bond lengths of 80 and 100 mm, respectively. With a bond length of 80 mm in both cases of the EM and EBM, it can be seen that the OF sensor bonded by CN adhesive showed the best TSP; meanwhile, the OF sensor bonded by epoxy resin combined with CN adhesive showed the worst TSP (see Figure 19). The effectiveness of CN adhesive can be observed clearly under sustained stress from $0.4f_u$.

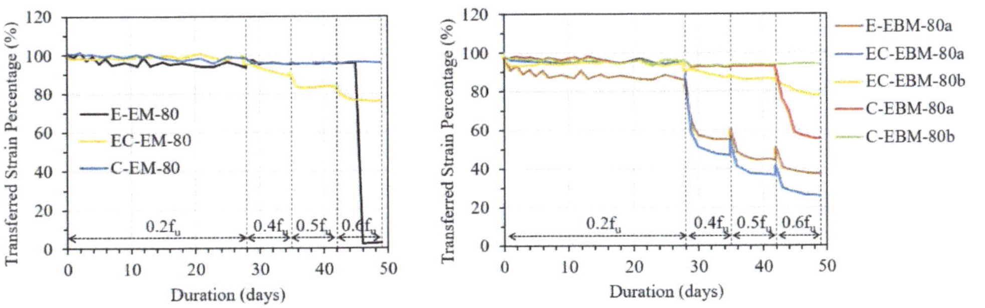


(a) (b)

**Figure 19.** Short-term creep effect on shear transfer with different types of adhesives: (**a**) EM with bond length of 80 mm and (**b**) EBM with bond length of 80 mm.

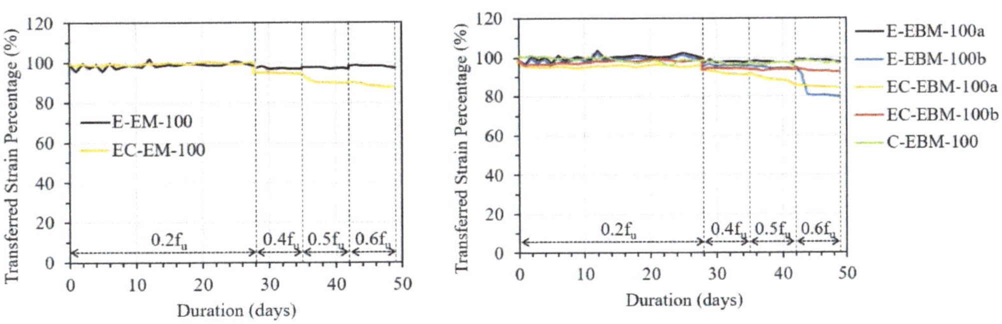


(a) (b)

**Figure 20.** Short-term creep effect on shear transfer with different types of adhesives: (**a**) EM with bond length of 100 mm and (**b**) EBM with bond length of 100 mm.

Under sustained stress from $0.4f_u$, the TSP of the OF sensor bonded with epoxy resin could be significantly decreased due to debonding when the bond length was insufficient to maintain its strain transfer ability. According to the comparison between different adhesive types, it is believed that the debonding occurred at the interface between the OF sensor and the adhesive. In this case, the debonding occurred at the interface between the secondary coating layer of the OF sensor and the epoxy resin.

Regarding the longer bond length of 100 mm, the difference in the TSP caused by the various adhesive types decreased. In detail, in the case of the EM, the OF sensor bonded by epoxy resin showed a TSP 8% higher than the OF sensor bonded by epoxy resin combined with CN adhesive (see Figure 20a). Meanwhile, the OF sensor bonded by CN adhesive revealed a TSP that was approximately 17% higher than the OF sensor bonded by the other type of adhesive, which showed the worst TSP (see Figure 20b).

### 3.2.4. Scanning Electron Microscope

In order to obtain further insight into the interfaces between OF sensors and epoxy resin, as well as the component layers of the OF sensors, a scanning electron microscope (SEM) analysis was performed after finishing the creep test. The cross-section samples were prepared using a Cross Section Polisher (IB-19510CP, JEOL Ltd., Tokyo, Japan). Then, a Field Emission Scanning Electron Microscope (JSM-7610, JEOL Ltd., Japan) was used for the SEM analysis.

Figure 21a,b show the SEM images of the OF sensors in the E specimen with bond lengths of 80 and 150 representing the cases with and without debonding, respectively. It can be seen that the core and cladding parts were broken after the preparation procedure of the samples due to their brittle characteristics. Meanwhile, the primary and secondary coatings were well preserved. Figure 21a shows the SEM image of E-EBM-80a, which showed a severe reduction in TSP due to debonding. It can be seen that clear gaps appeared due to debonding. Meanwhile, no gaps were observed between the component layers of the OF sensor, such as the primary and secondary coatings or the primary coating and core. This is consistent with the assumption that the bond strength between these layers is higher due to the long bond length compared to the bond strength between the OF sensor and the epoxy layer. In contrast, no gaps were observed in the SEM image of E-EBM-150a due to an unremarkable reduction in TSP (see Figure 21b).

(a) (b)

**Figure 21.** Scanning electron microscope images of FBG sensors in the E specimen: (**a**) Bond length of 80 mm and (**b**) Bond length of 150 mm.

## 4. Conclusions

In this study, the short-term creep effect on strain transfer from the FRP strip to the FBG-OF sensor was investigated under sustained stress levels of 20, 40, 50, and 60% of the ultimate tensile strength of the FRP strip. The three main investigated parameters were bond length, adhesive type, and bonding method. From the test results, the main conclusions were obtained as follows:

(1) According to previous studies regarding the FBG-OF sensor bonded by epoxy resin, it was found that a bond length of 40 mm was sufficient to control the shear lag effect for simple tensile loading. However, under short-term creep conditions to high sustained stress, such as 60% of the ultimate strength of the FRP strip, to maintain a strain

carrying capacity of more than 95%, minimum bond lengths of 100 mm and 120 mm are required for the embedded and external bonding methods, respectively.

(2) Regarding the OF sensor bonded with epoxy resin, when the bonding length was 80 mm or less, the strain transferred from the FRP strip to the OF sensor can be greatly reduced, even at stress levels of 40% of the ultimate strength of the FRP strip. This is due to adhesion damage at the interface between the secondary coating of the OF sensor and the adhesive; this phenomenon can be observed at the early stage of the loading duration.

(3) The OF sensor bonded with CN adhesive had better ability in strain transfer compared with the other adhesive types with the same bond length. In particular, in the case of the embedded method with a bond length of 80 mm, more than 95% of the TSP can be obtained at a stress level of 60% of the ultimate strength of the FRP strip.

(4) Regarding the OF sensor bonded with epoxy resin combined with CN adhesive, it is not recommended due to incompatibility in the mechanical properties.

(5) Sustained stress can cause additional creep deformation of FRP strips, even at low stress levels. Furthermore, it is recommended that an additional strain increase of at least 5% be considered due to creep effect when the FRP strip is exposed to conditions of sustained tensile stress.

**Author Contributions:** Conceptualization, H.V.T. and S.-Y.S.; data analysis, H.V.T. and S.-Y.S.; writing—original draft preparation, H.V.T.; writing—review and editing, H.V.T. and S.-Y.S. All authors have read and agreed to the published version of the manuscript.

**Funding:** This research was supported by the National Research Foundation of Korea (NRF) under a grant funded by the Korean government (MSIT) (No. 2021R1A4A2001964, No.2022R1A2C2004460).

**Institutional Review Board Statement:** Not applicable.

**Informed Consent Statement:** Not applicable.

**Data Availability Statement:** Not applicable.

**Conflicts of Interest:** The authors declare no conflict of interest.

## Abbreviations and Acronyms

| | |
|---|---|
| CN | Cyanoacrylate |
| DFOS | Distributed fiber optic sensors |
| EB | Externally bonded |
| EBM | External bonding method |
| EM | Embedded method |
| ESG | Electrical strain gauge |
| ESTC | Error in strain transfer coefficient |
| FBG | Fiber Bragg grating |
| FRP | Fiber-reinforced polymer |
| NSM | Near surface-mounted |
| OF | Optical fiber |
| PMMA | Polymethyl methacrylate |
| RC | Reinforced concrete |
| SEM | Scanning electron microscope |
| SHM | Structural health monitoring |
| STCE | Short-term creep effect |
| TSP | Transferred strain percentage |
| A | Additional |
| C | CN adhesive |
| E | Epoxy resin |
| EC | Epoxy resin and CN adhesive |

## References


1. Teng, J.G.; Chen, J.F.; Yu, T. *FRP-Strengthened RC Structures*; Wiley: Hoboken, NJ, USA, 2002.
2. De Lorenzis, L.; Teng, J.-G. Near-surface mounted FRP reinforcement: An emerging technique for strengthening structures. *Compos. Part B Eng.* **2007**, *38*, 119–143. [CrossRef]
3. Seo, S.-Y.; Feo, L.; Hui, D. Bond strength of near surface-mounted FRP plate for retrofit of concrete structures. *Compos. Struct.* **2013**, *95*, 719–727. [CrossRef]
4. Seo, S.-Y.; Choi, K.-B.; Kwon, Y.-S.; Lee, K.-S. Flexural Strength of RC Beam Strengthened by Partially De-bonded Near Surface-Mounted FRP Strip. *Int. J. Concr. Struct. Mater.* **2016**, *10*, 149–161. [CrossRef]
5. Seo, S.-Y.; Lee, M.S.; Feo, L. Flexural analysis of RC beam strengthened by partially de-bonded NSM FRP strip. *Compos. Part B Eng.* **2016**, *101*, 21–30. [CrossRef]
6. Feng, C.-W.; Keong, C.-W.; Hsueh, Y.-P.; Wang, Y.-Y.; Sue, H. Modeling of long-term creep behavior of structural epoxy adhesives. *Int. J. Adhes. Adhes.* **2005**, *25*, 427–436. [CrossRef]
7. Fernando, G.F.; Hameed, A.; Winter, D.; Tetlow, J.; Leng, J.; Barnes, R.; Mays, G.; Kister, G. Structural Integrity Monitoring of Concrete Structures via Optical Fiber Sensors: Sensor Protection Systems. *Struct. Health Monit.* **2003**, *2*, 123–135. [CrossRef]
8. Moyo, P.; Brownjohn, J.; Suresh, R.; Tjin, S. Development of fiber Bragg grating sensors for monitoring civil infrastructure. *Eng. Struct.* **2005**, *27*, 1828–1834. [CrossRef]
9. Sahota, J.; Gupta, N.; Dhawan, D. Fiber Bragg grating sensors for monitoring of physical parameters: A comprehensive review. *Opt. Eng.* **2020**, *59*, 060901. [CrossRef]
10. Hill, K.O.; Fujii, Y.; Johnson, D.C.; Kawasaki, B.S. Photosensitivity in optical fiber waveguides: Application to reflection filter fabrication. *Appl. Phys. Lett.* **1978**, *32*, 647–649. [CrossRef]
11. Meltz, G.; Morey, W.W.; Glenn, W.H. Formation of Bragg gratings in optical fibers by a transverse holographic method. *Opt. Lett.* **1989**, *14*, 823–825. [CrossRef]
12. Ansari, F. State-of-the-art in the applications of fiber-optic sensors to cementitious composites. *Cem. Concr. Compos.* **1997**, *19*, 3–19. [CrossRef]
13. López-Higuera, J.M.; Cobo, L.R.; Incera, A.Q.; Cobo, A. Fiber Optic Sensors in Structural Health Monitoring. *J. Light. Technol.* **2011**, *29*, 587–608. [CrossRef]
14. Majumder, M.; Gangopadhyay, T.K.; Chakraborty, A.K.; Dasgupta, K.; Bhattacharya, D. Fibre Bragg gratings in structural health monitoring—Present status and applications. *Sens. Actuators A Phys.* **2008**, *147*, 150–164. [CrossRef]
15. Zhang, Z.; Ansari, F. Crack tip opening displacement in micro-cracked concrete by an embedded optical fiber sensor. *Eng. Fract. Mech.* **2005**, *72*, 2505–2518. [CrossRef]
16. Zhang, Z.; Ansari, F. Fiber-optic laser speckle-intensity crack sensor for embedment in concrete. *Sens. Actuators A Phys.* **2006**, *126*, 107–111. [CrossRef]
17. Gu, X.; Chen, Z.; Ansari, F. Embedded fiber optic crack sensor for reinforced concrete structures. *Struct. J.* **2000**, *97*, 468–476.
18. Zhao, Y.; Ansari, F. Embedded fiber optic sensor for characterization of interface strains in FRP composite. *Sens. Actuators A Phys.* **2002**, *100*, 247–251. [CrossRef]
19. Siwowski, T.; Rajchel, M.; Howiacki, T.; Sieńko, R.; Bednarski, Ł. Distributed fibre optic sensors in FRP composite bridge monitoring: Validation through proof load tests. *Eng. Struct.* **2021**, *246*, 113057. [CrossRef]
20. Woods, J.E.; Lau, D.T.; Bao, X.; Li, W. Measuring strain fields in FRP strengthened RC shear walls using a distributed fiber optic sensor. *Eng. Struct.* **2017**, *152*, 359–369. [CrossRef]
21. Wang, B.; Teng, J.; De Lorenzis, L.; Zhou, L.-M.; Ou, J.; Jin, W.; Lau, K. Strain monitoring of RC members strengthened with smart NSM FRP bars. *Constr. Build. Mater.* **2009**, *23*, 1698–1711. [CrossRef]
22. Seo, S.-Y.; Park, J.-H.; Yun, H.-D.; Kim, K.-S.; Lee, G.-C.; Hong, S. Strain Transfer of Fiber Bragg Grating Sensor Externally Bonded to FRP Strip for Structural Monitoring after Reinforcement. *Materials* **2021**, *14*, 4382. [CrossRef] [PubMed]
23. Seo, S.-Y.; Park, J.-H.; Yun, H.-D.; Kim, K.-S. Installation Technique of Fiber Optic Sensor into FRP Used as NSM Structural Strengthening System. *Sustainability* **2020**, *12*, 8501. [CrossRef]
24. Tam, L.-H.; Jiang, J.; Yu, Z.; Orr, J.; Wu, C. Molecular dynamics investigation on the interfacial shear creep between carbon fiber and epoxy matrix. *Appl. Surf. Sci.* **2021**, *537*, 148013. [CrossRef]
25. *ASTM D3039/DM 3039*; Standard Test Method for Tensile Properties of Polymer Matrix Composite Materials. ASTM International: West Conshohocken, PA, USA, 2008.
26. Motwani, P.; Perogamvros, N.; Taylor, S.; Sonebi, M.; Laskar, A.; Murphy, A. Experimental investigation of strain sensitivity for surface bonded fibre optic sensors. *Sens. Actuators A Phys.* **2020**, *303*, 111833. [CrossRef]
27. ACI Committee. *State-of-the-Art Report on Fiber Reinforced Plastic (FRP) Reinforcement for Concrete Structures*; ACI Committee 440; American Concrete Institute: Farmington Hills, MI, USA, 1996.

28. Werneck, M.M.; Allil, R.C.S.B.; Ribeiro, B.A.; de Nazaré, F.V.B. *A Guide to Fiber Bragg Grating Sensors: Current Trends in Short-and Long-Period Fiber Gratings*; INTECH: Vienna, Austria, 2013.
29. Ansari, F.; Libo, Y. Mechanics of Bond and Interface Shear Transfer in Optical Fiber Sensors. *J. Eng. Mech.* **1998**, *124*, 385–394. [CrossRef]

sensors

Communication

# Algorithm of FBG Spectrum Distortion Correction for Optical Spectra Analyzers with CCD Elements

**Vladimir Anfinogentov [1], Kamil Karimov [1], Artem Kuznetsov [1], Oleg G. Morozov [1], Ilnur Nureev [1], Airat Sakhabutdinov [1,*], Konstantin Lipatnikov [1], Safaa M. R. H. Hussein [2] and Mustafa H. Ali [3]**

1 Department of Radiophotonics and Microwave Technologies, Kazan National Research Technical University Named after A.N. Tupolev-KAI, K. Marx Str. 10, 420111 Kazan, Russia; v.anfinogentov@yandex.ru (V.A.); mail12kamil2000@mail.ru (K.K.); AAKuznetsov@kai.ru (A.K.); microoil@mail.ru (O.G.M.); n2i2@mail.ru (I.N.); klipatnikov87@mail.ru (K.L.)

2 Department of physics, College of Education for Pure Sciences, University of Karbala, Karbala 56001, Iraq; safaa.mohammed@uokerbala.edu.iq

3 College of Dentistry, University of Mustansiriyah, Baghdad 14022, Iraq; mustafa.h@uomustansiriyah.edu.iq

* Correspondence: azhsakhabutdinov@kai.ru

**Abstract:** Nonlinear spectrum distortions are caused by the peculiarities of the operation of charge-coupled device elements (CCD), in which the signal exposition time (Time of INTegration–TINT) is one of the significant parameters. A change of TINT on a CCD leads to a nonlinear distortion of the resulting spectrum. A nonlinear distortion of the spectrum, in turn, leads to errors in determining the central wavelength of fiber Bragg gratings (FBGs) and spectrally sensitive sensors, which, in general, negatively affects the accuracy of the measuring systems. This paper proposes an algorithm for correcting the nonlinear distortions of the spectrum obtained on a spectrum analyzer using CCD as a receiver. It is shown that preliminary calibration of the optical spectrum analyzer with subsequent mathematical processing of the signal makes it possible to make corrections in the resulting spectrum, thereby leveling the errors caused by measurements at different TINT.

**Keywords:** nonlinear spectrum distortions; signal exposition time; fiber Bragg grating; fiber Bragg sensors; fiber optic sensors; fiber optic interrogator; optical spectrum analyzer; charge-coupled device elements; CCD

check for updates

**Citation:** Anfinogentov, V.; Karimov, K.; Kuznetsov, A.; Morozov, O.G.; Nureev, I.; Sakhabutdinov, A.; Lipatnikov, K.; Hussein, S.M.R.H.; Ali, M.H. Algorithm of FBG Spectrum Distortion Correction for Optical Spectra Analyzers with CCD Elements. *Sensors* **2021**, *21*, 2817. https://doi.org/10.3390/s21082817

Academic Editor: Damien Kinet

Received: 24 February 2021
Accepted: 15 April 2021
Published: 16 April 2021



## 1. Introduction

It is widely known that in recent years, fiber-optic measuring systems have become increasingly relevant. In fiber-optic sensor systems, various technologies of interrogation and multiplexing are used [1–5]. Different technologies are applied for dividing fiber-optic spectral-sensitive sensors: by wavelength [1], by time response [2], by frequency [3], by polarization [4], and by spatial [5] multiplexing. To determine the average wavelength of the sensors, optical analyzers, such as spectrometers with tunable Fabry-Perot system interferometers, or diffraction gratings with CCD arrays, are used. The complexity of multiplexing technologies is also related to the fact that any spectrum overlaps of FBG spectra leads to significant errors in measurements of their central wavelengths [6–8]. The multiplexing technologies and microwave-photon interrogation methods of spectrally-encoded and addressed Bragg gratings have been developed; they allow separating, spectrally, the responses of sensors in the same frequency range [9–12]. Some researchers worked on sensor detection and tracking using Slepyan codes [13–16]; thus, a measurement of temperature and deformation in the case of sensors spectra overlapping became possible. Recently, a more convenient technology, based on addressable fiber Bragg structures, was proposed [17]. It made it possible to design distributed sensor systems with a large number of sensors without complicated optoelectronic schemes [17–19].

Despite significant progress in this direction, the classical optical spectrum analyzers, using diffraction gratings with CCD, have yet to develop their full potential. The attractiveness is in the lower cost of such devices, compared to interrogators on tunable filters (with comparable measurement errors). The source of optical radiation is superluminescent laser diodes with a spectral width that overlaps the working spectral range of the analyzer.

The information parameter of the FBG sensor is the shift of its central wavelength. The data obtained from the CCD array allow measurements with an accuracy of ~160 pm, which is not acceptable. Therefore, to improve the accuracy, an approximation is used. Various types of approximation have been investigated: Gaussian curve; second-order parabola using three upper points [20]; a parabola using the least-squares method; approximating the position of the central wavelength using the center of mass method; and others [21,22]. All investigated methods make it possible to determine the central wavelength FBG with a margin of error; however, the center of mass method gives the best accuracy.

## 2. Problem Explanation

For an experimental study of the FBG sensor interrogating system, its prototype was assembled. The IBSEN I-MON 512 USB as a spectrometer, a SLD-761-HP1-DIL as a broadband light source, and an FS62WSS (HBM) as a bore-type temperature sensor were used. The sensor response was measured at several constant temperature values maintained by a thermostat with an accuracy of $\pm 0.1$ °C. The dependence "shift of the central wavelength-temperature" was plotted based on the measurement results. The second series of measurements were made at the facility, where the sensor was connected to the device through an existing long optical line. The decrease in the optical signal power was compensated by the increase in the exposition time (TINT) of the CCD array, which made it possible to "scale" the spectrum to the required signal-to-noise ratio. The spectrometer manufacturer (IBSEN) uses the term TINT as the variable and command in the software, and the term means "Time of INTegration". However, this approach led to the fact that the previously obtained correspondence "central wavelength–temperature shift" began to be violated: at the same temperature, the value of the central wavelength differed by tens of picometers (equivalent to an error of several degrees Celsius), from which a hypothesis was put forward about the nonlinear deformation of the spectrum with a change in the accumulation time.

It should be noted that there is no mention of this phenomenon in the user's manual of the spectrometer, which is, in our opinion, due to the fact that the manufacturer considers the spectrometer as an end product; in this regard, its use as an element of a more complex system is not within their goals.

In the works of other authors, it is shown that CCD-based spectrometers have a number of spectrum distortions, such as nonlinear pixel sensitivity to incident light [23] and chromatic distortions [24–26]. In these works, the sources of these errors are considered, as well as the methods of their compensation. In our work, we consider the spectrometer as a "black box", so we are not interested in the physical mechanisms of the spectrum distortion. We only operate with the data that it is capable of producing (taking into account the built-in mechanisms for picking up and converting the analog output of the CCD to the digital output) in response to standard commands–requests. It is also worth noting the work [27], where a modification of the algorithm for calculating the information characteristics of the FBG spectrum with its significant distortions is proposed.

Nevertheless, the interrogation system must consider the spectral characteristics both of spectrum analyzers and broadband light sources. It is necessary to calibrate spectrum analyzers and light sources jointly if we want to use their combination as an FBG interrogation system to avoid the FBG spectrum distortion due to different exposition times.

Thus, the task is to study the influence of the TINT parameter on distortions of the shape of the response spectrum of the FBG sensor, and advise methods to improve them.

## 3. Collecting of Initial Data

In order to solve this problem, a described experimental set-up was used. Here, the same fiber-optic sensor based on FBG was interrogated from two ends of the fiber. The optical line length connected to the fiber sensor at one end was equal to 10 m, and at the other end was equal to 10 km. The maximum amplitude of the spectral response from the sensor was set at ~40,000 quantization levels of the analog-to-digital converter to ensure an acceptable signal-to-noise ratio. This, in turn, led to the fact that the sensor was interrogated from the short end with TINT equal to 20 ms, and from the long end equal to 135 ms. The central wavelength was determined by the center of mass method for the same wavelength range of this sensor. It was found that the error in determining the central wavelength value of FBG in this experimental setup could reach 10 pm, which caused an error equal to ~1 K in temperature determination. Figure 1 shows the spectral range of FBG, measured at three different values of TINT, the ordinate axis is shown in convenient units, normed at 40,000 quantization levels. The used spectrometer (Ibsen I-MON) has a built-in temperature sensor, the data from which are used to compensate the temperature fluctuations of the spectrometer elements (optical and electronic paths) and associated errors in the interpretation of the FBG spectrum. The method of this compensation is described in the documentation on Ibsen I-MON.

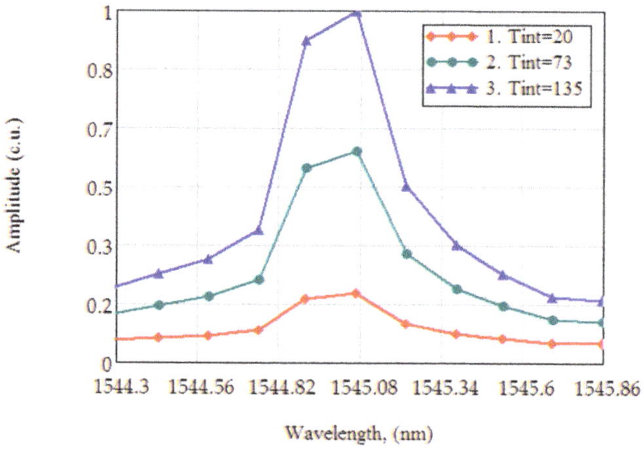


**Figure 1.** The FBG spectrum, measured at different TINT values (20, 73, 135 ms).

The central wavelength value, calculated for the same spectrum of FBG at different TINT values, can differ significantly in practice. The difference in determining the FBG central wavelength depends on the chosen method and can reach 40 pm. For example, as shown in Figure 1, the differences between the central wavelength values for FBG at different values of TINT (20, 73, and 135 ms), are 7.6, 9.5, and 1.9 pm. This accuracy cannot be satisfactory, when the requirements for the temperature determining the accuracy is less than 1 K.

For further investigation of the hypothesis, the initial broadband spectra were obtained at different values of TINT on the spectrum analyzer. Figure 2 shows the spectral characteristics of broadband laser radiation, obtained at different values of TINT (curve 1–20 ms, 2–48 ms, 3–77 ms, 4–106 ms, 5–135 ms).

For eleven different characteristic points of the spectrum, the dependence of their amplitudes on TINT was received; the dependences for these points with different initial values of the amplitude on TINT are shown in different colors in Figure 3. The dependence of the amplitude on TINT was plotted for the TINT range from 20 to 135 ms with the step equal to 1 ms, which provided 116 measurements.

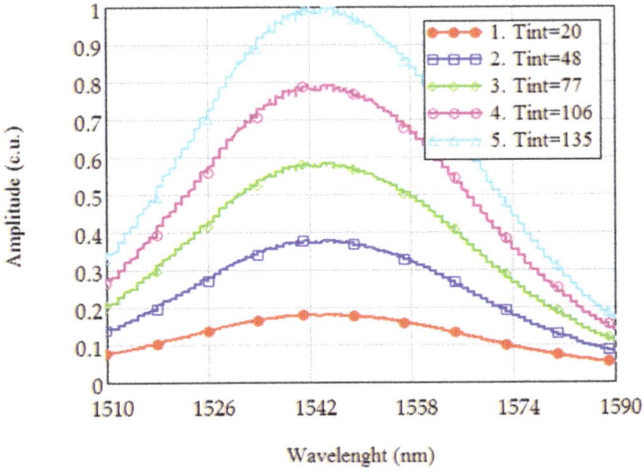


**Figure 2.** The broadband radiation spectra, obtained at different TINT values (1–20 ms, 2–48 ms, 3–77 ms, 4–106 ms, 5–135 ms).

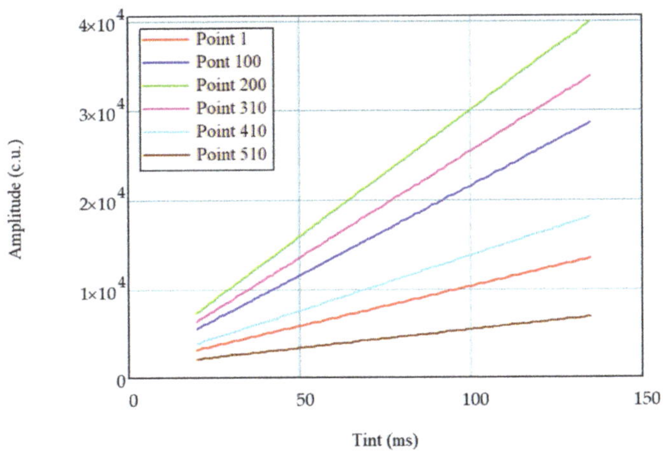


**Figure 3.** The dependence of the amplitude of different points of the spectrum on TINT.

From the dependences shown in Figure 3, it can be seen that, with an increase in TINT, a linear increase in the amplitude occurs; however, the coefficient of this linear dependence (the slope of the straight line) itself depends on the amplitude initial value, which is measured at the initial TINT value. In addition, it can be seen that the higher the amplitude, which is measured at the initial TINT value (at 20 ms), the faster it grows with the TINT increase. This leads to nonlinear distortion of the spectrum and, ultimately, to errors in determining the FBG central wavelength.

The dependence of the slope angle on the initial value of the amplitude obtained at the initial value of TINT was received. This dependence is shown in Figure 4. As can be seen from Figure 4, the slope of the linear dependence has a linear dependence on the initial amplitude.

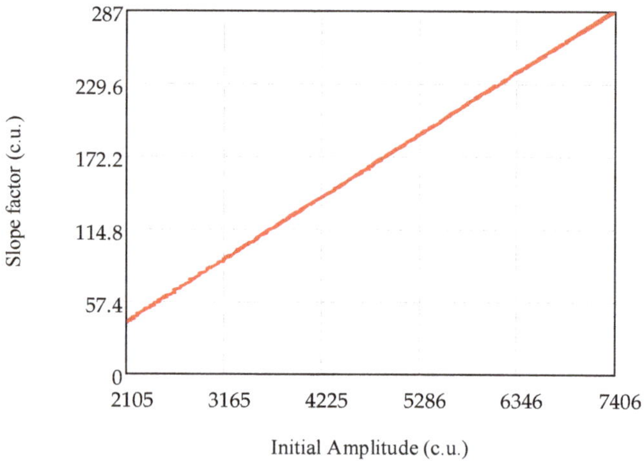


**Figure 4.** Dependence of the slope on the initial amplitude.

## 4. Mathematical Processing

The spectral response of the broadband source radiation for initial TINT value ($t_0$) can be obtained:

$$A_{i,0},\ i = \overline{1, N}, \tag{1}$$

where $N$ is the number of sampling points of the spectrum, $i = 1$ and $i = N$ correspond to minimal and maximum wavelengths of range, respectively.

The amplitude increases linearly with TINT increasing, which allows approximating the dependence by a straight line:

$$A(t) = A_0 + k(A_0) \cdot (t - t_0), \tag{2}$$

where $t_0$ is the initial TINT value, at which the calibration is performed; $t$ is the arbitrary TINT value; $k(A_0)$ is the linear dependence slope; and $A_0$ is the amplitude, obtained at the initial TINT ($t_0$).

Based on the fact that the slope of the linear dependence (Figure 4) also increases linearly with the increasing amplitude, measured at the initial TINT, the slope of the linear dependence in Equation (2) can be represented as its linear dependence on the initial value of the amplitude, obtained at initial TINT. This statement can be formulated as follows:

$$k(A_0) = \alpha \cdot A_0 + \beta. \tag{3}$$

Substituting Equation (3) into Equation (2), we obtain the dependence of the amplitude on TINT and the initial value of the amplitude, obtained at initial TINT value. A field of amplitude values depending on the TINT and the initial value of the amplitude, measured at initial TINT, can be obtained:

$$A(t) = A_0 + (\alpha \cdot A_0 + \beta) \cdot (t - t_0), \tag{4}$$

where $\alpha$ and $\beta$ are the coefficients of the linear dependence, $t_0$ is initial TINT, at which the amplitudes $A_0$ are measured.

It is possible to obtain a field of amplitude values depending on TINT and the initial value of the amplitude, measured at the initial TINT value, for the entire spectrum of the broadband radiation source:

$$\{A_{i,j}, t_j\},\ i = \overline{1, N}, j = \overline{0, M}, \tag{5}$$

where $N$ is the number of the spectrum sampling points, $M$ is the number of points for the TINT, changing in the range from $t_0$ to $t_M$, namely $t_j = t_0 + j(t_M - t_0)/M$.

Therefore, the measured amplitude $A_{i,j}(t_j, A_{i,0})$ for $i$-th sampling point can be written as dependence on TINT ($t_j$) and on the initial amplitude value $A_{i,0}$ for each:

$$A_{i,j}(t_j, A_{i,0}) = A_{i,0} + (t_j - t_0)(\alpha \cdot A_{i,0} + \beta),\ \ i = \overline{1, N}, j = \overline{1, M}. \tag{6}$$

The measured field of values Equation (5) makes it possible to determine the value of the slope of the linear dependence $k(A_{i,0})$ in Equation (2) for each initial value of the amplitude in each $i$-th spectrum point by calculating the characteristic points $\{A_{i,0}, k(A_{i,0})\}$, $i = 1{,}N$ using the least square method by formula:

$$K_i = k(A_{i,0}) = \sum_{j=1}^{M} (A_{j,i} - A_{0,i})(t_j - t_0) / \sum_{j=1}^{N} (t_j - t_0)^2,\ \ i = \overline{1, N}. \tag{7}$$

It gives a set of values $\{A_{i,0}, K_i\}$, $i = 1, N$, which, in turn, allow determining the $\alpha$ and $\beta$ values–the linear dependence coefficients of the slope on $A_0$ in Equation (3), by the linear equations system solving the following:

$$\begin{pmatrix} \sum_{i=1}^{N} A_{i,0}^2 & \sum_{i=1}^{N} A_{i,0} \\ \sum_{i=1}^{N} A_{i,0} & N \end{pmatrix} \cdot \begin{pmatrix} \alpha \\ \beta \end{pmatrix} = \begin{pmatrix} \sum_{i=1}^{N} A_{i,0} K_i \\ \sum_{i=1}^{N} K_i \end{pmatrix}. \tag{8}$$

Thus, after calculating the set of values $\{A_{i,0}, K_i\}$, $i = 1, N$, and the coefficients of the linear dependence of the slope in Equation (3)–$\alpha$ and $\beta$, in relation to Equation (6), the entire right-hand side becomes known. It allows to recalculate the initial amplitude of the arbitrary spectrum point, measured at initial TINT, depending on the amplitude, measured at arbitrary TINT ($t$):

$$A_{i,0} = \frac{A(t, A_{i,0})_i - \beta(t - t_0)}{1 + \alpha \cdot (t - t_0)}. \tag{9}$$

In Equation (9), $A(t, A_{i,0})$ is the amplitude, measured at arbitrary TINT, $\alpha$ and $\beta$ are the linear dependence coefficients of the slope in Equation (3), obtained as the solution of the linear equation system in Equation (8), and $A_{i,0}$ is the amplitude, measured at initial TINT.

Thus, knowing the current value of the amplitude at a point in the spectrum, measured at an arbitrary TINT, providing a comfortable signal-to-noise ratio, allows determining the amplitude value, measured at initial TINT, which is used in the calibration process for this sensor.

## 5. Experimental Setup

The experimental setup is shown in Figure 5a. The experimental setup consists of the thermostat, interrogator (Figure 5b), fiber sensors under calibration, and controlling system. The thermostat is shown at the left side of Figure 5a; it has its own temperature sensor which is connected to the interrogator by the RS-232 interface. The sensors are placed in the thermostat and held at a constant temperature. The thermostat's temperature is varied in the range with the given step. The interrogator is placed on the right side of the table in Figure 5a. The experimental setup is assembled so the sensors are affected only by temperature.

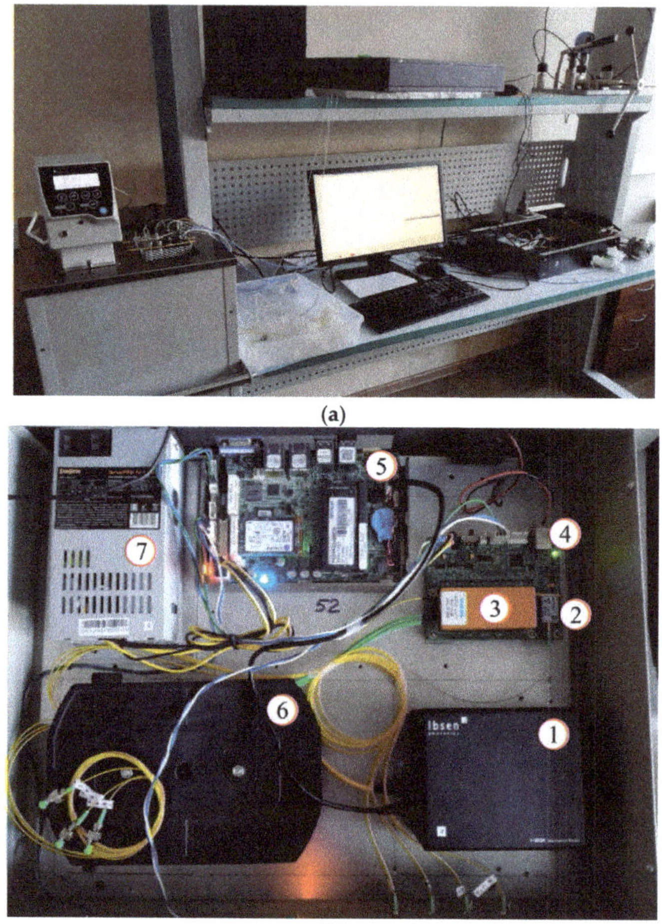


**Figure 5.** Experimental setup (**a**); FBG interrogator (**b**): 1–IBSEN I-MON 512 USB; 2–laser source SLD-761-HP1-DIL; 3–optic channel switch Sercalo MEMS switch rSC $1 \times 8$; 4–laser source and optical switch maintaining module; 5–computer Wafer ULT-3; 6–optical cross; 7–power supply.

The FBG interrogator (Figure 5b) includes the optical spectrum analyzer IBSEN I-MON 512 with a USB interface–1. The laser source SLD-761-HP1-DIL–2 is used as the broadband optical light source. The optical channel switch Sercalo MEMS switch rSC $1 \times 8$–3 is used to switch channels to have a possibility to interrogate several optical channels consistently with the given interval (~100 ÷ 200 ms). In this interrogator model, we used eight independent channels with eight sensors in each channel. The optical light source with the channel switch is maintained by our own designed control board–4. All measured data are collected on the on-board computer, Wafer ULT-3–5. The optical cross–6 is used for internal optical cabling. The common power supply–7 is used.

## 6. Calibration Data

The calibration curves received without and with the spectrum calibration algorithm are presented in Figure 6. All measurements were made for the bore-type temperature sensor FS62WSS (HBM) from two ends. The measurements made from the first end are marked by red rhombuses, and from the second end are marked by blue rhombuses. All measurements were made in the temperature range from 15 to 90 °C with a discrete step of 5 °C. Fifty independent measurements of the FBG central wavelength were made in each temperature point; thus, each rhombus in the figure means one measurement.

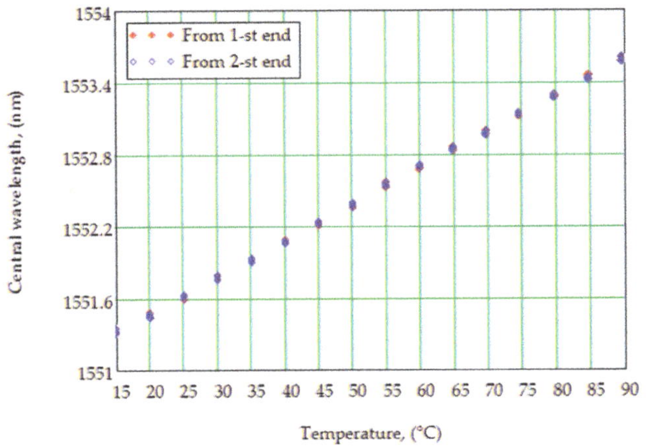


(**a**)

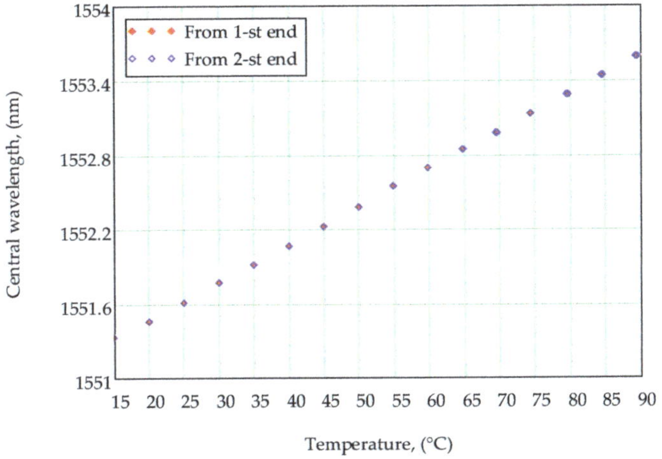


(**b**)

**Figure 6.** Calibration curves–FBG central wavelength dependences on temperature. The measured data: (**a**) without spectrum correction; (**b**) with spectrum correction.

As one can see, there is dispersion in obtaining the FBG central wavelength in the case when the spectrum calibration algorithm is not used (Figure 6a). The average error of the central wavelength approximation is 50 pm, which is equivalent to 5 K. While the calibration dependence received with the spectrum correction algorithm gives an error approximation of the FBG central wavelength of less than 0.1 pm, that allows approximating the temperature with 0.01 K accuracy (Figure 5b).

The received results approve the necessity of using the preliminary mandatory calibration of the light source with the CCD element and using these calibration results to correct the FBG spectra from the spectrum analyzer.

## 7. Conclusions

The proposed algorithm allows excluding nonlinear distortions of the fiber Bragg grating spectrum response, caused by different signal integration times on charge-coupled devices.

The spectrum correction, performed according to the proposed algorithm, allows excluding limitations associated with the fact that the calibration of all sensors in the fiber-optic measurement system, based on CCD, must be carried out at the same signal integration time values, which will be used in exploitation.

As a result, the compensation of nonlinear distortions allows calibrating sensors independently from the measuring system. In addition, it removes restrictions on strictly fixed optical lengths from an interrogator to the sensor. Moreover, it gives the opportunity to simplify the array of sensors formed in an optical channel, and to replace sensors with one another arbitrarily during exploitation.

We are free to discuss this problem and our results with scientists and manufacturers.

**Author Contributions:** Conceptualization, V.A., O.G.M. and A.S.; methodology, A.S. and V.A.; software, K.K. and K.L.; validation, V.A., K.L., I.N., A.S. and O.G.M.; formal analysis, A.S., K.L., M.H.A., and S.M.R.H.H.; investigation, I.N., V.A., A.S., M.H.A., and K.K.; resources, A.K.; data curation, A.K., K.L., S.M.R.H.H., and M.H.A.; writing—original draft preparation, A.S. and A.K.; writing—review and editing, A.K. and A.S.; visualization, K.K. and K.L.; supervision, O.G.M.; project administration, O.G.M.; funding acquisition, A.K. and O.G.M. All authors have read and agreed to the published version of the manuscript.

**Funding:** A.S. and O.G.M. were funded by the Ministry of Science and Higher Education of the Russian Federation (Agreement No. 075-03-2020-051, Topic No. fzsu-2020-0020), A.K. was funded by a grant from the President of the Russian Federation for the state support of young Russian scientists—candidates of sciences MK-3421.2019.8 (Agreement No. 075-15-2019-309).

**Institutional Review Board Statement:** Not applicable.

**Informed Consent Statement:** Not applicable.

**Data Availability Statement:** The data presented in this study are available on request from the corresponding author. The data are not publicly available due to rules of our contract conditions with our customer.

**Conflicts of Interest:** The authors declare no conflict of interest.

## References

1. Measures, R.M.; Melle, S.; Liu, K. Wavelength demodulated Bragg grating fiber optic sensing systems for addressing smart structure critical issues. *Smart Mater. Struct.* **1992**, *1*, 36–44. [CrossRef]
2. Davis, M.A.; Bellemore, D.G.; Kersey, A.D. Structural strain mapping using a wavelength/time division addressed fiber Bragg grating array. *Smart Struct. Mater. Second Eur. Conf.* **1994**, *2361*, 342–346. [CrossRef]
3. Matveenko, V.; Shardakov, I.; Voronkov, A.; Kosheleva, N.; Lobanov, D.; Serovaev, G.; Spaskova, E.; Shipunov, G. Measurement of strains by optical fiber Bragg grating sensors embedded into polymer composite material. *Struct. Control. Health Monit.* **2017**, *25*, e2118. [CrossRef]
4. Qiao, X.; Shao, Z.; Bao, W.; Rong, Q. Fiber Bragg Grating Sensors for the Oil Industry. *Sensors* **2017**, *17*, 429. [CrossRef] [PubMed]
5. Ma, Z.; Chen, X. Fiber Bragg gratings sensors for aircraft wing shape measurement: Recent applications and technical analysis. *Sensors* **2019**, *19*, 1–25. [CrossRef] [PubMed]
6. Karim, F. *Full Matlab Code for Synthesis and Optimization of Bragg Gratings*; Cambridge Scholars Publishing: Newcastle upon Tyne, UK, 2019; p. 24.
7. Cormier, G.; Boudreau, R.; Thériault, S. Real-coded genetic algorithm for Bragg grating parameter synthesis. *J. Opt. Soc. Am. B* **2001**, *18*, 1771–1776. [CrossRef]
8. Li, K. Review of the Strain Modulation Methods Used in Fiber Bragg Grating Sensors. *J. Sensors* **2016**, *2016*, 1–8. [CrossRef]
9. Koo, K.; Leblanc, M.; Tsai, T.; Vohra, S. Fiber-chirped grating Fabry-Perot sensor with multiple-wavelength-addressable free-spectral ranges. *IEEE Photon. Technol. Lett.* **1998**, *10*, 1006–1008. [CrossRef]
10. Wei, Z.; Ghafouri-Shiraz, H.; Shalaby, H. New code families for fiber-Bragg-grating-based spectral-amplitude-coding optical CDMA systems. *IEEE Photon. Technol. Lett.* **2001**, *13*, 890–892. [CrossRef]
11. Dai, B.; Gao, Z.; Wang, X.; Kataoka, N.; Wada, N. Performance comparison of $0/\pi$- and $\pm\ \pi/2$-phase-shifted superstructured Fiber Bragg grating en/decoder. *Opt. Express* **2011**, *19*, 12248–12260. [CrossRef]

12. Triana, C.A.; Pastor, D.; Varon, M. Optical code division multiplexing in the design of encoded fiber Bragg grating sensors. *Opt. Pura Apl.* **2016**, *49*, 17–28. [CrossRef]
13. Triana, A.; Pastor, D. Interrogation of super-structured FBG sensors based on discrete prolate spheroidal sequences. *Opt. Sens.* **2017**, *10231*, 102310. [CrossRef]
14. Djordjevic, I.B.; Saleh, A.H.; Küppers, F. Design of DPSS based fiber bragg gratings and their application in all-optical encryption, OCDMA, optical steganography, and orthogonal-division multiplexing. *Opt. Express* **2014**, *22*, 10882–10897. [CrossRef]
15. Kim, Y.; Jeon, S.-W.; Kwon, W.-B.; Park, C.-S. A Wide Dynamics and Fast Scan Interrogating Method for a Fiber Bragg Grating Sensor Network Implemented Using Code Division Multiple Access. *Sensors* **2012**, *12*, 5888–5895. [CrossRef]
16. Triana, A.; Pastor, D.; Varón, M. A Code Division Design Strategy for Multiplexing Fiber Bragg Grating Sensing Networks. *Sensors* **2017**, *17*, 2508. [CrossRef]
17. Morozov, O.; Sakhabutdinov, A.; Anfinogentov, V.; Misbakhov, R.; Kuznetsov, A.; Agliullin, T. Multi-Addressed Fiber Bragg Structures for Microwave-Photonic Sensor Systems. *Sensors* **2020**, *20*, 2693. [CrossRef]
18. Agliullin, T.; Gubaidullin, R.; Sakhabutdinov, A.; Morozov, O.; Kuznetsov, A.; Ivanov, V. Addressed Fiber Bragg Structures in Load-Sensing Wheel Hub Bearings. *Sensors* **2020**, *20*, 6191. [CrossRef]
19. Gubaidullin, R.; Agliullin, T.; Ivanov, V.; Sakhabutdinov, A. Tire dynamic monitoring setup based on microwave photonic sensors. *Opt. Technol. Telecommun.* **2019**, *11146*, 111461J. [CrossRef]
20. Jiang, J.; Liu, T.; Liu, K.; Zhang, Y. Investigation of peak wavelength detection of fiber Bragg grating with sparse spectral data. *Opt. Eng.* **2012**, *51*, 034403. [CrossRef]
21. Morozov, O.G.; Kuznetsov, A.A.; Morozov, G.A.; Nureev, I.I.; Sakhabutdinov, A.Z.; Faskhutdinov, L.M.; Artemev, V.I. Smart Photonic Carbon Brush. In Proceedings of the International Society for Optical Engineering, Ufa, Russia, 26 March 2016; Sultanov, A.H., Morozov, O.G., Burdin, V.A., Bourdine, A.V., Andreev, V.A., Eds.; SPIE: Ufa, Russia, 2016; Volume 9807, p. 98070M. [CrossRef]
22. Gubaidullin, R.R.; Agliullin, T.A.; Ivanov, V.; Morozov, O.G.; Sakhabutdinov, A.Z. Tire Dynamic Monitoring Setup Based on Microwave Photonic Sensors. In Proceedings of the Optical Technologies for Telecommunications 2018, Ufa, Russia, 24 June 2019; Volume 11146, p. 111461J. [CrossRef]
23. Xia, G.; Liu, Q.; Zhou, H.; Yu, F. A non-linearity correction method of charge-coupled device array spectrometer. In Proceedings of the AOPC 2015: Optical Test, Measurement, and Equipment, Beijing, China, 1 October 2015; Volume 9677, p. 96770J. [CrossRef]
24. Fu, H.-K.; Liu, Y.-L.; Chen, T.-T.; Wang, C.-P.; Chou, P.-T. The Study of Spectral Correction Algorithm of Charge-Coupled Device Array Spectrometer. *IEEE Trans. Electron Devices* **2014**, *61*, 3796–3802. [CrossRef]
25. Czech, E.; Fryc, I. Spectral correction of the measurement CCD array. *Opt. Eng.* **2002**, *41*, 2402–2406. [CrossRef]
26. Sun, Y.C.; Huang, C.; Xia, G.; Jin, S.Q.; Lu, H.B. Accurate wavelength calibration method for compact CCD spectrometer. *J. Opt. Soc. Am. A* **2017**, *34*, 498–505. [CrossRef]
27. Lamberti, A.; Vanlanduit, S.; De Pauw, B.; Berghmans, F. Influence of Fiber Bragg Grating Spectrum Degradation on the Performance of Sensor Interrogation Algorithms. *Sensors* **2014**, *14*, 24258–24277. [CrossRef]

*sensors*

MDPI

*Article*

# On the Advantages of Microwave Photonic Interrogation of Fiber-Based Sensors: A Noise Analysis

**Ulrich Nordmeyer [1,*], Torsten Thiel [2], Konstantin Kojucharow [3] and Niels Neumann [1]**

1 Institute of Electrical Information Technology, Clausthal University of Technology, 38678 Clausthal-Zellerfeld, Germany
2 Advanced Optics Solutions (AOS) GmbH, Overbeckstr. 39a, 01139 Dresden, Germany
3 Kojucharow Microwave Development and Components (KMDC), Zur Bleiche 15, 01279 Dresden, Germany
* Correspondence: ulrich.nordmeyer@tu-clausthal.de

**Abstract:** Although microwave photonic approaches have been used for fiber sensing applications before, most contributions in the past dealt with evaluating the sensor signal's amplitude. Carrying this topic on, the authors previously presented a scheme for the interrogation of fiber sensors that was based on a fiber Bragg grating's phase response for the electrical signal. However, neither has the measurement setup been analyzed nor have the amplitude and phase-based approaches been compared in detail before. Hence, this paper picks up the previously proposed setup, which relies on an amplitude modulation of the optical signal and investigates for sources of signal degradation, an aspect that has not been considered before. Following the incorporation of the microwave signal, the setup is suitable not only for an amplitude-based evaluation of fiber Bragg gratings but also for a phase-based evaluation. In this context, the signal-to-noise ratios are studied for the conventional amplitude-based evaluation approach and for the recently developed phase-based approach. The findings indicate a strong advantage for the signal-to-noise ratio of the phase response evaluation; an 11 dB improvement at the least has been found for the examined setup. Further studies may investigate the consequences and additional benefits of this approach for radio-over-fiber sensing systems or general performance aspects such as achievable sensitivity and sampling rates.

**Keywords:** electrical readout; fiber sensor; fiber bragg grating; microwave photonics; phase response; radio-over-fiber; SNR; wireless sensing

check for updates

**Citation:** Nordmeyer, U.; Thiel, T.; Kojucharow, K.; Neumann, N. On the Advantages of Microwave Photonic Interrogation of Fiber-Based Sensors: A Noise Analysis. *Sensors* **2023**, *23*, 3746. https://doi.org/10.3390/s23073746

Academic Editor: Oleg G. Morozov

Received: 3 March 2023
Revised: 30 March 2023
Accepted: 3 April 2023
Published: 4 April 2023



## 1. Introduction

The field of microwave photonics (MWP) has been subject to research interests for many years. Among a very large overall count of publications associated with MWP, some are dedicated to presenting an overview of this research field [1–5]. Reviewing the latest advances at the time of their writing, they cover different application areas such as communication technology, signal generation, signal processing and integrated photonics. At the same time, the sub-field of sensors and measurement tasks is still underrepresented. However, the publication history of the past decade shows increasing research work in this area. A partial overview is given by Yao [6], while other studies are linked to a variety of specific problem formulations such as distributed fast fiber-optic sensing [7], multiplexing sensors [8] and measuring properties of microwave signals [9]. Further work concentrates on the microwave-based interrogation of fiber grating sensors, e.g., aiming for resolution and speed improvements [10], high-resolution multi-point sensing [11], antigen biosensing [12] and wireless sensing schemes [13], all of which are incorporating Fiber Bragg Gratings (FBGs) as sensing elements. Earlier, more general studies on the microwave interrogation of fiber-optic devices began investigating the use of MWP for optical network analysis. The scheme was based on single-sideband (SSB) modulation [14,15] and afterwards adapted for measuring and reconstructing the phase response of FBGs [16,17]. The same basic principle was utilized for an improved evaluation of quasi-distributed sensing elements [18,19]. A similar concept involves double-sideband (DSB) modulation, and its applicability has been reviewed along with system modeling considerations and functionality examples [20].

In addition to these publications, the authors of this paper have also contributed to the topic of MWP-based FBG characterization and fiber-optic sensing in their previous studies. Starting with a presentation of the general wireless evaluation approach for FBGs [21] and a demonstration of pH sensing in such a radio-over-fiber (RoF) setup [22], the work was continued by researching a novel scheme of electrical read-out in the form of utilizing the group-delay characteristic as an alternative sensor characteristic to the conventionally used amplitude response [23]. The applicability of this evaluation scheme to thermometry has been evaluated [24] and the effects resulting from the DSB modulation have been studied and modeled in detail [25].

In contrast to the conventional approach, the authors' motivation was to make use of the electrical phase information that is present in MWP sensing setups but remained unused before. That means, other aspects that have been investigated before, such as improving vector analyzing of optical networks, precisely reconstructing the actual dispersion characteristic of FBGs or dealing with multi-point or quasi-distributed sensing of a large number of fiber sensors, are not in focus. An optimization of the novel approach led to simplified experimental setups and finally to a direct evaluation of the electrical phase, which for FBGs possesses a dependency on the measurand. After demonstrating the general suitability of this concept for sensing applications [26], the present paper is focused on analyzing the interrogation scheme for possible signal impairments and their consequences regarding measurement accuracy. The setup used previously for the general suitability study is picked up and the possible noise performance is analyzed in detail for the first time. Based on these considerations, the conventional amplitude characteristic read-out method is compared to the phase evaluation method, and both advantages and disadvantages are discussed. The study starts with the explanation of the basic concept of microwave photonic FBG interrogation and continues with the presentation of the experimental setup and the measurement results in Section 2. A detailed analysis of signal degradation sources and their consequences for the signal-to-noise ratio (SNR) of the respective signal follows in Section 3. Finally, the work is summarized, the core results are presented and starting points for further investigations are pointed out in Section 4.

## 2. Sensor Evaluation in the Electrical Domain

The evaluation of FBGs that are employed as optical sensors is conventionally carried out in the optical domain. Light, e.g., from a laser source, is fed into the sensor, and the power of the reflected or transmitted light is analyzed. This can be realized in a variety of ways. A source with a broad spectrum can be used, and the changes introduced by the FBG, which depend on the measurand and the FBG itself, can be analyzed with an optical spectrum analyzer (OSA). Alternatively, a tuneable laser source (TLS) can be used as the optical source and an optical power meter as the evaluation unit. Then, through a wavelength sweep of the TLS, the spectrum of the amplitude characteristic can be monitored. Both methods are costly in terms of hardware expenses and sampling time. A simpler but in general less accurate method employs a fixed wavelength laser source and a power meter. Provided that an initial state of the sensor is known or that the sensor characteristic is fully unambiguous, this scheme allows for the possible relative measurements to be translated into absolute ones.

In the following, the alternative idea of an MWP-supported evaluation of fiber-based sensors is presented.

### 2.1. Basic Concept

The fundamental concept of evaluating a fiber sensor in the electrical domain is based on the introduction of a microwave signal, as discussed by the authors before in detail [21–26]. This microwave signal modulates the light source, and the modulated light is fed into the sensor, which changes not only the signal's amplitude but also its phase. Depending on the configuration, the transmitted or the reflected signal is then opto-electrically converted by a photodiode (PD). Finally, the resulting analog electrical microwave signal is digitized and processed.

### 2.2. Experimental Setup

Figure 1 shows the block diagram of the experimental setup used for the investigations in this paper. For manufacturers and models of the used components, please refer to this diagram. Further parameters and the overall signal flow are explained in the following.

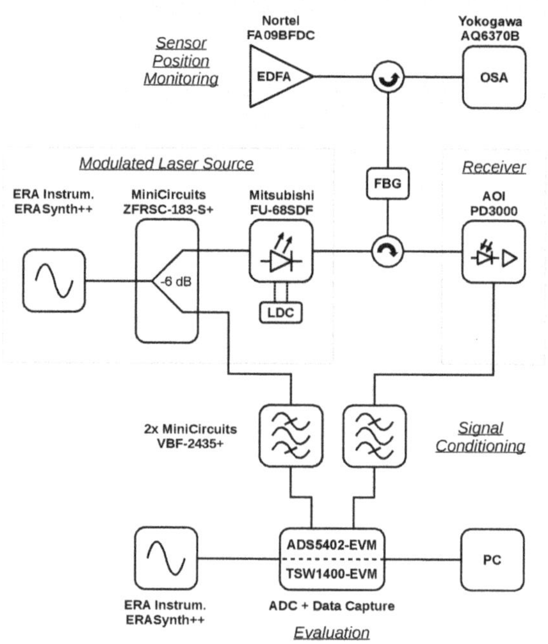


**Figure 1.** Block diagram of the experimental setup.

A chirped FBG is used as the sensor. It is designed for a full width at half minimum (FWHM) transmission bandwidth of 8 nm around a central wavelength of 1536.5 nm. Further characteristics are its length of 7 mm, its chirp $\frac{d\Lambda}{dz}$ of 1 nm mm$^{-1}$ and its apodization with a Gaussian cosine function for providing a delay characteristic as linear as possible across the FWHM bandwidth. Tuning is realized by a fine-threaded adjusting screw that is coupled to a cantilever's free tail. The FBG is bonded to the surface of the cantilever that bends convexly while tuning. The same FBG and tuning mechanism has been used in previous studies [26].

Due to the absence of a scale on the adjustment mechanism, the sensor position is monitored in the optical spectrum by a broadband optical source and an OSA via an optical circulator in the reflection configuration. An erbium-doped fiber amplifier (EDFA) is used for the broadband optical source. The OSA monitors a range from 1520 nm to 1560 nm at a resolution of 0.2 nm and 1001 points; the EDFA is set to an output power of 10.7 dBm.

A distributed feedback (DFB) laser diode (LD) provides the optical measurement signal. It is driven by a custom laser diode controller (LDC) at a forward current of 60 mA and a temperature of 21 °C. The DFB-LD is modulated by a microwave sine signal that is generated by a signal generator. Its output is configured for a frequency of 2.45 GHz and an output power of 16 dBm. The microwave signal is split into two paths in order to calculate a phase relation, which is crucial for the evaluation scheme.

The modulated optical signal is fed via a circulator into the second port of the FBG and then opto-electrically converted. This is carried out by a custom module that consists of a biased PD and a two-stage radio frequency (RF) amplifier with a gain of 9 dB.

Both signal paths—the directly branched off reference path and the sensor path—contain anti-aliasing filters as the last element before the signals are digitized by an analog-to-digital converter (ADC) and finally processed on a personal computer (PC). The filters are necessary due to the operation of the ADC in the 17th Nyquist zone at a sampling rate of 300 MHz in single-shot mode, collecting 65,536 samples per channel for each trigger event.

Previous results of the authors [26] have compared this approach with a benchmark setup that is based on an electrical vector network analyzer (EVNA) and proven its validity. Therefore, this step does not need to be repeated here. Besides that, the focus of this paper is on the noise analysis of the ADC-based setup as some performance drawbacks come along with the advantage of lower hardware expenses compared with the EVNA-based setup.

### 2.3. Derivation of the Characteristic Curves

Furthermore, in [26], the authors have derived in detail how the characteristic curves of FBG-based sensors can be determined. To summarize, the FBG is tuned, and its spectrum is thereby swept over the fixed, small bandwidth optical signal of the modulated laser source. A width of 0.5 nm and a range from 1533 nm to 1540 nm have been chosen for the tuning parameters of the FBG's center wavelength. After a settling time of 1 min to account for transient effects of the laser diode, five 20 s delayed single-shot measurements are triggered at each of the approached positions. The gathered data contain a time series of amplitude values for each ADC channel, i.e., for the reference signal and for the sensor signal. A sine wave is fitted to slices of both time discrete signals, resulting in parameters for frequency, amplitude, phase and offset. The fitting is carried out by an optimizer and repeated for 10 slices per single-shot measurement, with each slice having a length of 6000 samples. Finally, the mean and standard deviation for the resulting total of 50 values per sampling position (10 slices $\times$ 5 single shot measurements) are calculated for the amplitude and the phase. From the results, the amplitude characteristic can be derived directly from the sensor signal. For the phase characteristic, the sensor signal's phase has to be related to the reference signal's phase by subtracting the latter from the former. Finally, both characteristic curves are normalized to the center wavelength of the interval under study.

### 2.4. Experimental Outcomes

Figures 2 and 3 show the measurement results for the amplitude characteristic and for the phase characteristic, respectively. The curves follow the known amplitude and phase characteristics of the sensor, and the standard deviations are small. Judging by the available resolution, the phase characteristic appears to be an injective function, which is beneficial in contrast to the clearly non-injective amplitude characteristic. Between the measured positions of the used FBG, local extrema could exist, making the phase response a non-injective function as well, which limits the achievable sensor resolution. However, the final characteristic can be specially tailored as part of the sensor design. Furthermore, the macroscopically approximate linear progression of the phase's course provides an almost constant sensitivity for signal deviations. Possible sources for signal deviations that are inherent to the measurement setup and their impacts on the evaluation scheme are analyzed in the subsequent section. For this analysis, minimum and maximum absolute peak values of the signal magnitudes are relevant. These can be extracted from the raw data, resulting in the minimum peak voltage $V_{\text{Signal,peak,min}} = 41.50\,\text{mV}$ and the maximum peak voltage $V_{\text{Signal,peak,max}} = 72.33\,\text{mV}$.

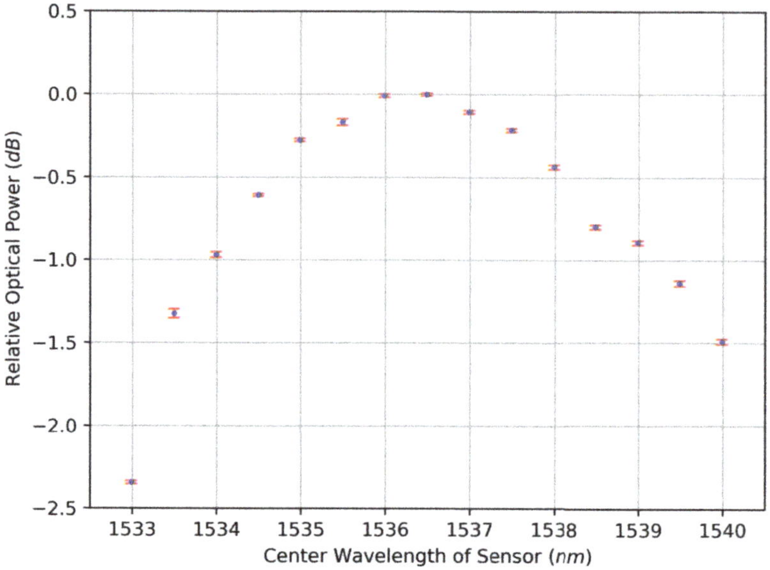


**Figure 2.** Measured dependency of the optical power from the center wavelength of the sensor. Mean (blue) and standard deviation (red) of 50 readings per sensor position.

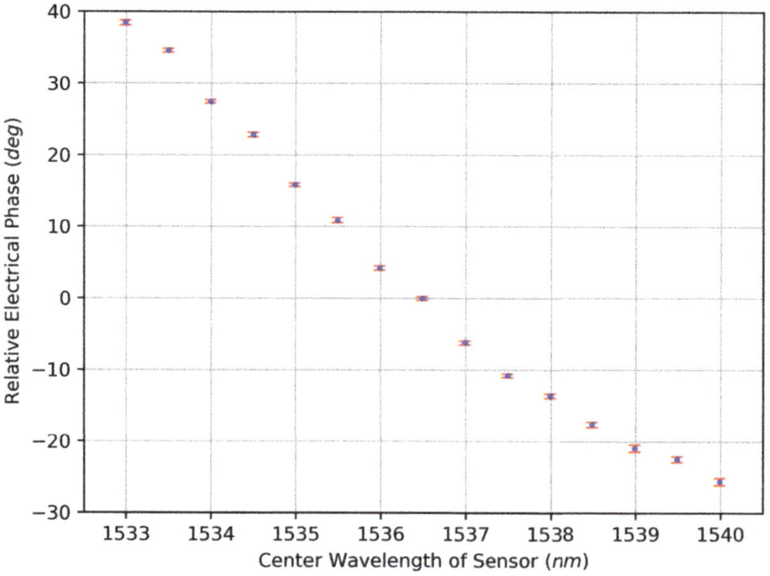


**Figure 3.** Measured dependency of the electrical phase from the center wavelength of the sensor. Mean (blue) and standard deviation (red) of 50 readings per sensor position.

## 3. Noise Analysis

Following the fact that utilizing the phase response of an FBG as a sensor characteristic is a novel and, up to this point, not well-researched approach, studying the advantages and disadvantages in comparison to prevalent evaluation schemes is of great interest. Therefore, sources of signal noise are identified, their impact on the measurement signal is quantified and the findings are discussed in this section.

### 3.1. Sources of Signal Degradation

The noise sources listed below refer to the setup depicted in Figure 1 and are based on the technical performance of the used devices.

#### 3.1.1. Phase Noise of the Signal Generators

The phase noise of the modulating signal generator has no effect on the amplitude evaluation provided that the amplitude characteristic is approximately linear in the range of values covered by the noise. This is because the lower sideband (LSB) and the upper sideband (USB) in this modulation regime contribute in equal shares to the sensor signal and both sidebands will jitter mirror symmetrically around the optical carrier. At the same time, phase noise of the modulation signal can influence the phase-based measurement if the reference path and the signal path have a length difference greater than the coherence length of the modulation signal. Without matching the path lengths, the coherence length of the signal has to be taken into account, which is known for electrical signal sources to be significantly longer than relevant to the setup. The ADC is sensitive to clock jitter, which translates to a sensitivity to the phase noise of the second signal generator. According to the datasheet of the ADC [27], the SNR due to clock jitter is calculated after

$$SNR_{\text{Jitter,Amplitude}} = -20 \cdot \log_{10}(2\pi \cdot f_{\text{In}} \cdot t_{\text{Jitter}})\,. \tag{1}$$

For the configured input frequency $f_{\text{In}}$ of 2.45 GHz and a jitter $t_{\text{Jitter}}$ according to

$$t_{\text{Jitter}} = \sqrt{t_{\text{Jitter,Clock}} + t_{\text{Jitter,ADC Aperture}}}\,, \tag{2}$$

with the jitter of the clock $t_{\text{Jitter,Clock}}$ being circa 350 fs and the internal aperture jitter of the ADC $t_{\text{Jitter,ADC Aperture}}$ being 100 fs, the SNR due to clock jitter $SNR_{\text{Jitter,Amplitude}}$ is calculated to 45.0 dB, corresponding to clock-jitter-induced noise power of $P_{\text{Jitter,Amplitude}} = -62.7$ dBm for the weakest signal of the experiment.

#### 3.1.2. Phase Noise of the ADC

The internal aperture jitter $t_{\text{Jitter,ADC Aperture}}$ of the ADC generates phase noise for the quantized signal and therefore defines the SNR of the phase evaluation $SNR_{\text{Jitter,Phase}}$. After quantization, the sensor signal frequency is 50 MHz, which gives a period $T$ of 20 ns. One period equals 360°, but from Figure 3, an effective phase range $\phi_{\text{Sensor}}$ of 65° becomes apparent for the used sensor. The SNR for the phase evaluation is therefore calculated after

$$SNR_{\text{Jitter,Phase}} = 10 \cdot \log_{10}\left(\frac{\phi_{\text{Sensor}} \cdot T}{360° \cdot t_{\text{Jitter,ADC Aperture}}}\right) \tag{3}$$

and yields the result $SNR_{\text{Jitter,Phase}} = 45.6$ dB. For an optimal sensor covering the full range of 360°, the result is $SNR_{\text{Jitter,Phase}} = 53.0$ dB, marking the best possible SNR for a phase evaluation based on the used ADC.

#### 3.1.3. Quantization Noise

An analog-to-digital conversion introduces quantization noise to the converted signal. The calculation of the root-mean-square (RMS) quantization noise voltage

$$V_{\text{QN}} = \frac{q}{\sqrt{12}} = \frac{V_{\text{Ref}}}{\sqrt{12} \cdot 2^N} \tag{4}$$

of an ideal ADC with the weight of a least significant bit $q$, which is the result of the ADC's reference voltage $V_{\text{Ref}}$ and its number of bits $N$, has been established before [28].

The general definition of an SNR is given by the signal power $P_{\text{Signal}}$ and the noise power $P_{\text{Noise}}$ according to

$$SNR = \frac{P_{\text{Signal}}}{P_{\text{Noise}}} = \frac{V_{\text{Signal}}^2}{V_{\text{Noise}}^2} = \frac{I_{\text{Signal}}^2}{I_{\text{Noise}}^2}. \tag{5}$$

Hence, the SNR due to quantization noise is calculated using Equations (4) and (5) according to

$$SNR_{\text{QN}} = \left( \frac{V_{\text{Signal,eff.}}}{\frac{V_{\text{Ref}}}{\sqrt{12 \cdot 2^N}}} \right)^2. \tag{6}$$

For a sine wave as the signal, its peak signal voltage $V_{\text{Signal,peak.}} = \sqrt{2} \cdot V_{\text{Signal,eff.}}$ leads, when applied to Equation (6), to the SNR from quantization noise in decibel:

$$SNR_{\text{QN}} = 20 \cdot \log_{10}\left( 2^N \cdot \sqrt{6} \cdot \frac{V_{\text{Signal,peak}}}{V_{\text{Ref}}} \right). \tag{7}$$

Two significant SNRs can be calculated, one for the lowest input signal level and one for the highest. For peak voltages of $V_{\text{Signal,peak,min}} = 41.50\,\text{mV}$ and $V_{\text{Signal,peak,max}} = 72.33\,\text{mV}$, an effective number of bits $ENOB = 9.8$ [27] inserted for $N$ and a reference voltage level $V_{\text{Ref}} = 1.2\,\text{V}$, the two SNRs result in $SNR_{\text{QN,min}} = 37.6\,\text{dB}$ and $SNR_{\text{QN,max}} = 42.4\,\text{dB}$, corresponding to the quantization noise power of $P_{\text{QN}} = -55.2\,\text{dBm}$.

#### 3.1.4. Johnson–Nyquist Noise

The power of the Johnson–Nyquist noise $P_{\text{JNN}}$ depends on the Boltzmann constant $k_{\text{B}}$, the measurement bandwidth $BW$ and the temperature $T$. Following the Nyquist–Shannon sampling theorem and taking all bandwidth-restricting components of the setup into account, the measurement bandwidth is 150 MHz. Because the measurements were conducted at a temperature of 21 °C, calculating the noise power using

$$P_{\text{JNN}} = 10 \cdot \log_{10}\left( \frac{k_{\text{B}} \cdot T \cdot BW}{1 \cdot 10^{-3}\text{W}} \right) \tag{8}$$

leads to a value of $-92.2\,\text{dBm}$.

#### 3.1.5. Relative Intensity Noise

The relative intensity noise (RIN) of the used DFB-LD is specified with better than $-145\,\text{dB}\,\text{Hz}^{-1}$ for the applied operating conditions. Considering the optical power of 10 dBm and the bandwidth of 150 MHz results in a RIN power of $P_{\text{RIN}} = -53.2\,\text{dBm}$.

#### 3.1.6. Shot Noise

Two sources of shot noise can be identified from the setup: the laser diode and the PD. The shot noise of the laser diode is part of its RIN, which has already been discussed. The shot noise of the PD depends on the photocurrent $I_{\text{Photo}}$. The definition of the RMS shot noise current is

$$\sqrt{\overline{i^2}} = \sqrt{2 \cdot e \cdot I_{\text{Photo}} \cdot BW}. \tag{9}$$

With Equation (5), the SNR due to shot noise is calculated by

$$SNR_{\text{SN}} = \left( \frac{I_{\text{Signal,eff.}}}{\sqrt{\overline{i^2}}} \right)^2. \tag{10}$$

The effective signal current, which is only one part of the total photocurrent, is calculated according to

$$I_{\text{Signal,eff.}} = \frac{V_{\text{Signal,peak}}}{Z_0 \cdot \sqrt{2G}}, \tag{11}$$

with characteristic impedance of $Z_0 = 50\,\Omega$ and gain of $G = 9\,\text{dB}$ of the RF amplifier in the receiver block of the setup. Combining Equations (9)–(11) and transferring the finding to a decibel scale results in

$$SNR_{\text{SN}} = 10 \cdot \log_{10}\left(\frac{V^2_{\text{Signal,peak}}}{4 \cdot e \cdot {Z_0}^2 \cdot G \cdot BW \cdot I_{\text{Photo}}}\right). \tag{12}$$

Again, minimum and maximum values for the peak signal voltage have to be taken into account. The photocurrent equals the product of the PD's sensitivity $S = 0.95\,\text{mA}\,\text{mW}^{-1}$ and the power of the incident light $P_{\text{Optical}}$:

$$I_{\text{Photo}} = S \cdot P_{\text{Optical}}. \tag{13}$$

It can be seen from the setup that $P_{\text{Optical}}$ is the sum of the LD's and EDFA's output powers, which is less than the losses of the optical path.Altogether, $P_{\text{Optical}}$ totals to not more than 7.8 dBm, which generates a photocurrent of 5.7 mA. With these values inserted, the $SNR_{\text{SN}}$ ranges from $SNR_{\text{SN,min}} = 52.0\,\text{dB}$ to $SNR_{\text{SN,max}} = 56.8\,\text{dB}$, corresponding to the shot noise power $P_{\text{SN}} = -69.6\,\text{dBm}$.

#### 3.1.7. Further Influences

Some more characteristics influence the SNR. Additional amplitude noise can be introduced to the DFB-LD by the LDC. Measurements have shown an amplitude noise of the used custom LDC is not greater than the noise of typical commercial LDCs. Existing chirp of the DFB-LD has no effect on the measurement as the optical phase is not evaluated. The 6 dB splitter could have a temperature dependence, resulting in imbalanced amplitudes and phases for the branches. The evaluation scheme would tolerate such changes for the expected small gradients, which make the changes much slower than the measurement speed. Further on, the RF amplifier's noise figure *NF* impacts the SNR. This affects only the SNR resulting from the Johnson–Nyquist noise, the relative intensity noise of the LD and the shot noise of the PD, which is following from the RF amplifier's position in the signal path. Due to a high integration level of the receiver module, the noise figure of the used RF amplifier could not be specified as part of the present work. In addition, the ADC can theoretically introduce phase jitter between the two channels to the measurement. However, this is an unlikely mechanism as the datasheet indicates a fully parallel architecture. Finally, the sensor sensitivity has to be considered. Any relative shift in the FBG's spectrum relative to the LD's spectrum, independent from its cause (e.g., temperature changes, mechanical impacts), will deteriorate the measurement.

### *3.2. Discussion*

Several noise sources have been identified and quantified for the observed setup up to this point. The total amplitude noise power $P_{\text{AN}}$ resulting from the known contributions can be calculated according to

$$P_{\text{AN}} = \sqrt{P^2_{\text{Jitter,ADC}} + P^2_{\text{QN}} + NF^2 \cdot \left(P^2_{\text{JNN}} + P^2_{\text{RIN}} + P^2_{\text{SN}}\right)}. \tag{14}$$

As the *NF* of the used RF amplifier is unknown, a typical value of 6 dB is assumed. Inserting this along with the previously calculated results into Equation (14) sums up to $P_{\text{AN}} = -47.1\,\text{dBm}$, resulting in a range of SNRs from $SNR_{\text{AN,min}} = 29.5\,\text{dB}$ to $SNR_{\text{AN,max}} = 34.3\,\text{dB}$. It can be seen that the dominating noise contributions are the quantization noise and the RIN of the LD. Accordingly, some room for improvement is observable. An LD with a better RIN can be used. More advantages can be gained from matching the ADC input level better to the reference voltage, which improves the quantization noise, and from reducing the input frequency or using an ADC with a suitable analog bandwidth, leading to lower noise due to clock jitter. Finally, the shot noise can be reduced by roughly 67 % solely by removing the sensor position monitoring because the EDFA contributes double the power to the total optical power compared to the LD.

As a consequence of the phase measurement being a relative evaluation, phase jitter of the modulating RF signal is inherently eliminated as long as the coherence length of the RF signal is not exceeded by the setup. The only degradation of the phase information is introduced by the ADC aperture jitter, leading to a signal-to-phase noise ratio of $SNR_{\mathrm{Jitter,Phase}} = 45.6\,\mathrm{dB}$. This can be improved by using an ADC with less internal aperture jitter or by optimizing the phase response characteristic of the FBG to cover a greater phase range, whereby an SNR improvement potential of more than 7 dB has been found for the latter.

It is clear from the previous considerations that the SNR of the amplitude evaluation greatly depends on the performance of many core components in the setup. An improvement in the SNR can be achieved only by costly changes to these components, especially to the LD and ADC but also to the RF amplifier stage. At the same time, using a reasonably selected sensor, the phase evaluation approach depends solely on the ADC performance. This involves not only a systematically better SNR but also offers the potential to select lower-quality components for the rest of the setup without degrading the SNR. For the investigated setup, the SNR of the phase evaluation approach is expected to be at least $SNR_{\mathrm{AN,max}} - SNR_{\mathrm{Jitter,Phase}} = 11\,\mathrm{dB}$ and at best more than $SNR_{\mathrm{AN,min}} - SNR_{\mathrm{Jitter,Phase}} = 16\,\mathrm{dB}$ better than for the amplitude evaluation approach. Albeit these values are specific to the used components, they highlight the finding that the phase evaluation is particularly advantageous over the conventional amplitude-based evaluation. Another benefit of the phase interrogation is the overall course of its characteristic curve due its broad linearity, which makes it generally less sensitive to relative shifts between sensor and laser sources.

## 4. Conclusions

This paper has explained the idea of basing the read-out of fiber sensors on the electrical evaluation of a microwave signal, which is introduced to modulate an optical source with a fixed wavelength, and distinguished this approach from the prior art of microwave photonics in sensing applications. An experimental setup and the measurement results have been presented. Two different characteristic curves of the same sensor can be interrogated with the approach: amplitude and phase. The setup has been analyzed for noise sources that degrade the microwave signal, and the consequences for either of the evaluation variants have been discussed.

The results indicate that strong advantages can be gained from using the phase-based approach because its only signal deterioration source is the ADC aperture jitter. An SNR gain of at least 11 dB has been achieved with the examined setup compared to the amplitude-based evaluation, which is considerably affected by more noise contributions. At the same time, it has been found that specific adjustments of the setup can improve the SNR for the amplitude-based measurement. These open the opportunity for further studies. Among the possible starting points are the quantization noise and the RIN of the optical source.

Furthermore, as has been addressed by the authors before [24], the implementation of this scheme in wireless sensing systems is of particular interest, as the already present microwave signal makes it predestined to adapt the concept for RoF measurement applications. After using the phase evaluation as the core of an RoF sensing system, even more advantages can be expected. This is because the phase of a wireless signal typically gets considerably less degraded during transmission than the amplitude of the same signal. Finally, the combined evaluation of phase and amplitude and accompanied potential information gain could be of interest for future studies. Further subjects of interest are the scheme's sensitivity to the measurand, its response time and the maximum achievable sampling rate, among others.

**Author Contributions:** Conceptualization, U.N., T.T., K.K. and N.N.; methodology, U.N. and N.N.; software, U.N.; validation, U.N.; formal analysis, U.N. and N.N.; investigation, U.N., T.T. and K.K.; resources, U.N., T.T., K.K. and N.N.; data curation, U.N.; writing—original draft preparation, U.N.; writing—review and editing, U.N. and N.N.; visualization, U.N.; supervision, N.N.; project administration, N.N.; funding acquisition, T.T., K.K. and N.N. All authors have read and agreed to the published version of the manuscript.

**Funding:** We acknowledge financial support by the Open Access Publishing Fund of the Clausthal University of Technology.

**Institutional Review Board Statement:** Not applicable.

**Informed Consent Statement:** Not applicable.

**Data Availability Statement:** The data presented in this study are available on request from the corresponding author. The raw data are not publicly available due to limited relevance to the study.

**Conflicts of Interest:** The authors declare no conflict of interest.

## Abbreviations

The following abbreviations are used in this manuscript:

| | |
|---|---|
| ADC | analog to digital converter |
| DFB | distributed feedback |
| DSB | double-sideband |
| EDFA | erbium-doped fiber amplifier |
| EVNA | electrical vector network analyzer |
| FBG | fiber Bragg grating |
| FWHM | full width at half minimum |
| LD | laser diode |
| LDC | laser diode controller |
| LSB | lower sideband |
| MWP | microwave photonics |
| OSA | optical spectrum analyzer |
| PC | personal computer |
| PD | photodiode |
| RF | radio frequency |
| RIN | relative intensity noise |
| RMS | root-mean-square |
| RoF | radio-over-fiber |
| SNR | signal-to-noise ratio |
| SSB | single-sideband |
| TLS | tuneable laser source |
| USB | upper sideband |

## References

1. Seeds, A.J.; Williams, K.J. Microwave Photonics. *IEEE/OSA J. Light. Technol.* **2006**, *24*, 4628–4641. [CrossRef]
2. Yao, J. Microwave Photonics. *IEEE/OSA J. Light. Technol.* **2009**, *27*, 314–335. [CrossRef]
3. Capmany, J.; Mora, J.; Gasulla, I.; Sancho, J.; Lloret, J.; Sales, S. Microwave Photonic Signal Processing. *IEEE/OSA J. Light. Technol.* **2013**, *31*, 571–586. [CrossRef]
4. Xu, K.; Wang, R.; Dai, Y.; Yin, F.; Li, J.; Ji, Y.; Lin, J. Microwave photonics: Radio-over-fiber links, systems, and applications. *Photonics Res.* **2014**, *2*, B54–B63. [CrossRef]
5. Marpaung, D.; Yao, J.; Capmany, J. Integrated microwave photonics. *Nat. Photonics* **2019**, *13*, 80–90. [CrossRef]
6. Yao, J. Microwave Photonic Sensors. *IEEE/OSA J. Light. Technol.* **2021**, *39*, 3626–3637. [CrossRef]
7. Zhou, D.; Dong, Y.; Yao, J. Truly Distributed and Ultra-Fast Microwave Photonic Fiber-Optic Sensor. *IEEE/OSA J. Light. Technol.* **2020**, *38*, 4150–4159. [CrossRef]
8. Hervás, J.; Madrigal, J.; Barrera, D.; Sales, S. Multiplexing FBG sensors combining microwave photonics and phase modulation. In Proceedings of the 2017 19th International Conference on Transparent Optical Networks (ICTON), Girona, Spain, 2–6 July 2017; pp. 1–4. [CrossRef]
9. Pan, S.; Yao, J. Photonics-Based Broadband Microwave Measurement. *IEEE/OSA J. Light. Technol.* **2017**, *35*, 3498–3513. [CrossRef]

10. Yao, J. Microwave Photonics for High-Resolution and High-Speed Interrogation of Fiber Bragg Grating Sensors. *Fiber Integr. Opt.* **2015**, *34*, 204–216. [CrossRef]
11. Wang, G.; Liao, B.; Cao, Y.; Guo, T.; Feng, X.; Guan, B.O.; Yao, J. Microwave Photonic Interrogation of a High-Speed and High-Resolution Multipoint Refractive Index Sensor. *IEEE/OSA J. Light. Technol.* **2022**, *40*, 1245–1251. [CrossRef]
12. Cao, Y.; Wang, X.; Guo, T.; Ran, Y.; Feng, X.; Guan, B.O.; Yao, J. High-resolution and temperature-compensational HER2 antigen detection based on microwave photonic interrogation. *Sens. Actuators B Chem.* **2017**, *245*, 583–589. [CrossRef]
13. Liu, W.; Fu, H.; Zhang, A.P.; He, S. Fiber Bragg Grating Based Wireless Sensor Module With Modulated Radio-Frequency Signal. *IEEE Microw. Wirel. Compon. Lett.* **2010**, *20*, 358–360. [CrossRef]
14. Hernández, R.; Loayssa, A.; Benito, D. On the use of single-sideband modulation for optical vector network analysis. In Proceedings of the MWP 2003 Proceedings: International Topical Meeting on Microwave Photonics, Budapest, Hungary, 10–12 September 2003; pp. 173–176. [CrossRef]
15. Hernández, R.; Loayssa, A.; Benito, D. Optical vector network analysis based on single-sideband modulation. In Proceedings of the 16th Annual Meeting of the IEEE Lasers and Electro-Optics Society, Tucson, AZ, USA, 27–28 October 2003; Volume 2, pp. 909–910. [CrossRef]
16. Erro, M.J.; Loayssa, A.; Taínta, S.; Hernández, R.; Benito, D.; Garde, M.J.; Muriel, M.A. Phase Reconstruction for the Frequency Response Measurement of FBGs. In Proceedings of the 2007 IEEE International Symposium on Intelligent Signal Processing, Alcala De Henares, Spain, 3–5 October 2007; pp. 1–3. [CrossRef]
17. Erro, M.J.; Loayssa, A.; Taínta, S.; Hernández, R.; Benito, D.; Garde, M.J.; Muriel, M.A. On the Measurement of Fiber Bragg Grating's Phase Responses and the Applicability of Phase Reconstruction Methods. *IEEE Trans. Instrum. Meas.* **2011**, *60*, 1416–1422. [CrossRef]
18. Sartiano, D.; Hervás, J.; Madrigal, J.M.; Pérez-Galacho, D.; Sales, S. On the Use of Microwave Photonics Techniques for Novel Sensing Applications. In Proceedings of the 2019 21st International Conference on Transparent Optical Networks (ICTON), Angers, France, 9–13 July 2019; pp. 1–4. [CrossRef]
19. Sales, S.; Barrea, D.; Hervás, J.; Madrigal, J. Microwave Photonics for Optical Fiber Sensors. In Proceedings of the 2019 Optical Fiber Communications Conference and Exhibition (OFC), San Diego, CA, USA, 3–7 March 2019; pp. 1–3. [CrossRef]
20. Hervás, J.; Ricchiuti, A.L.; Li, W.; Zhu, N.H.; Fernández-Pousa, C.R.; Sales, S.; Li, M.; Capmany, J. Microwave Photonics for Optical Sensors. *IEEE J. Sel. Top. Quantum Electron.* **2017**, *23*, 327–339. [CrossRef]
21. Neumann, N.; Schuster, T.; Plettemeier, D. Novel approach for simultaneous wireless transmission and evaluation of optical sensors. In *Advanced Sensor Systems and Applications VI*; Liu, T., Jiang, S., Neumann, N., Eds.; International Society for Optics and Photonics, SPIE: Bellingham, WA, USA, 2014; Volume 9274. [CrossRef]
22. Schuster, T.; Neumann, N.; Plettemeier, D.; Körbitz, R.; Richter, A. A fiber-optic pH sensor with wireless Radio over Fiber read-out. In Proceedings of the 2015 IEEE SENSORS, Busan, Republic of Korea, 1–4 November 2015; pp. 1–4. [CrossRef]
23. Nordmeyer, U.; Neumann, N.; Wang, X.; Thiel, T.; Kojucharow, K.; Plettemeier, D. Microwave based electrical read-out of optical sensors. In Proceedings of the 20. GMA/ITG-Fachtagung Sensoren und Messsysteme, Nürnberg, Germany, 25–26 June 2019; pp. 119–124. [CrossRef]
24. Nordmeyer, U.; Neumann, N.; Wang, X.; Thiel, T.; Kojucharow, K.; Plettemeier, D. Application of microwave based electrical read-out of fiber Bragg gratings in thermometry. In Proceedings of the SMSI 2020—Sensors and Measurement Science International, Nürnberg, Germany, 20–25 June 2020; AMA Service: Wunstorf, Germany, 2020; pp. 181–182. [CrossRef]
25. Nordmeyer, U.; Neumann, N.; Wang, X.; Thiel, T.; Kojucharow, K.; Plettemeier, D. Evaluation of optical fibre sensors in the electrical domain. *J. Sens. Sens. Syst.* **2020**, *9*, 199–208. [CrossRef]
26. Nordmeyer, U.; Neumann, N.; Thiel, T.; Kojucharow, K.; Plettemeier, D. Using the Phase Response of Fiber Bragg Gratings for Measurement Applications. In Proceedings of the Sensors and Measuring Systems, 21. ITG/GMA-Symposium, Nuremberg, Germany, 10–11 May 2022; pp. 200–203.
27. Texas Instruments Inc. *Dual 12-Bit 800Msps Analog-to-Digital Converter, ADS5402 Datasheet (Rev. B)*; Texas Instruments Inc.: Dallas, TX, USA, 2014.
28. Kester, W.; Analog Devices Inc. *Data Conversion Handbook*; Analog Devices Series; Elsevier Science: Amsterdam, The Netherlands, 2005.

sensors

Article

# Noise Cancellation of Helicopter Blade Deformations Measurement by Fiber Bragg Gratings

Raoul R. Nigmatullin [1,*], Timur Agliullin [2], Sergey Mikhailov [3], Oleg Morozov [2], Airat Sakhabutdinov [2], Maxim Ledyankin [3] and Kamil Karimov [2]

1 Department of Radio-Electronic and Information-Measuring Tools, Kazan National Research Technical University Named after A.N. Tupolev-KAI, K. Marx Str. 10, 420111 Kazan, Russia
2 Department of Radiophotonics and Microwave Technologies, Kazan National Research Technical University Named after A.N. Tupolev-KAI, K. Marx Str. 10, 420111 Kazan, Russia; taagliullin@mail.ru (T.A.); microoil@mail.ru (O.M.); azhsakhabutdinov@kai.ru (A.S.); mail12kamil2000@mail.ru (K.K.)
3 Department of Aero-Hydrodynamics, Kazan National Research Technical University Named after A.N. Tupolev-KAI, K. Marx Str. 10, 420111 Kazan, Russia; samikhailov@kai.ru (S.M.); maledyankin@kai.ru (M.L.)
* Correspondence: renigmat@gmail.com

**Abstract:** The work presents data treatment methods aimed at eliminating the noise in the strain sensor data induced by vibrations of the helicopter blade in flight conditions. The methods can be applied in order to enhance the metrological performance of the helicopter weight estimation system based on the deformation measurement of the main rotor blades. The experimental setup included a composite plate fixed to the vibrating stand on the one end, with six fiber-optic strain sensors attached to its surface. In this work, the procedure of the optimal linear smoothing (POLS) and 3D-invariant methods were used to obtain monotone calibration curves for each detector, thereby making it possible to distinguish the increase of load applied to the free end of the plate with an increment of 10 g. The second method associated with 3D invariants took into account 13 quantitative parameters defined as the combination of different moments and their intercorrelations up to the fourth-order inclusive. These 13 parameters allowed the calculation of the 3D surface that can serve as a specific fingerprint, differentiating one set of initial data from another one. The combination of the two data treatment methods used in this work can be applied successfully in a wide variety of applications.

**Keywords:** fiber Bragg grating; noise cancellation; helicopter blade deformation measurement; procedure of the optimal linear smoothing; 3D-invariant method



**Citation:** Nigmatullin, R.R.; Agliullin, T.; Mikhailov, S.; Morozov, O.; Sakhabutdinov, A.; Ledyankin, M.; Karimov, K. Noise Cancellation of Helicopter Blade Deformations Measurement by Fiber Bragg Gratings. *Sensors* **2021**, *21*, 4028. https://doi.org/10.3390/s21124028

Academic Editor: Damien Kinet

Received: 26 April 2021
Accepted: 9 June 2021
Published: 11 June 2021



## 1. Introduction

Helicopter main rotor blades are essential components that define the flight performance and safety of the helicopter. Therefore, monitoring of their state is of the utmost importance during helicopter exploitation. Moreover, a sophisticated aeroelastic analysis of the rotor blades is necessary at the early stages of helicopter development, since the blades are subjected to significant displacements during flight and the aerodynamic conditions around the rotor are highly unsteady and complex [1]. In order to perform the abovementioned tasks, real-time blade deformation measurements are required, which enable the blade wear or damage detection during helicopter operation as well as the evaluation of blade displacements during development testing procedures.

The existing approaches to helicopter blade deformation measurement can be divided into four main categories: the systems based on the detection of reflected light [2–4], the systems based on laser units [5], the systems based on electrical resistance strain gauges [6–8], and the fiber-optic strain measurement systems [9–11].

The methods of the first group capture the light reflected from the rotor blades in order to determine their deformation. Thus, the system presented in [2] uses light from the external source that is modulated by the rotating blades to define the amplitude of blade flapping, its velocity, bending, etc. The system described in [3] comprises CCD detectors installed in the helicopter fuselage and is able to determine the blade pitch angle and flapping. The work [4] presents an approach to blade deformation monitoring based on the digital image correlation (DIC) technique, in which the surface of the blade is painted with a high-contrast random dot pattern. Two high-resolution digital cameras are separated by a certain distance and view the same area on the blade so that the parallax is used to measure three-dimensional displacements of the blade surface. In general, the main advantage of the systems based on the reflected light detection is that they do not require the structural modification of the blade. On the other hand, the performance of such systems depends on the conditions of the ambient environment, such as weather conditions, altitude, time of day, disturbances from external light sources.

In order to increase the measurement accuracy of blade deformation, in comparison with the aforesaid methods, systems based on laser units have been proposed. Thus, in [5], Projection Moiré Interferometry (PMI) with infrared laser source is applied to measure dynamic rotor blade deflections. Similarly, to the previous group of approaches, the laser units-based methods are realized without blade structural modifications, while not being affected by disturbances from external visible light sources. However, the accuracy of such approaches also depends on the weather conditions and optical density of the environment.

Electrical resistance strain gauges have been applied for the development of rotor blade strain measurement systems for more than four decades [6]. The NASA/Army UH-60A Airloads Program was completed in 1994 [7,8] and performed over 200 flights, collecting data from the instrumentation system installed on the main rotor of the helicopter. The system included 25 strain gauges, accelerometers, pressure transducers, temperature sensors, and acquired a vast amount of data, including deformation to estimate the airloads. Unlike the methods that detect optical radiation reflected from the blades, the systems based on electrical resistance strain gauges are immune to the influence of weather conditions and generally provide higher accuracy and resolution of measurements. However, the implementation of electrical resistance strain gauges requires the corresponding modifications of the blade structure. In addition, despite the low weight and compact dimensions of the gauges, the installation of a high number of such sensors may lead to a significant increase in the overall blade weight due to the metal wires connecting the gauges.

At present, a considerable amount of research activity is aimed at the development of fiber-optic sensor systems for blade deformation measurement. The main advantages of FBG-based techniques are low weight, immunity to electromagnetic interference, small dimensions (the optical fiber diameter is 125 μm), and the absence of an electrical power supply to the sensors [9]. Moreover, optical fibers can be embedded into composite structures during fabrication [10] or attached to the surface of the blade [11]. The experimental setup presented in [12] includes 60 fiber-optic sensors based on fiber Bragg gratings (FBG) nonuniformly allocated along four optical fibers so that one-third of the sensors are positioned close to the blade root where the strain gradients are higher. The system is used to measure blade deformation in vertical and lateral directions. The system discussed in [13] uses 16 FBGs interrogated by optical frequency domain reflectometry (OFDR) and is capable of retrieving the first and second bending modes of the blade.

The application of blade deformation measurements to aeroelastic analysis requires the detection of blade oscillatory deformations during aerodynamic testing procedures. However, an alternative application for such systems can be found in the vertical displacement definition of the blades in order to estimate the gross vehicle weight of a helicopter. The development of such systems significantly contributes to the safety of helicopters, since, at present, the weight is generally determined by manually tracking the weight of cargo, passengers, fuel, etc., which does not guarantee adequate accuracy. The work [14] proposes a system that calculates the weight of a helicopter by measuring and averaging the distances between the fuselage and the underside of each blade of the main rotor while in hover or level unaccelerated forward flight. The system comprises two laser distance measuring devices at the front and rear of the fuselage. Predetermined calibration data are used to compensate for the effects of vertical and horizontal drag forces on the fuselage due to the spinning rotor and forward velocity. The task of blade vertical displacement measurement can be also accomplished using strain sensors installed on the blades [14], similarly to the previously mentioned developments. In this application, the oscillatory behavior of the blade vertical displacements caused by aeroelastic effects may deteriorate the accuracy and resolution of the weight measurement. The problem is illustrated in Figure 1, in which Figure 1a represents a qualitative graph of a signal from the strain sensor located on a steady blade under the load that increased over time in five fixed steps, while Figure 1b shows the qualitative graph of the signal from the same sensor on the oscillating blade under the same sequence of loads. As it can be seen from this example, the oscillations significantly complicate the task of distinguishing the loads acting on the blade.

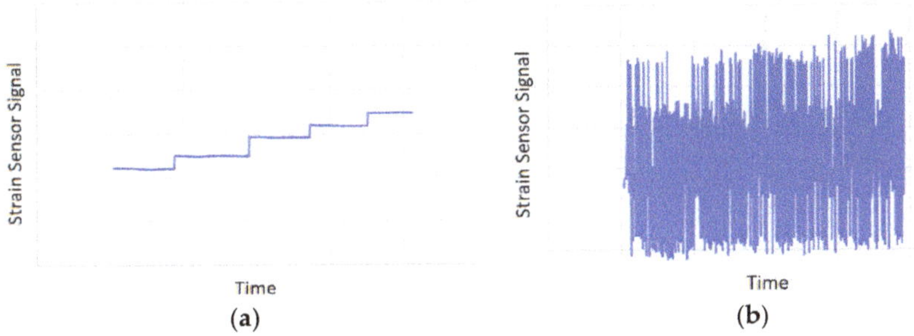


**Figure 1.** Example of data obtained by a strain sensor located on a blade for five cases of vertical loads: (**a**) without oscillations; (**b**) with oscillations.

The aim of the current work is to propose the method for the elimination of noise in the strain measurement data caused by the blade oscillations, thereby enhancing the metrological performance of the helicopter weight measurement system.

## 2. Deformation Measurement Using Fiber Bragg Gratings

A fiber Bragg grating (FBG) is a segment of an optical fiber that reflects particular wavelengths of optical radiation and transmits the others. FBGs are manufactured by inscribing a systematic variation of the refractive index of the fiber core. The central wavelength of the reflected light is called the Bragg wavelength and is defined according to the following generally accepted formulation:

$$\lambda_B = 2n_{eff}\Lambda, \quad (1)$$

where $n_{eff}$ is the effective refractive index of the grating, and $\Lambda$ is the grating period, i.e., the period of the refractive index variation.

The working principle of FBG is based on the change of either the effective refractive index ($n_{eff}$) or the grating period ($\Lambda$) and the resulting shift of its central wavelength ($\lambda_B$) when the sensor is subjected to physical fields, such as strain, temperature, pressure, humidity, etc. [15]. Figure 2a shows the schematic representation of an FBG, and Figure 2b presents the typical spectral response of an FBG for transmitted and reflected light.

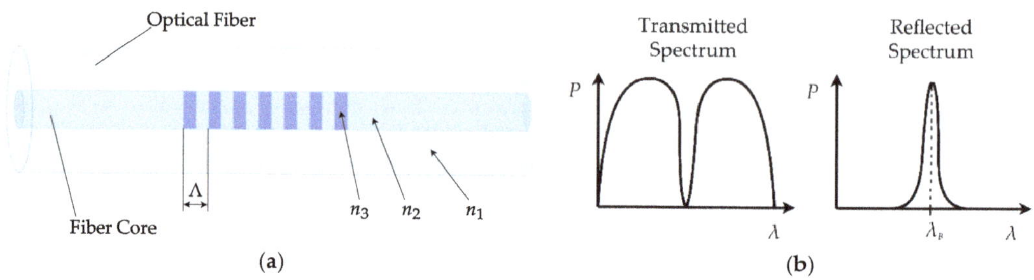


**Figure 2.** (**a**) Schematic representation of a fiber Bragg grating: Λ is the grating period, $n_1$ is the refractive index of the fiber cladding, $n_2$ is the refractive index of the fiber core, and $n_3$ is the refractive index of periodic variations forming the FBG; (**b**) spectra of transmitted and reflected light of an FBG.

The most common techniques for FBG multiplexing and interrogation, such as wavelength [16], time [17], frequency [18], polarization [19], and spatial [20] division are implemented using complex electro-optical devices, including spectrum analyzers, tunable Fabry–Perot interferometers, diffraction gratings, etc. The common disadvantage of the mentioned interrogation techniques is the complexity and high cost of devices required for their implementation. In order to simplify the FBG interrogation, the usage of addressed fiber Bragg structures (AFBSs) has been proposed [21,22]. An AFBS is a type of FBG, the spectral response of which has two ultra-narrowband components. The frequency spacing between the components is called the address frequency of the AFBS, it is in the microwave range and unique for each sensor in the system, and does not change when the AFBS is subjected to strain or temperature variations. The microwave-photonic approach for AFBS interrogation utilizes a broadband light source, an optical filter with predefined linear slope spectral response, and a photodetector. The central wavelength shift of AFBS is defined using the amplitude of the beating signal generated at the photodetector at the address frequencies of the sensors. This approach is further developed with the introduction of multiaddressed fiber Bragg structures (MAFBSs) [23], which have three or more ultra-narrowband components in their spectral response. The combination of address frequencies expands the sensor capacity of the measurement system and increases the accuracy of the MAFBS central wavelength definition.

The FBG interrogation device used in the current work implemented conventional wavelength-division multiplexing of FBG sensors, where each grating had a different Bragg wavelength so that their spectral responses did not superimpose on each other during measurements. The device was based on the Ibsen® I-MON spectrum analyzer and was operated by custom software.

## 3. Noise Cancellation Methods

### *3.1. Description of the POLS Method*

In order to decrease the noise evoked by vibro-oscillations, we use the procedure of the optimal linear smoothing (POLS) that was successfully applied earlier in papers [24–29] covering different regions of physics.

The basic formula is given by the following expression:

$$Ysm_j(w) = \frac{\sum_{i=1}^{N} K\left(\frac{x_j - x_i}{w}\right) y_i}{\sum_{i=1}^{N} K\left(\frac{x_j - x_i}{w}\right)}, \quad K(t) = \exp\left(-t^2\right), \; j = 1, 2, \ldots, N \tag{2}$$

Here, *K*(*t*) is a smoothing kernel given in the form of the Gaussian function. The value *w* determines the width of the smoothing window. We want to remark the following attractive features of this expression:

(a) This expression is linear with respect to the function $y_i$ subjected to the smoothing procedure. Therefore, it does not contain any additional distortions evoked by the possible treatment of a function $y_i$;
(b) If w >> 1, then (as it is easily seen from Expression (2)) the smoothed function $Ysm_j$ coincides with its mean value;
(c) If $w$ becomes close to the zero value, then the smoothing kernel $K(t)$ coincides practically with delta function $\delta(x_j - x_i)$ and therefore, in this case, $Ysm_j\,(0) \cong y_j$.

These attractive features allow the selection of high-frequency oscillations and leave only low-frequency "trend" that is proved to be useful for practical applications.

In this paper, we apply the simple Expression (2) for the elimination of high-frequency oscillations, and the remaining trend allows us to select the desired parameters that differentiate the oscillations without load from oscillations with a load.

### *3.2. Description of 3D-DGI Method*

In this section, we describe the mathematical details associated with the derivation of the complete discrete geometrical invariant (DGI) in 3D space. We remind here that preliminary results based on the application of the incomplete DGI form of the fourth order in 3D space are outlined recently in [29]. Let us consider the complete form of the fourth-order power law.

$$\begin{aligned}
L_k^{(4)} &= A_{40}^{(1,0)}(y_1 - r_{1k})^4 + A_{40}^{(2,0)}(y_2 - r_{2k})^4 + A_{40}^{(3,0)}(y_3 - r_{3k})^4 - \\
&- B_{22}^{(1,2)}(y_1 - r_{1k})^2 \cdot (y_2 - r_{2k})^2 - B_{22}^{(1,3)}(y_1 - r_{1k})^2 \cdot (y_3 - r_{3k})^2 \\
&- B_{22}^{(2,3)}(y_2 - r_{2k})^2 \cdot (y_3 - r_{3k})^2 + C_{211}^{1(2,3)}(y_1 - r_{1k})^2 \cdot (y_2 - r_{2k}) \cdot (y_3 - r_{3k}) \\
&+ C_{211}^{2(1,3)}(y_2 - r_{2k})^2 \cdot (y_1 - r_{1k}) \cdot (y_3 - r_{3k}) + \\
&+ C_{211}^{3(1,2)}(y_3 - r_{3k})^2 \cdot (y_1 - r_{1k}) \cdot (y_2 - r_{2k}) \\
&- \tfrac{1}{2} D_{31}^{(1,2)}(y_1 - r_{1k}) \cdot (y_2 - r_{2k}) \left[ (y_1 - r_{1k})^2 + (y_2 - r_{2k})^2 \right] - \\
&- \tfrac{1}{2} D_{31}^{(1,3)}(y_1 - r_{1k}) \cdot (y_3 - r_{3k}) \left[ (y_1 - r_{1k})^2 + (y_3 - r_{3k})^2 \right] - \\
&- \tfrac{1}{2} D_{31}^{(2,3)}(y_2 - r_{2k}) \cdot (y_3 - r_{3k}) \left[ (y_2 - r_{2k})^2 + (y_3 - r_{3k})^2 \right].
\end{aligned} \tag{3}$$

In Expression (3), the upper indices define the combination of the variables $y_\alpha$ ($\alpha$ = 1, 2, 3) fixing the location of an arbitrary point $M(y_1, y_2, y_3)$ in 3D space, the low indices determine the values of the power law exponents that correspond to the algebraic form of the fourth order. The choice of the sign's combination ($\pm$) before the constants in (3) will be explained below. Three random sequences are determined by the values $r_{\alpha k}$ ($\alpha$ = 1, 2, 3; $k$ = 1, 2, ..., $N$). Expression (1) represents itself the complete form of the fourth order that contains the combination of three variables associated with an arbitrary point $M(y_1, y_2, y_3)$ and three arbitrary sequences $r_{\alpha k}$. The desired DGI is obtained from the following requirement:

$$\frac{1}{N}\sum_{k=1}^{N} L_k^{(4)} = I_4 \tag{4}$$

In order to remove in Expression (4) the cubic terms, we introduce the variables

$$Y_\alpha = y_\alpha - \langle r_\alpha \rangle, \ \langle r_\alpha \rangle = \frac{1}{N}\sum_{k=1}^{N} r_{\alpha k} \tag{5}$$

and nullify the linear terms. This requirement helps us to separate the desired variables $Y_\alpha$ from each other and keep only the terms of the second and fourth orders, correspondingly. In order to decrease the number of constants in (3) and derive the DGI not depending on some additional constants, one defines three key ratio constants $R^{(\alpha,\beta)}$, with ($\alpha$,$\beta$) = (1,2), (1,3), (2,3),

$$
\begin{aligned}
&R^{(\alpha,\beta)} = \frac{B^{(\alpha,\beta)}}{A} = \frac{C^{\gamma(\alpha,\beta)}}{A} = \frac{D^{(\alpha,\beta)}}{A}, \\
&A_{40}^{(\alpha)} = A_{40}^{(\beta)} = A_{40}^{(\gamma)} \equiv A, \ \ \alpha,\beta,\gamma = 1,2,3.
\end{aligned} \tag{6}
$$

It is convenient also to introduce the following notations for the integer moments and their intercorrelations and present them as

$$
\begin{aligned}
&Q_{\alpha^n\beta^m\gamma^l} = \frac{1}{N}\sum_{k=1}^{N}\left((\Delta r_{3k})^m(\Delta r_{2k})^n(\Delta r_{1k})^l\right) \equiv \left\langle(\Delta r_\alpha)^m(\Delta r_\beta)^n(\Delta r_\gamma)^l\right\rangle, \\
&\alpha \geq \beta \geq \gamma, \ (\alpha,\beta,\gamma) = 1,2,3.
\end{aligned} \tag{7}
$$

In the result of the introduced notations (6) and (7), the system of linear equations for the finding of unknown ratios $R^{(\alpha,\beta)}$ from the nullification requirement of the entering linear terms accepts the following form:

$$
\begin{aligned}
&\left[2Q_{221} - Q_{332} + \tfrac{3}{2}Q_{211} + \tfrac{1}{2}Q_{222}\right] \cdot R^{(1,2)} + \\
&+\left[2Q_{331} - Q_{322} + \tfrac{3}{2}Q_{311} + \tfrac{1}{2}Q_{333}\right] \cdot R^{(1,3)} - 2Q_{321} \cdot R^{(2,3)} = 4Q_{111}, \\
&\left[2Q_{211} - Q_{331} + \tfrac{3}{2}Q_{221} + \tfrac{1}{2}Q_{111}\right] \cdot R^{(1,2)} - 2Q_{321} \cdot R^{(1,3)} + \\
&+\left[2Q_{332} - Q_{311} + \tfrac{3}{2}Q_{322} + \tfrac{1}{2}Q_{333}\right] \cdot R^{(2,3)} = 4Q_{222}, \\
&-2Q_{321} \cdot R^{(1,2)} + \left[2Q_{311} - Q_{221} + \tfrac{3}{2}Q_{331} + \tfrac{1}{2}Q_{111}\right] \cdot R^{(1,3)} + \\
&+\left[2Q_{322} - Q_{211} + \tfrac{3}{2}Q_{332} + \tfrac{1}{2}Q_{222}\right] \cdot R^{(2,3)} = 4Q_{333}.
\end{aligned} \tag{8}
$$

The linear system of equations helps to reduce three moments ($Q_{333}$, $Q_{222}$, $Q_{111}$) and seven intercorrelations of the third order ($Q_{332}$, $Q_{322}$, $Q_{221}$, $Q_{211}$, $Q_{331}$, $Q_{311}$, $Q_{321}$) to the calculation of three unknown ratios $R^{(\alpha,\beta)}$ only. We should notice also that the combination of the algebraic signs in (3) is chosen in that way for the keeping of the partial solution $R = 1$ of system (8) in the case when all three random sequences $r_{\alpha k}$ are identical to each other, i.e., $r_{1k} = r_{2k} = r_{3k}$. It is natural to define it as the case of spherical symmetry. If only two sequences coincide with each other (for example, $r_{1k} = r_{2k} \neq r_{3k}$), then we deal with the case of the cylindrical symmetry. In this case, the linear system (8) is reduced to a couple of linear equations relatively the variables $R^{(1,2)} \neq R^{(1,3)} = R^{(2,3)}$. After averaging procedure applied to Expression (4), the structure of the fourth-order form can be rewritten as

$$
K_4(Y_1,Y_2,Y_3) + K_2(Y_1,Y_2,Y_3) = I_4 \tag{9}
$$

The fourth- and the second-order forms entering to the left-hand side can be presented as

$$
\begin{aligned}
&K_4(Y_1,Y_2,Y_3) = Y_1^4 + Y_2^4 + Y_3^4 + R^{(1,2)}Y_1Y_2\left[Y_3^2 - \tfrac{1}{2}(Y_1+Y_2)^2\right] + \\
&+R^{(1,3)}Y_1Y_3\left[Y_2^2 - \tfrac{1}{2}(Y_1+Y_3)^2\right] + R^{(2,3)}Y_2Y_3\left[Y_1^2 - \tfrac{1}{2}(Y_2+Y_3)^2\right].
\end{aligned} \tag{10a}
$$

$$
\begin{aligned}
K_2(Y_1,Y_2,Y_3) = {} & A_{11}Y_1^2 + A_{22}Y_2^2 + A_{33}Y_3^2 + \\
& + A_{12}Y_1Y_2 + A_{13}Y_1Y_3 + A_{23}Y_2Y_3.
\end{aligned} \tag{10b}
$$

The constants $A_{\alpha\beta}$ figuring in Expression (10b) are defined as

$$
\begin{aligned}
&A_{11} = 6Q_{11} - \left(Q_{22} + \tfrac{3}{2}Q_{21}\right)R^{(1,2)} - \left(Q_{33} + \tfrac{3}{2}Q_{31}\right)R^{(1,3)} + Q_{32}R^{(2,3)}, \\
&A_{22} = 6Q_{22} - \left(Q_{11} + \tfrac{3}{2}Q_{21}\right)R^{(1,2)} + Q_{31}R^{(1,3)} - \left(Q_{33} + \tfrac{3}{2}Q_{32}\right)R^{(2,3)}, \\
&A_{33} = 6Q_{33} + Q_{21}R^{(1,2)} - \left(Q_{11} + \tfrac{3}{2}Q_{31}\right)R^{(1,3)} - \left(Q_{22} + \tfrac{3}{2}Q_{32}\right)R^{(2,3)}, \\
&A_{12} = -\left(4Q_{21} + \tfrac{3}{2}Q_{11} + \tfrac{3}{2}Q_{22} - Q_{33}\right)R^{(1,2)} + 2Q_{32}R^{(1,3)} + 2Q_{31}R^{(2,3)}, \\
&A_{13} = 2Q_{32}R^{(1,2)} - \left(4Q_{31} + \tfrac{3}{2}Q_{11} + \tfrac{3}{2}Q_{33} - Q_{22}\right)R^{(1,3)} + 2Q_{12}R^{(2,3)}, \\
&A_{23} = 2Q_{31}R^{(1,2)} + 2Q_{21}R^{(1,3)} - \left(4Q_{32} + \tfrac{3}{2}Q_{22} + \tfrac{3}{2}Q_{33} - Q_{11}\right)R^{(2,3)}.
\end{aligned} \tag{11}
$$

The constant $I_4$ (defined by 3 moments and 12 intercorrelations of the fourth order) figuring in the right-hand side of (9) is defined as

$$\begin{aligned} I_4 = Q_{1111} + Q_{2222} + Q_{3333} - \left(Q_{2211} - Q_{3321} + \tfrac{1}{2}Q_{2111} + \tfrac{1}{2}Q_{2221}\right)R^{(1,2)} - \\ -\left(Q_{3311} - Q_{3221} + \tfrac{1}{2}Q_{3111} + \tfrac{1}{2}Q_{3331}\right)R^{(1,3)} - \\ -\left(Q_{3322} - Q_{3211} + \tfrac{1}{2}Q_{3222} + \tfrac{1}{2}Q_{3332}\right)R^{(2,3)}. \end{aligned} \tag{12}$$

It is interesting to notice that in the case of the spherical symmetry ($r_{1k} = r_{2k} = r_{3k}$), all correlations coincide with each other and the value of $I_4$ equals zero. The form of the fourth order (9) admits the separation of the variables in the spherical system of coordinates. If one accepts the following conventional notations:

$$\begin{aligned} y_1 &= \langle y_1 \rangle + R\sin\theta\cos\varphi, \\ y_2 &= \langle y_2 \rangle + R\sin\theta\sin\varphi, \\ y_3 &= \langle y_3 \rangle + R\cos\theta, \\ &0 \le \theta \le \pi,\ 0 \le \varphi \le 2\pi, \end{aligned} \tag{13}$$

then substitution of these variables into (9) leads to the following biquadratic equation relatively the unknown radius $R(\theta,\varphi)$:

$$[R(\theta,\varphi)]^4 + \left(\frac{P_2(\theta,\varphi)}{P_4(\theta,\varphi)}\right)[R(\theta,\varphi)]^2 - \frac{I_4}{P_4(\theta,\varphi)} = 0. \tag{14a}$$

The desired solution ($R(\theta, \varphi) > 0$) is written as

$$R(\theta,\varphi) = \left[\frac{\sqrt{P_2^2(\theta,\varphi) + 4I_4 \cdot P_4(\theta,\varphi)} - P_2(\theta,\varphi)}{2P_4(\theta,\varphi)}\right]^{\frac{1}{2}} \tag{14b}$$

The polynomials $P_{2,4}(\theta,\varphi)$ entering in (14) are defined by the following expressions:

$$\begin{aligned} P_4(\theta,\varphi) = \sin^4\theta \cdot \cos^4\varphi + \sin^4\theta \cdot \sin^4\varphi + \cos^4\theta + \\ +R^{(1,2)}\sin^2\theta\sin\varphi\cos\varphi\left[\cos^2\theta - \tfrac{\sin^2\theta}{2}(\sin\varphi + \cos\varphi)^2\right] + \\ +R^{(1,3)}\sin\theta\cos\theta\cos\varphi\left[\sin^2\theta\sin^2\varphi - \tfrac{1}{2}(\sin\theta\cos\varphi + \cos\theta)^2\right] + \\ +R^{(2,3)}\sin\theta\cos\theta\sin\varphi\left[\sin^2\theta\cos^2\varphi - \tfrac{1}{2}(\sin\theta\sin\varphi + \cos\theta)^2\right] \end{aligned} \tag{15a}$$

$$\begin{aligned} P_2(\theta,\varphi) = A_{11}\sin^2(\theta)\cos^2(\varphi) + A_{22}\sin^2(\theta)\sin^2(\varphi) + \\ +A_{33}\cos^2(\theta) + A_{12}\sin^2(\theta)\sin(\varphi)\cos(\varphi) + \\ +A_{13}\sin(\theta)\cos(\theta)\cos(\varphi) + A_{23}\sin(\theta)\cos(\theta)\sin(\varphi). \end{aligned} \tag{15b}$$

The last Expressions (13)–(15) determine the final form of the DGI in 3D space. It includes three surfaces determined by Expression (13). The further analysis shows that Expression (14b) equals zero (because $I_4 = 0$) in the case of the coincidence of three compared random sequences ($r_{1k} = r_{2k} = r_{3k}$). The radius $R(\theta,\varphi)$ can contain the complex expression when the integrand in (14b) becomes negative. It accepts the negative values when the constant $I_4$ (which in most cases is defined by Expression (12)) becomes negative. In this case, it is convenient to rewrite expression (13) in the form

$$\begin{aligned} y_1 &= \langle y_1 \rangle + |R(\theta,\varphi)|\sin\theta\cos\varphi, \\ y_2 &= \langle y_2 \rangle + |R(\theta,\varphi)|\sin\theta\sin\varphi, \\ y_3 &= \langle y_3 \rangle + |R(\theta,\varphi)|\cos\theta, \\ |R(\theta,\varphi)| &= \sqrt{[\mathrm{Re}(R(\theta,\varphi))]^2 + [\mathrm{Im}(R(\theta,\varphi))]^2}, \\ &0 \le \theta < \pi,\ 0 \le \varphi < 2\pi. \end{aligned} \tag{16}$$

The curves defined by Equation (16) facilitate considerably the numerical analysis of initial data. As it follows from this preliminary analysis, this 3D surface is determined by the combination of 13 parameters: three moments of the first order $<y_\alpha>$, $\alpha = 1, 2, 3$ from (3), six correlators of the second-order $A_{\alpha\beta}$ from (11), three reduced correlators $R^{(\alpha,\beta)}$ from (8), and invariant of the fourth-order $I_4$ from (12). We want to stress here again that the final Expressions (14) and (15) do not use any model and are determined completely by the measured data together with their measurement errors. Finishing this section, one can say that this method can be applied for the reduction of initial data. It is necessary to note that the dimension of the radius $[R(\theta,\varphi)]$ coincides with the dimension of initial data $y_\alpha$.

This reduction procedure can be divided into the following stages:

1. Initially, any available data can be written in the form of rectangle matrix $[N \times M]$, where number $N$ ($j = 1, 2, \ldots, N$—number of rows) determines the given data points and $M$ ($m = 1, 2, \ldots, M$—columns) determines the number of the repeated measurements forming in total the statistically significant sampling. As the result of the application of the 3D-DGI method, we obtain the reduced matrix $[M \times S]$, where each column of the reduced matrix ($Pr_{m,s}$: $<y_\alpha>(3)$, $R^{(\alpha,\beta)}(3)$, $A_{\alpha\beta}(6)$, $I_4(1)$; $\alpha,\beta = 1, 2, 3$) determines the complete combination of the moments and their intercorrelations (3 + 3 + 6 + 1 = 13) up to the fourth-order inclusive. In the result of the application of the 3D-DGI method, we obtain $s = 1, 2, \ldots, S$ ($S = 13$) distributions that demonstrate the variations of each statistical parameter $Pr_s(m)$ with respect to the number of repeated measurements ($m = 1, 2 \ldots, M$).

2. The further reduction is possible if one takes into account that each random function $y_s(m) \equiv Pr_s(m)$ is located inside the rectangle $M \times (\text{Range}[y_s(m)])$, where $Range(f) = \max(f) - \min(f)$. For comparison of one random function $y_{1,s}(m)$ with another $y_{2,s}(m)$ corresponding to the chosen parameter $s$ ($s = 1, 2, \ldots, S$), one can use the following simple formula:

$$Q_{1,2}(s) = \frac{Range(y_{1,s}) + Range(y_{2,s})}{\max(y_{1,s}, y_{2,s}) - \min(y_{1,s}, y_{2,s})} \tag{17}$$

This expression, in spite of its simplicity, is really effective for the comparison of the statistical closeness of a pair of random functions belonging to the given/another sampling participating in the comparison operation. Really, if the function $Q_{1,2}(s)$ is located in the interval [1,2], then the pair random functions are statistically close to each other. In the case when $Q_{1,2}(s) \in [0,1)$, one can conclude that the pair random functions compared are statistically different. Besides this important parameter (16), one can take into account the symmetry of the random function $y_1(m)$. Any random function located in the rectangle $M \times Range[y(m)]$ crosses the line $<Pr(m)>$, coinciding with its mean value. Therefore, for evaluation of the symmetry of a random function, one can introduce the value

$$Sm(y) = \frac{\text{mean}(y) - 0.5 \cdot (\max(y) + \min(y))}{Range(y)} \tag{18}$$

If the value $Sm(y)$ is located near zero ($Sm(y) \approx 0$), then the line $<y>$ divides the rectangle $M \times Range[y(m)]$ into two almost equal parts. In other cases, the value $Sm(y) \in [-0.5, 0.5]$ determines the measure of asymmetry. After the application of Expression (17) for comparing similar columns (belonging to the same parameter $Pr_s$), one can receive finally the vector of the length $S = 13$ that contains information about the statistical closeness of two matrices compared. It is interesting to notice that simple Expression (17) can be used also for comparison each successive measurement with another one in the given rectangle matrix $[N \times M]$. If one compares the vectors $y_m$ forming the columns of the initial matrix with each other, then in the result of application (17), one can obtain the symmetrical matrix $U(m_1,m_2)$ ($m_{1,2} = 1, 2, \ldots, M$) with elements located in the interval $0 \leq U(m_1,m_2) \leq 2$. Only elements located in interval $1 \leq U(m_1,m_2) \leq 2$ will correspond to a "good" experiment, while the elements from interval $0 \leq U(m_1,m_2) < 1$ should be considered as possible "outliers" and correspond to "bad" experiment.

3. How to find the parameters belonging only to one matrix in order to compare them with similar parameters of another tested matrix in cases when the reference matrix is absent? Initially, it is necessary to scale each column to the same interval

$$Prn_s = \frac{DPr_s}{Range(\mathrm{Pr}_s)} \equiv \frac{Pr_s - \langle Pr_s \rangle}{Range(\mathrm{Pr}_s)},$$
$$-\frac{1}{2} \leq Prn_s \leq \frac{1}{2},\ \langle Pr_s \rangle = \frac{1}{M}\sum_{m=1}^{M} Pr_{m,s}. \tag{19}$$

This normalization procedure makes all parameters $Pr_s$ statistically close to each other with mean value equaled zero mean($Prn_s$) = 0 and with the $Range(Prn_s)$ = 1. If one integrates Expression (19) for each parameter one can receive the statistically different curves $JP_s$ = Integral($Prn_s$) for each initial parameter ($s$ = 1, 2, … , $S$). The distributions of the ranges of these integral curves $P_1 = Range(JP_s)$, together with the distribution of asymmetries $P_2 = Range(Sm(JP_s))$ calculated with the help of Expression (18), give finally the matrix containing [$S \times 2$] containing ($S$ = 13) rows and two columns only. If we calculate the ranges of these two columns, we obtain finally two values only that can characterize the initial matrix [$N \times M$]. If we have a set of matrices $[N \times M]_q$ ($q$ = 1, 2, … , $Q$), then this simple and general procedure allows us to select the "best" one having minimal values of these two key parameters $P_{1,2}$. It will characterize the stability of the initial sequence and their minimal values will serve as a criterion for the selection of the "best" TLS among other TLS(s). We want to emphasize here that this final stage of treatment of "big" matrices differs from the procedure used in the paper [29]. Earlier, one of us (RRN) had the set of rectangle matrices that can be characterized as "normal/reference" ones and compared them with "strange/tested" matrices associated with defects. Data that will be analyzed below do not contain this information. Therefore, we propose another procedure described above for the selection of the "best" data expressed in the form of rectangle matrices.

Concluding this section, it is necessary to mention the following. The POLS is the universal tool that can be applicable for smoothing all available data. It possesses some remarkable features that make it really universal for many applications, which include the following:

1. The POLS is a linear tool, and it does not distort initial data;
2. When the value of the smoothing window tends to zero ($w \to 0$), then the smoothed replica coincides with the initial data. In another limiting case, when w >> 1, the smoothed replica coincides with its arithmetic mean.

Another new instrument as the 3D geometrical invariants can be defined as the "universal" tool as well, by virtue of the following features:

1. Thanks to 13 universal parameters defining the feature space, it allows the comparison of the different random sequences having different natures;
2. It can be applied to analyses of the TLS and, therefore, this tool forms a universal platform that cannot contain treatment errors;
3. Section 3, given above, gives an example for its application to real data.

We should also note that both methods keep the units of the initial data.

## 4. Experimental Results and Their Treatment

### 4.1. Experimental Setup

In order to imitate the process of helicopter blade deformation measurement, an experimental setup was designed that included a composite plate with six strain detectors based on fiber Bragg gratings attached to the upper surface of the plate using epoxy adhesive. The FBGs used in the experiment were manufactured at the Department of Radiophotonics and Microwave Technologies of KNRTU-KAI (Kazan) using a common fabrication technique based on the Lloyd interferometer. Figure 3a presents a photograph of the experimental plate, and Figure 3b shows a schematic representation of the detector arrangement on the plate (positions 1.1–1.3, 2.1–2.3), as well as the main dimensions.

During the experimental procedure, the plate was fixed on one end to the vibrating stand, while the opposite end remained free, mimicking the vibrations of a helicopter blade. To the free end of the plate, a load with predefined mass was attached at point *A* (see Figure 3b). The values of the Bragg wavelength shifts of all FBG detectors were acquired for seven different load cases ($L$ = 0, 50, 60, . . . , 100 g). It must be noted that in this preliminary research, the authors did not pursue the goal to compare the vibrations with the real-life reference, which will be the aim of the subsequent studies. The measurements were taken during a certain time interval (~8 s) so that 800 points of data were registered with the rate of ~100 Hz for each load case. The data were collected using an FBG interrogation device developed at KNRTU-KAI based on the Ibsen I-MON 512 interrogation monitor.

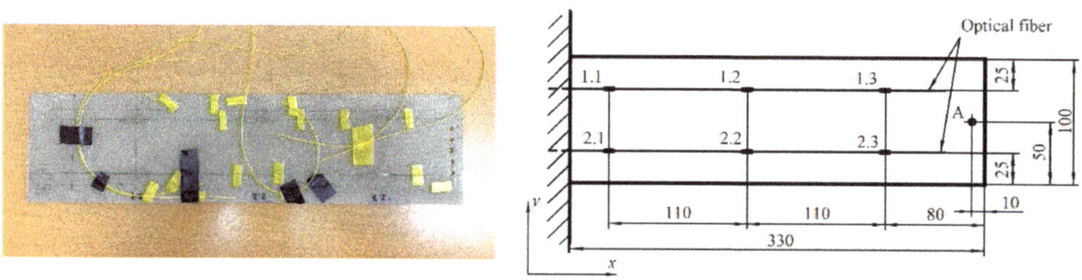


(a) (b)

**Figure 3.** Experimental plate with six FBG detectors: (**a**) photograph; (**b**) schematic representation of the plate with the positions of the detectors (1.1–1.3, 2.1–2.3), and the position of the load application (point A) (dimensions are in mm).

### 4.2. Treatment Procedure

Based on the treatment procedure explained in Section 3, we received seven matrices. Each matrix included 6 columns and 800 data points corresponding to different loads ($L$ = 0, 50, 60, . . . , 100 g), which were subjected to the same vibrations measured in nm as the wavelength shifts of the fiber-optic detectors. Each column inside the matrix 800 × 6 included the recorded vibrations associated with the fixed detector. Each detector shown in Figure 3b was tightly associated with a number of columns (their correspondence to each other shown below), and it was located on the plate imitating the helicopter blade.

In order to select the desired value of the smoothing window $w$ figuring in Expression (2), we chose a simple criterion that the low-frequency trends obtained after POLS should be almost uncorrelated with each other. If we fix the interval of correlations $[-0.5 \times 10^{-3}, 0.5 \times 10^{-3}]$ for the smoothed trends and require that

$$-0.0005 \le \frac{Ysm(0,w) \cdot Ysm(L,w)}{\sqrt{(Ysm(0,w))^2 (Ysm(L,w))^2}} \le 0.0005, \quad (20)$$

where $Ysm(L,w)$ is determined by Expression (2) and corresponds to the smoothed function for the given load $L$ = 0, 50, 60, . . . , 100 g, then from condition (20), one can find the desired value $w$ from the interval [100–200]. We chose the value $w$ = 150, corresponding to the middle of the found interval. The figures below demonstrate the low-frequency fluctuations that were obtained for detectors D-1.1 and D-2.2 for comparison with Figure 4a,b. We notice that after application of the POLS the smoothed curve associated with fluctuations with load is shifted up. It takes place because condition (20) allows decreasing the value of the initial correlations.

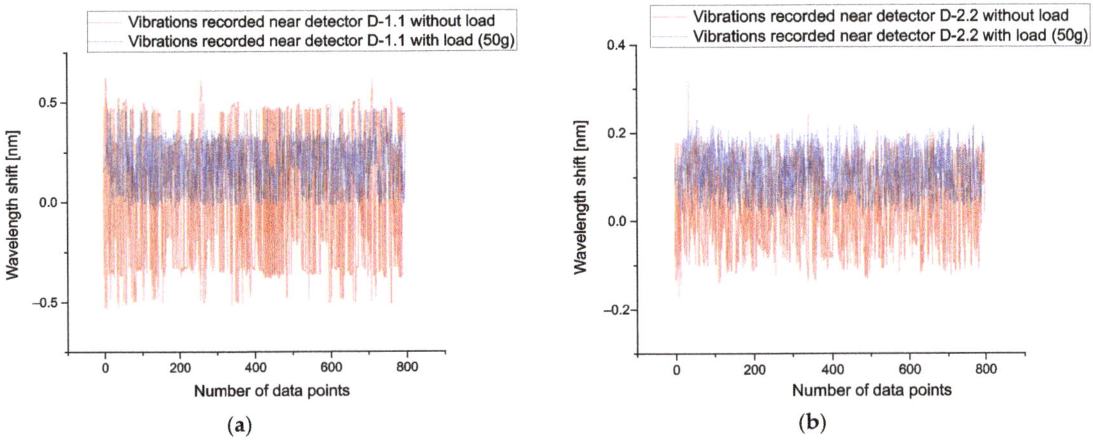


**Figure 4.** (**a**) This figure shows the level of vibrations without loads (red lines) and with minimal load corresponding to the load equal to 10 g (blue lines). This figure reflects the noise fluctuation for the first detector D-1.1. In order to stress their difference, the same picture Figure 4 (**b**) is placed on the right-hand side for detector D-2.2 (with load 10 g) and without load (red lines), correspondingly. For other detectors, the obtained pictures look the same and therefore they are omitted.

Figure 5a demonstrates the results of the application POLS for $w = 150$ and detector D-1.1. Figure 5b shows the same result for D-2.2 at $w = 150$, as well. Analysis of these curves shows that for construction of the calibration curves showing the dependence of each detector with respect to the applied load one can select the differences between mean values corresponding to these smoothed curves.

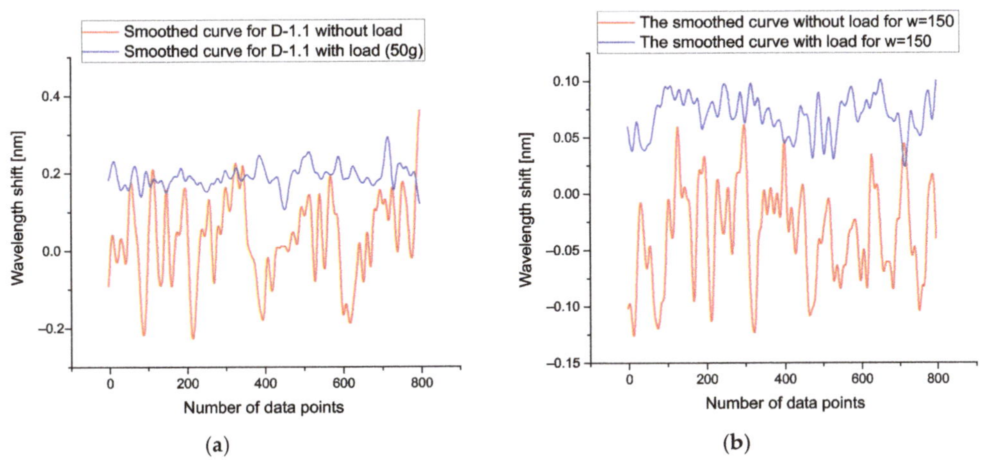


**Figure 5.** (**a**) Results of the application of POLS for $w = 150$ and detector D-1.1; (**b**) the same results for D-2.2 at $w = 150$.

Figure 6a demonstrates six calibration curves calculated for all types of detectors. All curves are monotone and we see the increase of these curves with respect to the applied load. Figure 6b comprises the data calculated with respect to Expression (17). It demonstrates the sensitivity of different detectors with respect to the applied load. Three detectors D-2.3, D-1.1, and D-2.2 are the most sensitive, while the detector D-2.1 has minimal sensitivity. The detectors D-1.2 and D-1.3 have an intermediate sensitivity. We highlight again that the mean values conserve the vibration units.

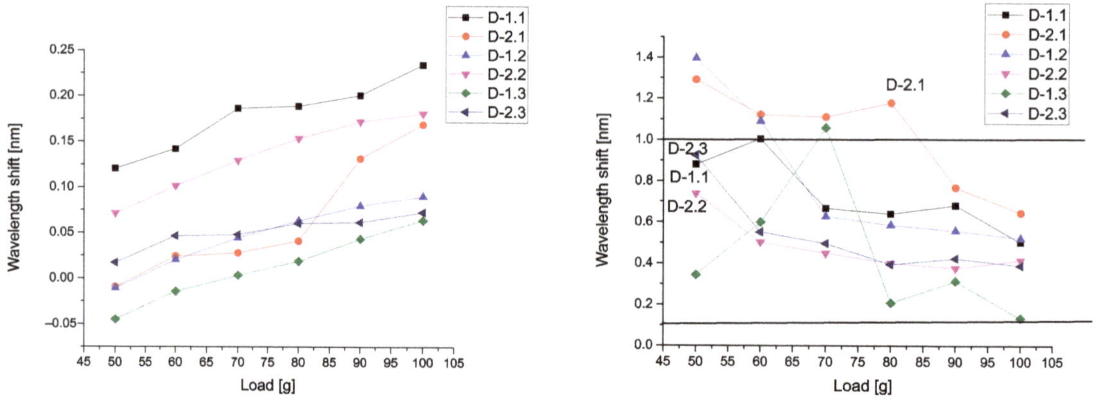


(a) (b)

**Figure 6.** (**a**) Six calibration curves (mean values) calculated for all types of detectors; (**b**) comparison of correlators calculated for different detectors.

It is interesting to analyze also 13 parameters (given by Expressions (8), (11), and (12)) that form the desired 3D surface. In order to compare these parameters with each other, we form the corresponding SRAs after applying the POLS procedure for different loads. Figure 7a demonstrates 3 curves calculated for the load L = 0 g. In order to notice the possible monotone behavior of these curves, we use the corresponding SRAs for different loads. On the right-hand side (Figure 7b), we place similar triple curves corresponding to L = 50 g. Other curves for the loads L = 60–100 g look similar and therefore are not shown.

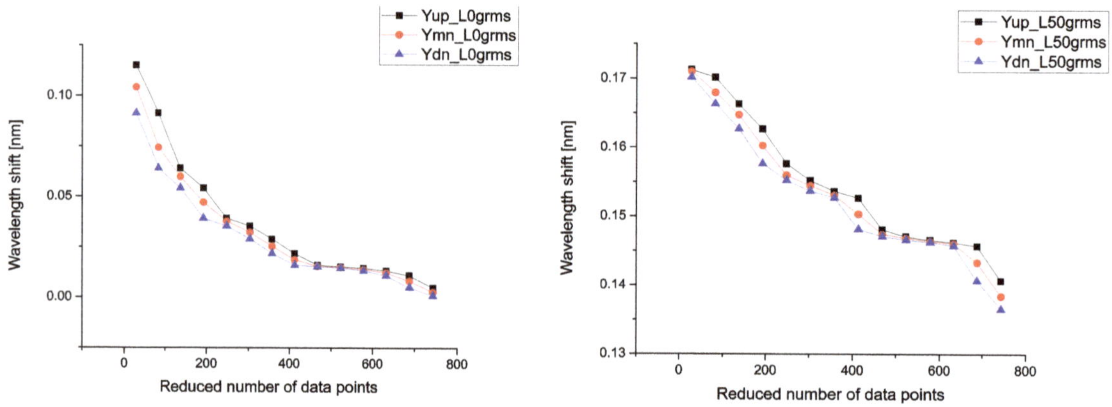


(a) (b)

**Figure 7.** (**a**) Sequences of the ranged amplitudes (SRAs) calculated for the load L = 0 g; (**b**) curves of SRAs corresponding to L = 50 g. Other curves for the loads L = 60–100 g look similar and therefore are not shown. Black markers (Yup) denote the distribution of maximum values, red markers (Ymn) represent the distribution of mean values, and blue markers (Ydn) show the distribution of minimum values.

In order to compare these parameters with each other, we take the mean values for the correlators of the first order (corresponding to the gravity center), then correlators of the second, third, and fourth orders, correspondingly. For us, it is important to detect a monotone behavior among these correlators. Figure 8a demonstrates six calibration curves calculated for the mean values of the first-order correlators. All curves are almost monotone, and we see an increase in these curves with respect to the applied load.

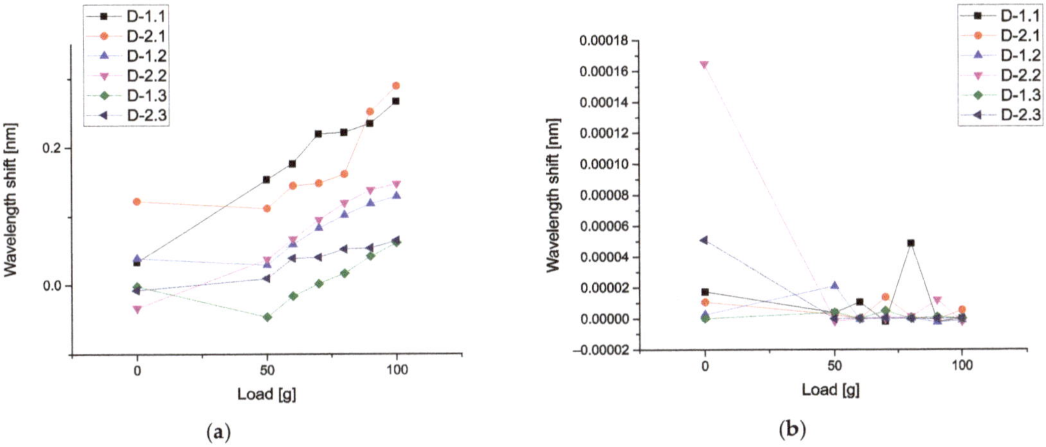


**Figure 8.** (**a**) Calibration curves calculated for mean values of the first-order correlators; (**b**) mean values of the second-order correlators.

However, the detectors D-2.1 D-1.2 and D-1.3 show nonmonotonic behavior in the limits of the load 0–50 g. We associate this phenomenon with the influence of the vibration waves amplitudes that are propagated over the plate. This phenomenon needs more detailed research. These curves are similar to Figure 6a. Figure 8b shows the nonmonotonic behavior of the correlators of the second order. In comparison with Figure 8a, the behavior of the correlators of the second order is different. As one can see, for this figure, the values of these correlators are small; only detector D-2.2 demonstrates the high values of the correlators of the second order when the applied load is absent.

Figure 9a demonstrates six calibration curves calculated for mean values of the third-order correlators. These curves have different behavior if we compare them with Figure 8. On the small figure inside, we show the correlators of the third order for detectors D-2.2 and D-2.3 that have a monotonic behavior. Figure 9b shows the different behavior of the correlators of the fourth order, in comparison with Figure 9a. As one can see, for this figure, the values of these correlators are small. However, only detectors D-2.2, D-1.1, and D-2.1 demonstrate the relatively high values of the correlators of the fourth order when the applied load is absent.

Analysis of these figures shows that they have different sensitivity to the applied load. Only the mean correlators of the first order having monotonic property can be used for calibration purposes. It presents an interesting possibility to give also the plots of 3D surfaces at different loads. It will reflect a unique combination of the 13 correlation parameters and this surface will serve as a *specific fingerprint* facilitating an initial analysis of the available data. Figure 10a shows the 3D surface for the L = 0 g, while Figure 10b presents the surface for L = 50 g. The 3D surface for the L = 90 g is demonstrated in Figure 11a. Figure 10b shows the surface for L = 100 g. In all these 3D surfaces, the unit of the OZ axis is conserved, and it is given in the units coinciding with initial data (values of vibrations in our case). As one can notice from the comparison of these figures, they are different in their values and have different forms.

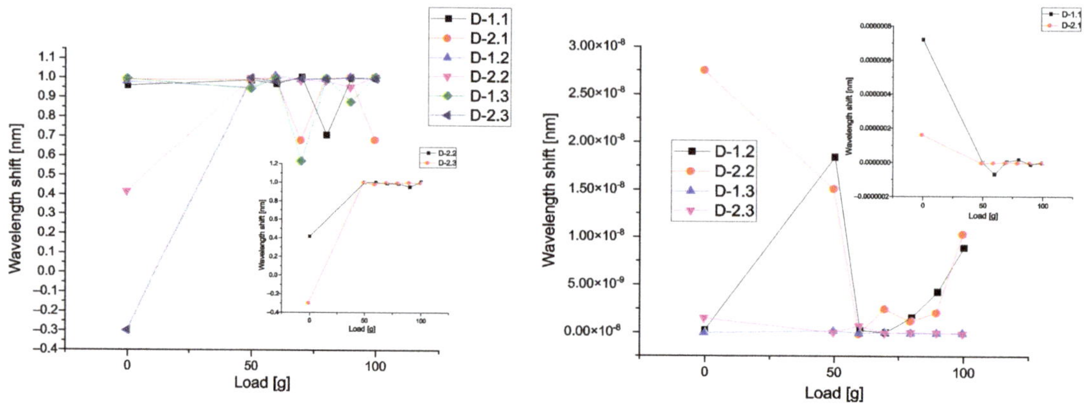


(a) (b)

**Figure 9.** (**a**) Calibration curves calculated for mean values of the third-order correlators; (**b**) calibration curves calculated for mean values of the fourth-order correlators.

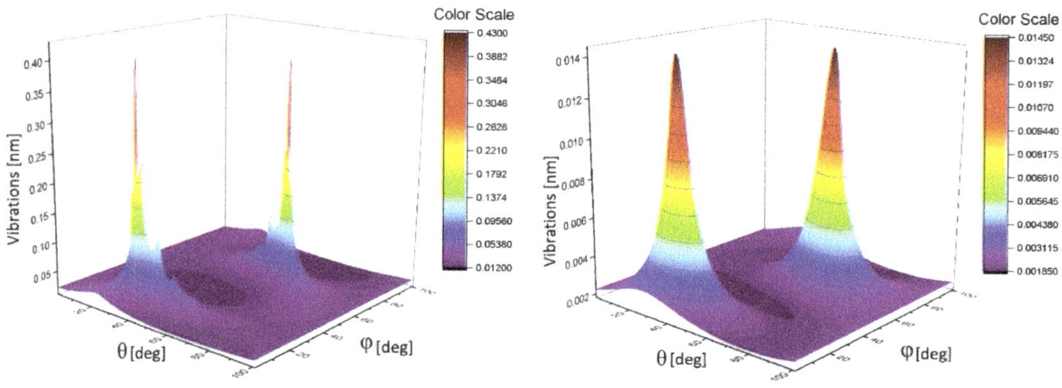


(a) (b)

**Figure 10.** (**a**) The 3D surface for the L = 0 g; (**b**) the 3D surface for L = 50 g.

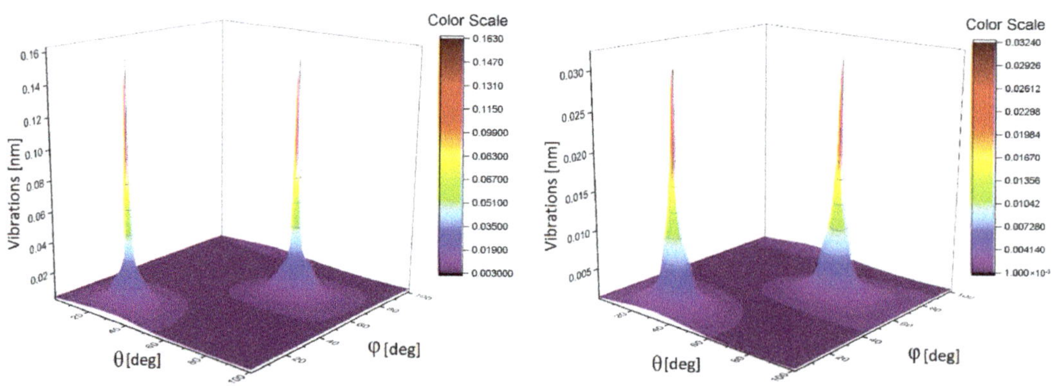


(a) (b)

**Figure 11.** (**a**) The 3D surface for the L = 90 g; (**b**) the 3D surface for L = 100 g.

## 5. Discussion and Conclusions

The proposed data treatment method based on the procedure of the optimal linear smoothing (POLS) allowed us to eliminate the noise induced by vibrations of the experimental plate, imitating the helicopter blade in flight conditions. As the result of the treatment procedure, monotonic calibration curves were obtained for each detector, making it possible to distinguish the increase of load applied to the plate with an increment of 10 g, which otherwise would be impracticable. It should be noted that the calibration curves comprising mean values of POLS-treated wavelength shift data of the detectors (Figure 6a) demonstrated higher mean deformation of the detector 2.2 (and even higher in case of the detector 2.3) than the one of the detector 2.1, which does not correspond to the a priori assumption that the detectors 2.1 and 1.1 (which are the nearest to the fixed end of the plate) should be subjected to the highest deformations, while 2.3 and 1.3 should experience the lowest strain. More detailed analysis based on a complete set of data allows concluding that detector 2.1 had lower sensitivity, in comparison with others, which can be caused by relatively weak adhesive bonding to the surface of the plate.

Highly likely, this peculiarity is related to the chosen material and formation in it the deformation waves.

However, the calibration curves calculated for mean values of the first-order correlators (Figure 8a) do agree with the mentioned assumption, showing the higher values at the detectors 1.1 and 2.1, and the lowest ones at the detectors 1.3 and 2.3.

As one can infer from the content given above, the second method associated with 3D invariants takes into account 13 quantitative parameters associated with a combination of different moments and their intercorrelations up to the fourth-order inclusive. They reflect more sophisticated behavior of the moments and their intercorrelations evoked by the applied load. Figures 8b and 9a,b show their peculiarities. These 13 parameters allow the calculation of the desired 3D surface that can serve as a specific fingerprint, differentiating one set of initial data from another one. Figure 10a,b and Figure 11a,b vividly demonstrate these peculiarities.

Therefore, the combination of these two methods that were used to analysis of real load data can be applied successfully to other data, because they are general, do not use a specific model, and are free from treatment errors.

**Author Contributions:** Conceptualization, R.R.N. and A.S.; methodology, R.R.N.; software, R.R.N. and K.K.; validation, A.S., T.A., K.K. and M.L.; formal analysis, R.R.N.; investigation, T.A., K.K. and M.L.; resources, S.M., M.L. and O.M.; data curation, R.R.N. and K.K.; writing—original draft preparation, R.R.N. and T.A.; writing—review and editing, R.R.N., A.S. and S.M.; visualization, R.R.N. and T.A.; supervision, O.M. and S.M.; project administration, R.R.N.; funding acquisition, O.M. and S.M. All authors have read and agreed to the published version of the manuscript.

**Funding:** R.R.N., A.S., T.A., O.M. and K.K. were funded by the Ministry of Science and Higher Education of the Russian Federation (Agreement No. 075-03-2020-051, Topic No. fzsu-2020-0020, FOKRAT), S.M. and M.L. were funded by the Ministry of Science and Higher Education of the Russian Federation in the framework of Grant FZSU-2020-00219, Project No. 075-03-2020-051/3 of 9 June 2020.

**Institutional Review Board Statement:** Not applicable.

**Informed Consent Statement:** Not applicable.

**Data Availability Statement:** The data presented in this study are available on request from the corresponding author. The data are not publicly available due to the rules of our contract conditions with our customer.

**Conflicts of Interest:** The authors declare no conflict of interest.

## References


1. Lee, H.; Viswamurthy, S.R.; Park, S.C.; Kim, T.; Shin, S.J. Helicopter Rotor Load Prediction Using a Geometrically Exact Beam with Multicomponent Model. *J. Aircr.* **2010**, *47*, 1382–1390. [CrossRef]
2. Maxwell, R.H. Optical Tracker System for Determining the Position of a Rotating Body. U.S. Patent 5929431A, 27 July 1999.
3. Christopher, I.M. Helicopter Blade Position Detector. U.S. Patent 8190393B2, 29 May 2012.
4. Sirohi, J.; Lawson, M. Measurement of helicopter rotor blade deformation using digital image correlation. *Opt. Eng.* **2012**, *51*, 3603. [CrossRef]
5. Fleming, G.A.; Gorton, S.A. Measurement of rotorcraft blade deformation using Projection Moire Interferometry. *Shock Vib.* **2000**, *7*, 149–165. [CrossRef]
6. Gaukroger, D.R.; Payen, D.B.; Walker, A.R. Application of Strain Gauge Pattern Analysis. In Proceedings of the Sixth European Rotorcraft and Powered Lift Aircraft Forum, Bristol, UK, 16–19 September 1980; p. 19.
7. Leung, J.G.M.; Wei, M.Y.; Aoyagi, M. UH-60A airloads data acquisition and processing system. In Proceedings of the AIAA/IEEE Digital Avionics Systems Conference, 13th DASC, Phoenix, AZ, USA, 30 October–3 November 1994; pp. 206–211. [CrossRef]
8. Datta, A.; Chopra, I. Validation of Structural and Aerodynamic Modeling Using UH-60A Airloads Program Data. *J. Am. Helicopter Soc.* **2006**, *51*, 43–58. [CrossRef]
9. Broadway, C.; Min, R.; Leal-Junior, A.G.; Marques, C.; Caucheteur, C. Toward Commercial Polymer Fiber Bragg Grating Sensors: Review and Applications. *J. Lightwave Technol.* **2019**, *37*, 2605–2615. [CrossRef]
10. Chehura, E.; Jarzebinska, R.; Da Costa, E.F.R.; Skordos, A.A.; James, S.W.; Partridge, I.K.; Tatam, R.P. Multiplexed fibre optic sensors for monitoring resin infusion, flow, and cure in composite material processing. In *Smart Sensor Phenomena, Technology, Networks, and Systems Integration 2013*; International Society for Optics and Photonics: Bellingham, WA, USA, 2013; Volume 8693, p. 86930F. [CrossRef]
11. Lawson, N.J.; Correia, R.; James, S.W.; Partridge, M.; Staines, S.E.; Gautrey, J.E.; Garry, K.P.; Holt, J.C.; Tatam, R.P. Development and application of optical fibre strain and pressure sensors for in-flight measurements. *Meas. Sci. Technol.* **2016**, *27*, 104001. [CrossRef]
12. Suesse, S.; Hajek, M. Rotor Blade Shape Estimation with Fiber-Optical Sensors for a Health and Usage Monitoring System. In Proceedings of the 42nd European Rotorcraft Forum, Lille, France, 5–8 September 2016.
13. Wada, D.; Igawa, H.; Kasai, T. Vibration monitoring of a helicopter blade model using the optical fiber distributed strain sensing technique. *Appl. Opt.* **2016**, *55*, 6953–6959. [CrossRef]
14. Daniell, J.G.B.; Molnar, G.A. Helicopter Weight Measurement. U.S. Patent 5,229,956, 20 July 1993.
15. Sahota, J.K.; Gupta, N.; Dhawan, D. Fiber Bragg grating sensors for monitoring of physical parameters: A comprehensive review. *Opt. Eng.* **2020**, *59*, 060901. [CrossRef]
16. Melle, S.; Liu, K. Wavelength demodulated Bragg grating fiber optic sensing systems for addressing smart structure critical issues. *Smart Mater. Struct.* **1992**, *1*, 36. [CrossRef]
17. Davis, M.A.; Bellemore, D.G.; Kersey, A.D. Structural strain mapping using a wavelength/time division addressed fiber Bragg grating array. *Proc. SPIE* **1994**, *2361*, 342–345. [CrossRef]
18. Matveenko, V.P.; Shardakov, I.N.; Voronkov, A.A.; Kosheleva, N.A.; Lobanov, D.S.; Serovaev, G.S.; Spaskova, E.M.; Shipunov, G.S. Measurement of strains by optical fiber Bragg grating sensors embedded into polymer composite material. *Struct. Control Health Monit.* **2017**, *25*, e2118. [CrossRef]
19. Qiao, X.; Shao, Z.; Bao, W.; Rong, Q. Fiber Bragg grating sensors for the oil industry. *Sensors* **2017**, *17*, 429. [CrossRef]
20. Ma, Z.; Chen, X. Fiber Bragg gratings sensors for aircraft wing shape measurement: Recent applications and technical analysis. *Sensors* **2019**, *19*, 55. [CrossRef]
21. Agliullin, T.; Gubaidullin, R.; Sakhabutdinov, A.; Morozov, O.; Kuznetsov, A.; Ivanov, V. Addressed Fiber Bragg Structures in Load-Sensing Wheel Hub Bearings. *Sensors* **2020**, *20*, 6191. [CrossRef]
22. Ledyankin, M.A.; Mikhailov, S.A.; Nedel'ko, D.V.; Agliullin, T.A. Implementation of the Radiophotonic Method for Measuring Blade Deformations of a Helicopter Main Rotor Model. *Russ. Aeronaut.* **2020**, *63*, 767–770. [CrossRef]
23. Morozov, O.; Sakhabutdinov, A.; Anfinogentov, V.; Misbakhov, R.; Kuznetsov, A.; Agliullin, T. Multi-Addressed Fiber Bragg Structures for Microwave-Photonic Sensor Systems. *Sensors* **2020**, *20*, 2693. [CrossRef] [PubMed]
24. Nigmatullin, R.R.; Osokin, S.I.; Toboev, V.A. NAFASS: Discrete spectroscopy of random signals. *Chaos Solitons Fractals* **2011**, *44*, 226–240. [CrossRef]
25. Pershin, S.M.; Bunkin, A.F.; Lukyanchenko, V.A.; Nigmatullin, R.R. Detection of the OH band fine structure in a liquid water by means of new treatment procedure based on the statistics of the fractional moments. *Lazer Phys. Lett.* **2007**, *4*, 809–812. [CrossRef]
26. Nigmatullin, R.R.; Ionescu, C.; Baleanu, D. NIMRAD: Novel technique for respiratory data treatment. *J. Signal Image Video Process.* **2014**, *8*, 1517–1532. [CrossRef]
27. Ciurea, M.L.; Lazanu, S.; Stavaracher, I.; Lepadatu, A.-M.; Iancu, V.; Mitroi, M.R.; Nigmatullin, R.R.; Baleanu, C.M. Stress-induced traps in multilayed structures. *J. Appl. Phys.* **2011**, *109*, 013717. [CrossRef]
28. Cetin, S.S.; Băleanu, C.M.; Nigmatullin, R.R.; Băleanu, D.; Ozcelik, S. Chemical Bonding Structure of TiO2 Thin Films grown on N-Type Si. *Thin Solid Film.* **2011**, *519*, 5712–5719. [CrossRef]
29. Băleanu, C.M.; Nigmatullin, R.R.; Cetin, S.S.; Băleanu, D.; Ozcelik, S. New method and treatment technique applied to interband transition in GaAs1-x Px ternary alloys. *Cent. Eur. J. Phys.* **2011**, *9*, 729–739. [CrossRef]

*sensors*

MDPI

*Article*

# Design and Measurement of a Dual FBG High-Precision Shape Sensor for Wing Shape Reconstruction

**Huifeng Wu [1,2], Lei Liang [1,*], Hui Wang [1], Shu Dai [1], Qiwei Xu [1] and Rui Dong [3]**

1 National Engineering Laboratory for Fiber Optic Sensing Technology, Wuhan University of Technology, Wuhan 430070, China; huizifeng_829@163.com (H.W.); wanghui1989@whut.edu.cn (H.W.); ds0567@whut.edu.cn (S.D.); xxdsgt2021@163.com (Q.X.)
2 School of Electronic Information and Automation, Guilin University of Aerospace Technology, Guilin 541004, China
3 Engineering Comprehensive Training Center, Guilin University of Aerospace Technology, Guilin 541004, China; dongrui87856@163.com
* Correspondence: 130130@126.com

**Abstract:** FBG shape sensors based on soft substrates are currently one of the research focuses of wing shape reconstruction, where soft substrates and torque are two important factors affecting the performance of shape sensors, but the related analysis is not common. A high-precision soft substrates shape sensor based on dual FBGs is designed. First, the FBG soft substrate shape sensor model is established to optimize the sensor size parameters and get the optimal solution. The two FBG cross-laying method is adopted to effectively reduce the influence of torque, the crossover angle between the FBGs is $2\alpha$, and $\alpha = 30°$ is selected as the most sensitive angle to the torquer response. Second, the calibration test platform of this shape sensor is built to obtain the linear relationship among the FBG wavelength drift and curvature, rotation radian loaded vertical force and torque. Finally, by using the test specimen shape reconstruction test, it is verified that this shape sensor can improve the shape reconstruction accuracy, and that its reconstruction error is 6.13%, which greatly improves the fit of shape reconstruction. The research results show that the dual FBG high-precision shape sensor successfully achieves high accuracy and reliability in shape reconstruction.

**Keywords:** fiber Bragg grating; curvature; torque; shape reconstruction



**Citation:** Wu, H.; Liang, L.; Wang, H.; Dai, S.; Xu, Q.; Dong, R. Design and Measurement of a Dual FBG High-Precision Shape Sensor for Wing Shape Reconstruction. *Sensors* **2022**, 22, 168. https://doi.org/10.3390/s22010168

Academic Editor: Oleg G. Morozov

Received: 7 December 2021
Accepted: 22 December 2021
Published: 28 December 2021



## 1. Introduction

Fiber grating shape sensing technology has been a new research direction in the field of fiber grating sensing in recent years, especially in three-dimensional shape reconstruction, and it is a widely studied and researched method that uses FBG shape sensors to obtain shape change information parameters [1], for example, in position tracing and positioning of probes, soft manipulators, endoscopes, etc., in the medical field [2,3]; deformation monitoring of key structures in the aerospace field, such as deformation monitoring of airplane skin, wings, etc. [4–6]; real-time monitoring of the motion status and spatial position of mechanical arms embedded in robot links and robot arms of continuum robots in industrial fields [7–9]; realization of joint positioning and posture monitoring during human movement by flexible smart wearable devices [10,11]; and the structural health monitoring of urban civil buildings [12] and bridges [13].

After MIT and NASA jointly designed a new digital deformable wing with polyimide film skin and composite lattice honeycomb structure, the development and application of FBG shape sensors are mainly concentrated in two types: Shape sensor based on a single core fiber grating, and shape sensor based on a multicore fiber.

A multicore fiber shape sensor is suitable for extremely large curvatures (bending radii up to 5 mm). As early as 2012, the US Luna set a multicore fiber shape sensor (total length of 30 m) on the surface of flexible structures of approximately 10 m, obtained the space deformation of the measured surface by reconstructing the multicore fiber space position, and had found that the reconstruction error was within 1.5% [5]. In recent years, distributed optical fiber shape sensors have developed rapidly. Chen, X.Y et al. established a distributed Brillouin optical time domain analysis based on a seven-core MCF to verify the linear relationship between fiber curvature and the Brillouin frequency shift of the outer core in the MCF and realized a curvature with a spatial resolution of 20 cm using a 1 km fiber length [14]. According to the temperature-sensitive characteristics of optical fibers, Zhao, Z et al. designed a high-performance temperature-insensitive shape sensor based on a differential pulse pair system, whose spatial resolution reached 10 cm, and its sensitivity was greatly improved [15]. Ba, D.X et al. proposed a high-sensitivity distributed shape sensor based on a seven-core optical fiber and phase sensitive optical time domain reflectometry, which can achieve high strain sensitivity as low as 0.3 $\mu\varepsilon$ on a 24 m-long MCF with a spatial resolution of 10 cm [16].

Some of the inevitable problems of using multicore optical fibers as shape sensors are twisting, eliminating the measurement error caused by distortion and improving the accuracy of shape measurement and reconstruction, which have become the focus of research for many scholars in recent years. Some scholars have proposed solutions for spiral multicore fibers [17], but the high fabrication cost and immature fabrication technology of spiral multicore fibers prevent their large-scale application.

In comparison, the moldability of a single-core fiber grating shape sensor is better, especially the combined sensing mode of a soft substrate material and FBG, which is more widely used [18]. Wang et al. proposed a sensing network and method-based FBG sensor for 3D shape detection of flexible robotic arms and proved high detection accuracy for complex 3D shapes by simulation [19]. Zhi, G.H et al. designed a new intensity modulated fiber optic sensor with an enhanced sensitivity region to measure curvature and proposed a wind speed monitoring system based on the fiber optic shape sensor, which monitored the frequency of the fiber optic sensor period signal to measure wind speed. The experimental results show that the measurement error and repeatability of the wind speed monitoring system are within $\pm$0.65 m/s and $\pm$2.9%, respectively [20]. Abro, A. Z et al. [21,22] developed and designed a smart sensing garment SSG and a FBG smart belt. The FBG was embedded in special silicone to make a shape sensor that was mounted on the palm, wrist and elbow joints to measure changes in body posture. The sensitivity of the FBG smart belt was between 0.018% and 0.021% for male subjects at speeds of 2 and 3 km/h, respectively. He, Y.L et al. [23–25] used silica gel and polyimide with FBG to design two different types of shape sensors; one was a soft actuator based on silica gel to reconstruct the 3D shape of the soft actuator in different bending states, and the other was to embed FBG into polyimide to make a morphing wing smart skin. The 3D shape reconstruction of the polyimide skin of the wing was completed. The test results show that the maximum error of FBG measurement is less than 5%. Wang, Q.L et al. [26] proposed a bending curvature sensor for a soft surgical manipulator, which buried a Bragg grating into polyimide to form a polyimide sensing layer and integrated the sensing layer into a soft manipulator model to achieve bending curvature monitoring of the soft surgical manipulator. The maximum error between the measured value and the actual bending curvature was less than 2.1%. Arnaldo G et al. [27] presented a 3D displacement sensor based on a single fiber Bragg grating in CYTOP fiber, to obtain the influence of each displacement condition, namely, axial strain, torsion and bending on the FBG reflection spectrum, and had relative errors below 5.5%, but the fiber Bragg grating in CYTOP was prone to breakage and not suitable for harsh environments.

For single-core fiber grating shape sensors, how to eliminate the measurement error caused by distortion to improve the accuracy of shape reconstruction was not considered in the previous literature. In addition, there is little literature on how to eliminate the influence of the properties of soft substrate materials and volume size on the performance of shape sensors to improve the measurement accuracy of sensors, and only a few papers have analyzed the effect of the depth of FBG preburial in soft substrates [28–30].

In this paper, we choose the more widely used single-core FBG and soft substrate combination mode to design a new FBG shape sensor, analyze in detail the influence of the length, thickness and width of the soft substrate on the accuracy of the shape sensor and choose the most reasonable soft substrate size. At the same time, considering the irresistible factor of torque in the use of FBG shape sensors, we adopt the cross double-grid method to measure the FBG center wavelength drift caused by torque. By a calibration test, the linear relationship between FBG wavelength drift and rotation radian loaded torque is obtained. The shape sensor designed in this paper can structurally improve the measurement accuracy.

## 2. FBG Sensing Principle and Structure Analysis

### 2.1. Sensory Principle of FBG

From Hooke's law, we know that after a uniform elastic solid material is stressed, there is a linear relationship between stress $\varepsilon$ and strain (unit deformation) in the material, as shown in Equation (1) [31].

$$\varepsilon = \frac{\Delta l}{l} \tag{1}$$

A microarc segment of the elastic material is intercepted, in which length L and thickness H are chosen, and the location of the preburied FBG is X from the upper plane, as shown in Figure 1. When the elastic material is subjected to axial stress, two cross-sections of the microarc segment relatively rotate by a microsmall angle $\theta$, as shown in Figure 1b. The length of the neutral plane does not change, the length of the upper plane is stretched by $\Delta l$, the length of the under plane is shortened by $\Delta l$ and the length of the plane on which the FBG is located is stretched by $\Delta x$, where $\Delta x < \Delta l$ and R is the radius of curvature.

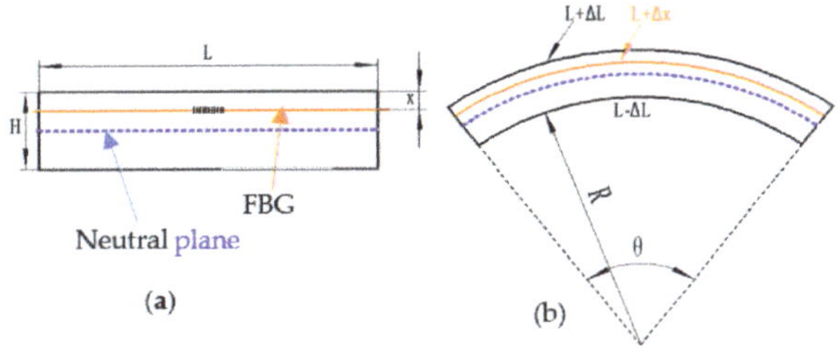


**Figure 1.** Bending mechanism of the elastic material: (**a**) Stress-free without bending, (**b**) stress and bend.

When the length of the microarc segment is small enough, $sin\theta = \theta$

$$\text{upper plane}: \ L + \Delta L = (R + H)\theta \tag{2}$$

$$\text{neutral plane}: \ L = (R + \frac{1}{2}H)\theta \tag{3}$$

$$\text{under plane}: \ -\Delta L = R\theta \tag{4}$$

From Equations (2) and (4), the following is obtained:

$$2\Delta L = H\theta \tag{5}$$

From Equations (3) and (5), the following is obtained:

$$\frac{\Delta L}{L} = \frac{\frac{1}{2}H\theta}{\left(R + \frac{1}{2}H\right)\theta} = \frac{\frac{1}{2}H}{R + \frac{H}{2}} \tag{6}$$

For a standard single–method optical fiber, the relationship between the center wavelength drift $\Delta\lambda$ and the center wavelength $\lambda$ and the changes in temperature $\Delta T$ and strain $\varepsilon$ can be obtained.

$$\frac{\Delta\lambda}{\lambda} = (1 - P_e)\varepsilon + (\alpha + \xi)\Delta T \tag{7}$$

where $P_e$-photoelastic parameter of the optical fiber core for a general single-mode fiber, where $P_e \approx 0.22$.

$\alpha$-thermal expansion of the optical fiber core.

$\xi$-thermal-optic coefficient of the optical fiber core.

Assume that the temperature change is zero, Equation (7) simplifies to

$$\frac{\Delta\lambda}{\lambda} = (1 - P_e)\varepsilon = (1 - P_e)\frac{\Delta x}{L} = (1 - P_e)\frac{\frac{1}{2}H}{R + \frac{H}{2}} \tag{8}$$

In the case where the center wavelength of the FBG is known, $H$ is the thickness of the silica gel and the FBG center wavelength drift $\Delta\lambda$ can be measured from Equation (8). It follows that R is linearly related to $\frac{\Delta\lambda}{\lambda}$:

$$R \propto a \cdot \frac{\lambda}{\Delta\lambda} \tag{9}$$

where $a$ is a constant associated with the thickness, size and material of the shape sensor, the wavelength drifts of the fiber grating, etc., and the specific value is determined by testing.

### 2.2. *Effect of Soft Substrate Material Size on Shape Sensors*

There are few reports on the influence of the size of soft substrate shape sensors, i.e., the length, width and thickness of the soft material, on the performance. Few papers have analyzed the influence of FBG preburial depth; however, choosing the appropriate soft substrate material size is also one of the focuses of FBG shape sensor design. To analyze the influence of the length, width and thickness of the soft material on the performance of the FBG shape sensor, we choose common silica gel as the soft substrate and use the experimental model shown in Figure 2 to perform the following simulation and analysis in the workbench in the case where the size of the stainless steel, the direction and the position of the force are unchanged.

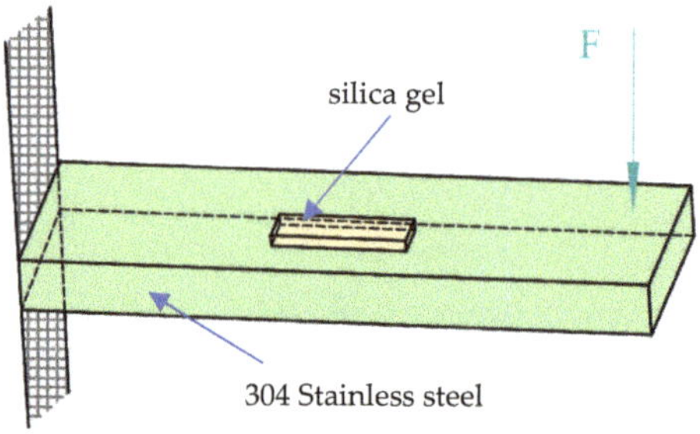


**Figure 2.** Sensor simulation model.

As shown in Figure 2, one end of the 304 stainless steel is fixed, the other free end is applied with a force F = 20 N, the direction is vertically downward and the silica gel is pasted on the stainless steel. The center always remains in the same line as the center of the stainless steel, regardless of the change in silica gel size. The dimensions of the stainless steel were as follows: Length of 100 mm, width of 25 mm, thickness of 10 mm. The length, width and thickness of the silica gel are indicated by L, D and H respectively.

2.2.1. Variation of Silica Gel Length

When the width and thickness of the silica gel are constant, the length is changed and the stress of the silica gel is analyzed. The size changes of silica gel are shown in Table 1.

**Table 1.** Variation of silica gel length.

| Size | a | b | c | d |
|---|---|---|---|---|
| length (mm) | 30 | 25 | 15 | 8 |
| width (mm) | 8 | 8 | 8 | 8 |
| thickness (mm) | 3 | 3 | 3 | 3 |

From Figure 3, we clearly find that the stress distribution of the silica gel varies significantly by length when the width and thickness of the silica gel are constant.

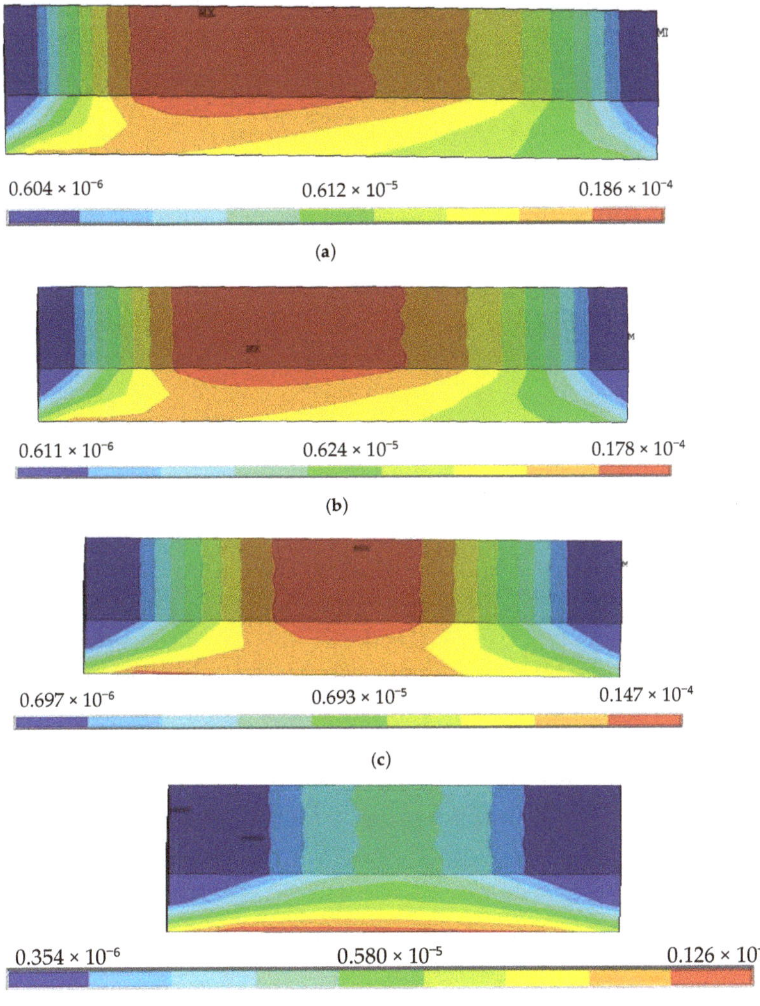


**Figure 3.** Stress distribution nephogram of silica gel with different lengths, (**a**) L = 30 mm, (**b**) L = 25 mm, (**c**) L = 15 mm, (**d**) L = 8 mm.

(1) If the length of the silica gel is longer, the stress area is larger and the maximum peak of the strain is greater, while the stress is concentrated in the upper layer of the neutral surface of the silica gel; if the length of the silica gel is smaller, the stress begins to shift from the upper layer to the lower layer of the neutral surface.
(2) Regardless of the length of the silica gel, the stress distribution in the silica gel has left-right axis symmetry.

2.2.2. Variation of Silica Gel Width

When the length and thickness of the silica gel are constant, the width of the silica gel is changed and the stress distribution of the silica gel is analyzed. The size changes of silica gel are shown in Table 2.

**Table 2.** Variation of the silica gel width.

| Size | a | b | c | d |
|---|---|---|---|---|
| Length (mm) | 30 | 30 | 30 | 30 |
| Width (mm) | 25 | 15 | 10 | 8 |
| Thickness (mm) | 3 | 3 | 3 | 3 |

From Figure 4, we can find that the difference in silica gel stress distribution is small for different widths. When the width of the silica gel is varied, the maximum peak value of the strain of the silica gel varies from 0.0189 to 0.0186 MPa, with a small change in value, so we think that the effect of silica gel width on its stress can be ignored.

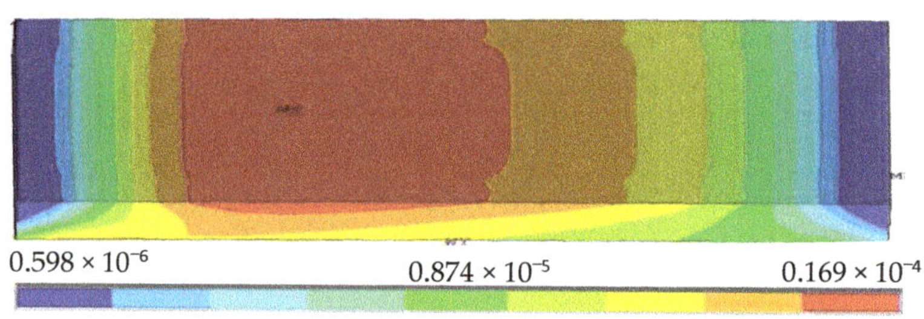


(**a**)

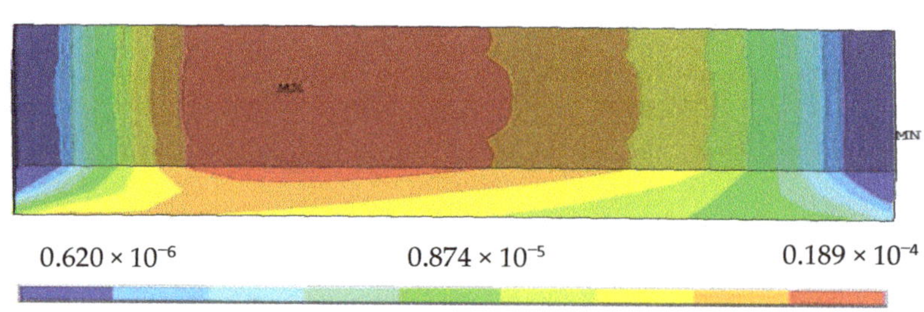


(**b**)

**Figure 4.** *Cont.*

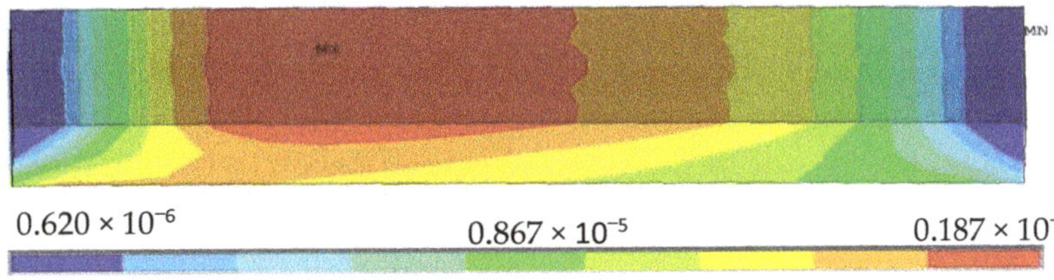


(**c**)

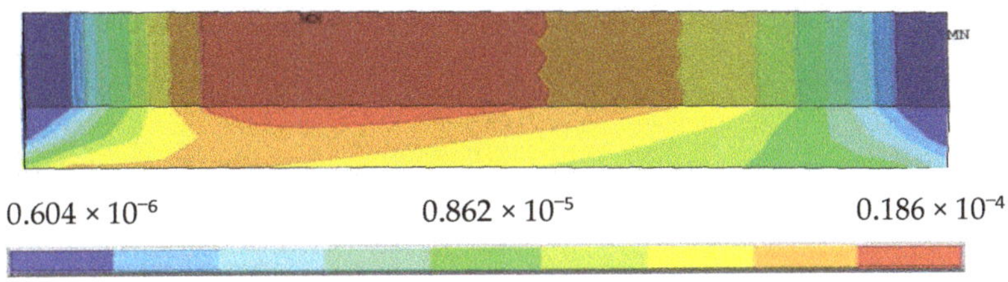


(**d**)

**Figure 4.** Stress distribution nephogram of silica gel with different widths: (**a**) D = 25 mm, (**b**) D = 15 mm, (**c**) D = 10 mm, (**d**) D = 8 mm.

### 2.2.3. Variation of Silica Gel Thickness

When the length and width of the silica gel are constant, the thickness of the silica gel is changed. Size changes of silica gel are shown in Table 3.

**Table 3.** Variation of the silica gel thickness.

| Size | a | b | c | d |
|---|---|---|---|---|
| length (mm) | 30 | 30 | 30 | 30 |
| width (mm) | 10 | 10 | 10 | 10 |
| thickness (mm) | 2 | 5 | 8 | 10 |

From Equation (9), we can deduce that if the thickness of the silica gel is larger, the measured radius of curvature is larger, but this is not actually the case, from Figure 5, we can find that the thickness changes and the silica gel stress distribution varies significantly when the silica gel length and width are constant.

(1) When the silica gel thickness changes, the maximum stress position of silica gel will also change, the thickness of the silica gel gradually increases, the area of the stress maximum gradually decreases, its position moves closer to the center of the silica gel, the gel thickness tends to the thickness of the stainless steel and the stress distribution is likely to be transformed from the upper layer to both the upper and lower layers.

Therefore, the thicker the silica gel thickness, the better the performance of the sensor; this method has not been established.

(2) The stress of silica gel reaches the maximum when the thickness is 5 mm, and the maximum stress increases with increasing thickness when the thickness is below 5 mm. The maximum stress decreases with increasing thickness when the thickness is above 5 mm.

In summary, selecting a reasonable silica gel size has a great role in improving the accuracy of the FBG shape sensor. The longer the length of the silica gel is, the larger the stress distribution area and the larger the maximum stress, while the width has less influence on the stress. Considering the practical application of shape sensors, the preparation process, the length of the grating of FBG and other factors, silica gel is selected as a rectangular structure with a length of 25–30 mm, width of 8–10 mm and thickness of 2–3 mm.

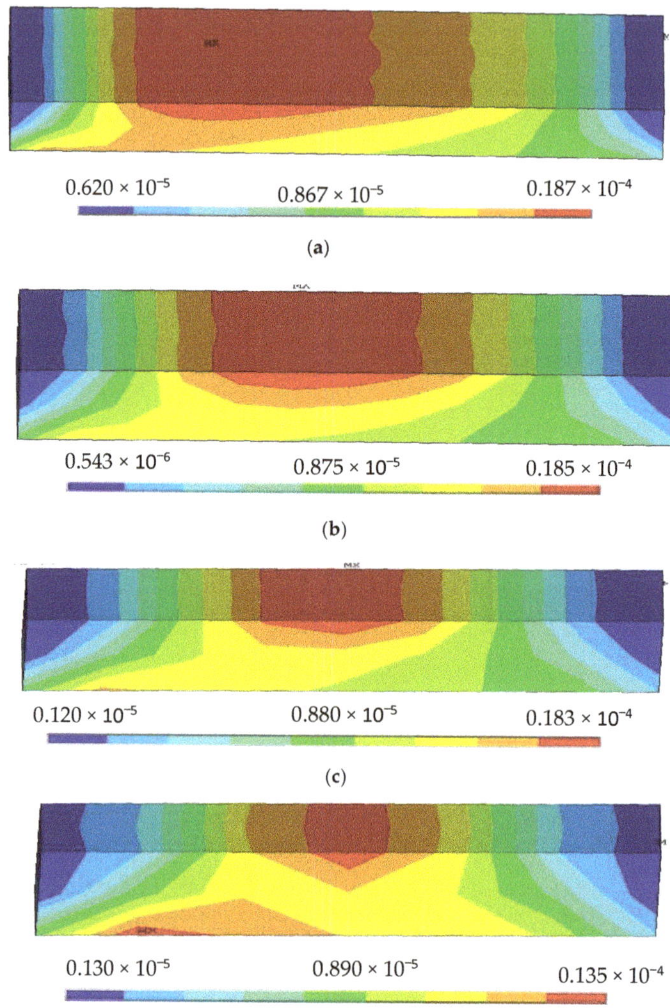


**Figure 5.** Stress distribution nephogram of silica gel with different thickness: (**a**) H = 2 mm, (**b**) H = 5 mm, (**c**) H = 8 mm, (**d**) H = 10 mm.

## 3. High-Precision FBG Shape Sensor Structure Design

Two main aspects should be considered in the design of high-precision FBG shape sensors based soft substrates: (1) The choice of the soft substrate—silica gel, as one of the most widely used and maturely processed soft materials, is suitable for the study of validation tests; (2) the layout problem of the grating in the silica gel. The influence of torque on the FBG soft substrate shape sensors cannot be ignored, and in order to minimize the influence of torque and improve the accuracy of the sensor, the FBG arrangement in Figure 6 is chosen.

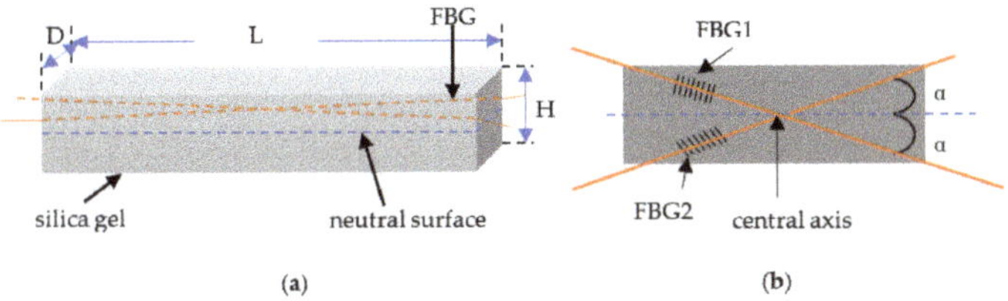


**Figure 6.** Structure diagram of the dual FBG high-precision shape sensor: (**a**) Dual FBG shape sensor structure Figure, (**b**) FBG layout Figure.

The angle between the grating and the silica gel central axis is $\alpha_1$ and $\alpha_2$. We choose the simulation model shown in Figure 7 to simulate the stress of the FBG shape sensor when it is loaded with three loading methods, i.e., bending, torque and bending plus torque, and discuss the variation in the equivalent force of the FBG when the angles $\alpha_1$ and $\alpha_2$ are varied. All simulations are completed by Mechanical APDL software, and the equivalent stress stated in this section is the maximum equivalent stress on the nodes where the FBG is located.

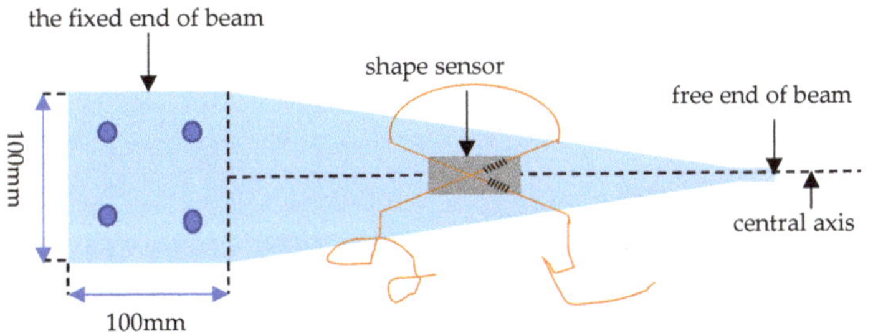


**Figure 7.** Simulation model of dual FBG high-precision shape sensor.

In the simulation model shown in Figure 7, we set the silica gel size to a length L of 23 mm, width D of 8 mm and thickness H of 2 mm, while the length $L_b$ and thickness $H_b$ of the cantilever beam size are variable. The applied load is shown in Table 4.

**Table 4.** Parameters of the applied load.

| Parameter | Force | Torques | Force + Torques | |
|---|---|---|---|---|
| value | 800 N | 800 N·m | 800 N | 800 N·m |
| direction | vertical down | clockwise | vertical down | clockwise |
| loading point | End of beam (free end) | Any position of beam | End of beam (free end) | Any position of beam |

*3.1. The Stress Relevance between the Three Loading Methods*

To ensure that the equivalent stress of the FBG is independent of the size of the cantilever beam and the location of the torque loaded, when three different loads are applied to the cantilever beam, we perform simulation and analysis for two different parameter settings, i.e., constant and variable beam size.

3.1.1. Constant Beam Size ($L_b$ = 500 mm $H_b$ = 4 mm)

The loaded torque points are located at 1/4, 1/2 and 3/4 of the beam from the fixed end, and the equivalent stress simulation analysis of the FBG is shown in Table 5.

**Table 5.** Absolute value of the maximum equivalent stress.

| Loaded Point | Force | | Torque | | Force + Torque | |
|---|---|---|---|---|---|---|
| | FBG1 (MPa) | FBG2 (MPa) | FBG1 (MPa) | FBG1 (MPa) | FBG2 (MPa) | FBG1 (MPa) |
| 1/4 | $5.78 \times 10^{-4}$ | $5.78 \times 10^{-4}$ | $-1.80 \times 10^{-11}$ | $1.82 \times 10^{-11}$ | $5.78 \times 10^{-4}$ | $5.78 \times 10^{-4}$ |
| 1/2 | $5.78 \times 10^{-4}$ | $5.78 \times 10^{-4}$ | $-1.04 \times 10^{-8}$ | $1.04 \times 10^{-8}$ | $5.78 \times 10^{-4}$ | $5.78 \times 10^{-4}$ |
| 3/4 | $5.78 \times 10^{-4}$ | $5.78 \times 10^{-4}$ | $-1.41 \times 10^{-4}$ | $1.41 \times 10^{-4}$ | $4.37 \times 10^{-4}$ | $7.19 \times 10^{-4}$ |

3.1.2. Variable Beam Size

The length $L_b$ and thickness $H_b$ of the cantilever beam are variable, and the loaded torque points are located at 1/3 and 2/3 of the beam from the fixed end. The equivalent stress simulation of the FBG is shown in Table 6.

**Table 6.** Absolute value of the maximum equivalent stress.

| Size/ Loaded Point | Force | | Torque | | Force +Torque | |
|---|---|---|---|---|---|---|
| | FBG1(MPa) | FBG2 (MPa) | FBG1 (MPa) | FBG2 (MPa) | FBG1 (MPa) | FBG2 (MPa) |
| $L_b$ = 300 mm, $H_b$ = 4 mm,1/3 | $1.12 \times 10^{-4}$ | $1.12 \times 10^{-4}$ | $-5.45 \times 10^{-8}$ | $5.45 \times 10^{-8}$ | $1.12 \times 10^{-4}$ | $1.12 \times 10^{-4}$ |
| $L_b$ = 400 mm, $H_b$ = 4 mm,1/3 | $1.86 \times 10^{-4}$ | $1.86 \times 10^{-4}$ | $-3.62 \times 10^{-10}$ | $3.62 \times 10^{-10}$ | $1.86 \times 10^{-4}$ | $1.86 \times 10^{-4}$ |
| $L_b$ = 500 mm, $H_b$ = 5 mm,2/3 | $1.13 \times 10^{-4}$ | $1.13 \times 10^{-4}$ | $-5.45 \times 10^{-8}$ | $5.45 \times 10^{-8}$ | $1.13 \times 10^{-4}$ | $1.13 \times 10^{-4}$ |
| $L_b$ = 600 mm, $H_b$ = 8 mm,2/3 | $2.49 \times 10^{-4}$ | $2.49 \times 10^{-4}$ | $-3.57 \times 10^{-7}$ | $3.57 \times 10^{-7}$ | $2.49 \times 10^{-4}$ | $2.49 \times 10^{-4}$ |

We analyze the tables of stress in Tables 5 and 6 and obtain the following conclusions:

(1) When the cantilever beam is subjected to a separate torque, the maximum equivalent stress at the FBG on the silica gel is $F_t$; when the cantilever beam is subjected to a separate force, the maximum equivalent stress at the FBG on the silica gel is $F_f$; and when the cantilever beam is subjected to a combination of torque and force, the maximum equivalent stress at the FBG on the silica gel is $F_m$; Equation (10) is always established, which is independent of the position of the loaded torque and independent of the size of the cantilever beam.

$$F_t + F_f = F_m \tag{10}$$

(2) The value of the maximum equivalent stress at the FBG on the silica gel $F_t$ is affected by the size of the beam, the location of the loaded torque and other factors.

### 3.2. *Influence of Angles $\alpha_1$ and $\alpha_2$ on Torque*

In the experimental model shown in Figure 7, we set the cantilever beam length as 500 mm and thickness as 6 mm and load a force of 800 N vertically downward at the end of the beam and a torque of 800 N·m at 1/4 of the beam from the fixed end. In Mechanical APDL, we simulate and analyze the maximum equivalent stress at the nodes on four coordinate lines, which are the X-axis and Y-axis of the center of the silica gel, and the positions of the two FBGs are laid out. Simulation of equivalent stress on the 4-coordinate line at different values of $\alpha$ are shown in Figure 8.

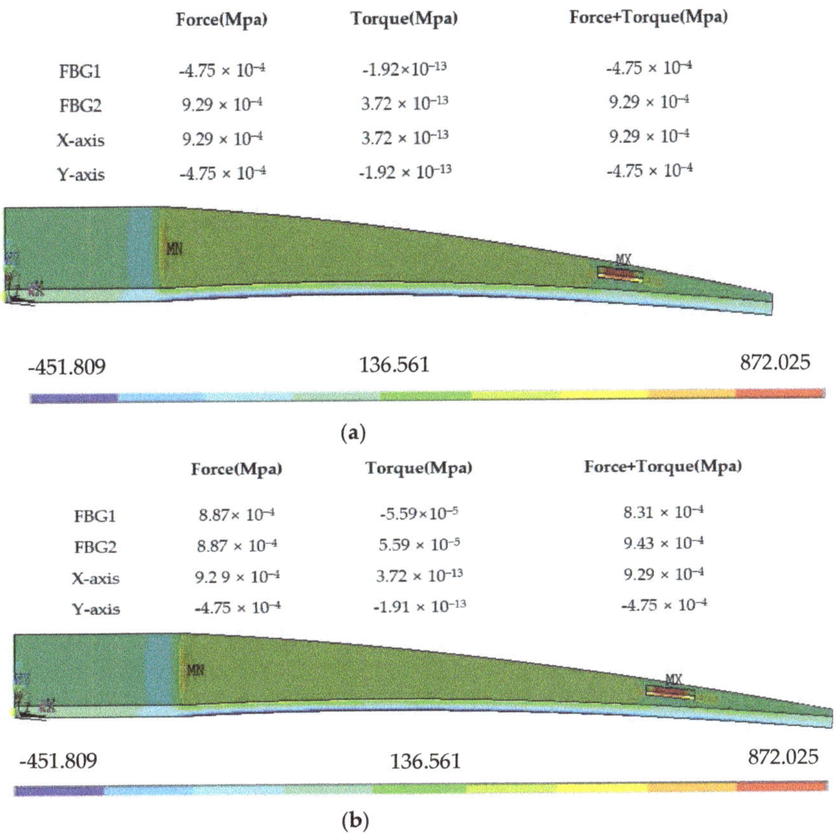


**Figure 8.** Simulation of equivalent stress on the 4-coordinate line at different values of $\alpha$, (**a**) $\alpha_1 = 90°\alpha_2 = 0°$, (**b**) $\alpha_1 = \alpha_2 = 10°$.

From the simulation results, we can find the following characteristics:

(1) $\alpha_1 = 90°$, $\alpha_2 = 0°$ : The two FBGs correspond to the X-axis and Y-axis of the center of the silica gel, and the two maximum equivalent stress values at the positions of the two FBGs are $3.72 \times 10^{-13}$ MPa and $1.92 \times 10^{-13}$ MPa, respectively, which are generated by torque. These results show that the stress generated by torque is not sensitive to the X-axis and the Y-axis

(2) $\alpha_1 = \alpha_2 \neq 0 \neq 90°$: The maximum equivalent stress at the nodes where FBGs is located are shown in Table 7.

**Table 7.** Absolute value of the maximum equivalent stress.

| $\alpha_1/\alpha_2$ (°) | Stress$_{\text{(torque)}}$ (MPa) | Stress$_{\text{(force)}}$ (MPa) | $\alpha_1/\alpha_2$ (°) | Stress$_{\text{(torque)}}$ (MPa) | Stress$_{\text{(force)}}$ (MPa) |
|---|---|---|---|---|---|
| $\alpha_1 = 0$ $\alpha_2 = 90$ | - - - - | - - - - | $\alpha_1 = \alpha_2 = 45$ | $1.63 \times 10^{-4}$ | $2.27 \times 10^{-4}$ |
| $\alpha_1 = \alpha_2 = 10$ | $5.59 \times 10^{-5}$ | $8.87 \times 10^{-4}$ | $\alpha_1 = \alpha_2 = 50$ | $1.61 \times 10^{-4}$ | $1.05 \times 10^{-4}$ |
| $\alpha_1 = \alpha_2 = 20$ | $1.05 \times 10^{-4}$ | $7.65 \times 10^{-4}$ | $\alpha_1 = \alpha_2 = 60$ | $1.41 \times 10^{-4}$ | $1.24 \times 10^{-4}$ |
| $\alpha_1 = \alpha_2 = 30$ | $1.41 \times 10^{-4}$ | $5.78 \times 10^{-4}$ | $\alpha_1 = \alpha_2 = 70$ | $1.05 \times 10^{-5}$ | $3.14 \times 10^{-4}$ |
| $\alpha_1 = \alpha_2 = 40$ | $1.61 \times 10^{-4}$ | $3.49 \times 10^{-4}$ | $\alpha_1 = \alpha_2 = 80$ | $5.59 \times 10^{-5}$ | $4.32 \times 10^{-4}$ |

In summary, the equivalent stress by torque has the following characteristics: When $\alpha_1$ is equal to 10° and 80°, 20° and 70°, 30° and 60°, or 40° and 50°, the maximum equivalent stress is equal for two different angles. With the increase of $\alpha_1$ and $\alpha_2$, the maximum equivalent stress increased and then decreases. The equivalent stress is the maximum when $\alpha_1$ and $\alpha_2$ are equal to 45°

The equivalent stress by force has the following characteristics: When $\alpha_1$ is equal to 0°, the FBG is the most sensitive, and with the increase of $\alpha_1$ and $\alpha_2$, the maximum equivalent stress decreases and then increased.

For a comprehensive comparison, when $\alpha_1 = \alpha_2 = 30°$, as far as the equivalent stresses by the three loads at the FBGs are concerned, the stress by force is large and the stress by torque is relatively pronounced. Therefore, the parameters of the dual FBG high-precision shape sensor are set as follows:

The length is 23 mm, the width is 8 mm, the thickness is 2 mm, angle $\alpha_1 = \alpha_2 = 30°$, the FBG length is 5 mm and the depth that the FBGs are preburied is 0.5 mm from the neutral plane in the silica gel.

## 4. Sensor Performance Testing

### *4.1. Fabrication of the Dual FBG High-Precision Shape Sensor*

The silica gel in this paper is a translucent silicone, liquid, and it must be mixed with a certain proportion of curing agent evenly to become a solid state as shown in Figure 9. The cured silica gel product is high temperature resistant, tough and tear resistant. After many tests, it is discovered that the optimal ratio of silica gel and curing agent is 15 g:0.1 g– ~ 10 g:0.1 g, which can not only avoid the silica gel being cured in the process of vacuuming, but also ensure that the curing time at normal temperature does not exceed 8 h, which is shorter in the 90 °C heating environment.

**Figure 9.** Liquid-like translucent silica gel.

A large number of bubbles will be generated during the mixing process of silica gel and curing agent. If there are a large number of bubbles in the curved shape sensor, it will affect the stress transfer between FBG and silica gel, and reduce the sensitivity of the sensor, so it is necessary to vacuum before the silica gel is injected into the mold, as shown in Figure 10.

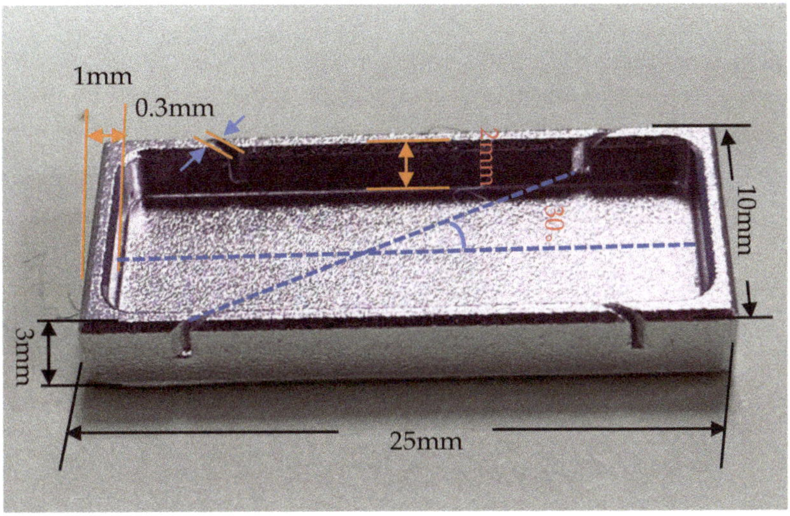


**Figure 10.** Mold of Sensor Fabrication.

The sensor curing process and the finished sensor after curing are shown in Figure 11.

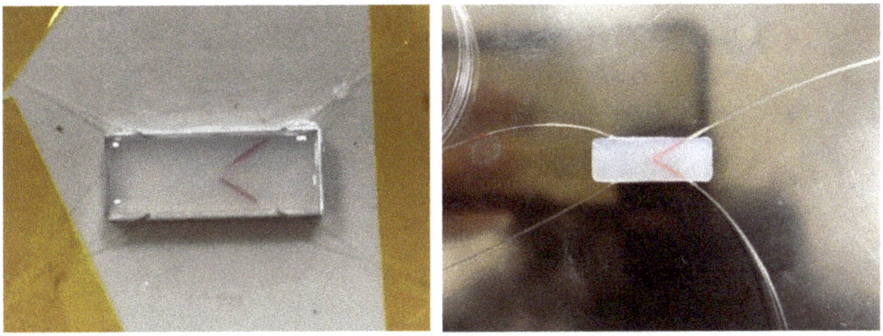

(a) (b)

**Figure 11.** Fabrication of Dual FBG High-Precision Shape sensor, (**a**) Sensor curing process, (**b**) Sensor sample.

*4.2. FBG Shape Sensor Test Platform*

The shape sensor designed in this paper is mainly used for shape reconstruction of objects similar to wing structures, so the cantilever beam is used for parameter calibration, especially when the influence of torque on the accuracy of the shape sensor is considered. The traditional cylinder calibration method is not used because in the reconstruction process, we mainly consider the relationship between the rotation angle of the reconstructed object and the FBG wavelength drift, rather than the relationship between the size of the object subjected to torque and wavelength drift. The performance parameters of the dual FBG high-precision shape sensor were tested by building the sensor test platform shown in Figure 12.

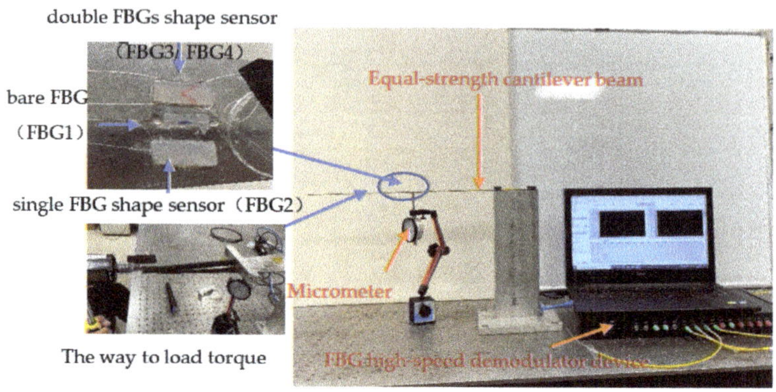


**Figure 12.** Dual FBG High-Precision Shape Sensor Test Platform.

The material of the beam is stainless steel, the length of the beam is 500 mm and three types of sensors are attached at 200 mm from the fixed end of the beam: Single FBG silica gel shape sensor (labeled FBG2) on the left, bare FBG (labeled FBG1) in the middle and dual FBG silica gel shape sensor (labeled FBG3/FBG4) on the right. The parameters of all FBGs are shown in Table 8

**Table 8.** The detailed parameters of the FBG sensors.

| Parameter | FBG1 (Bare FBG) | FBG2 (Single FBG Sensor) | FBG4 (Dual FBG Shape Sensor) | FBG3 |
|---|---|---|---|---|
| Fiber type | single mode | single mode | single mode | single mode |
| Wavelength (nm) | 1549.190 | 1536.141 | 1529.953 | 1536.185 |
| Grating length (mm) | 5 | 5 | 5 | 5 |
| Peak reflectivity (%) | 75 | 75 | 75 | 75 |

The test experiment was divided into several groups, and each group of experimental processes was repeated many times. By processing the data recorded in the experiment, various performance parameters of this sensor, such as sensitivity, repeatability error and hysteresis error, can be analyzed.

## *4.3. Sensor Parameter Calibration*

### 4.3.1. Curvature Calibration Loaded with Vertical Force

At the free end of the beam, weights of 101 g, 175 g, 175 g, 175 g, 500 g, 500 g and 500 g were added in turn, and the reading of the corresponding FBG high-speed demodulator wavelength for each weight was recorded one by one.

The relationship between the curvature $k$ of any point on the curve and the derivative of the curve equation w is shown in Equation (11).

$$k = \pm \frac{w''}{\left(1 + w'^2\right)^{\frac{3}{2}}} \tag{11}$$

For cantilever beams, the curve equation w is the deflection curve of the beam after the beam has been subjected to the force, which has the following physical relationship with the curvature $k$ and the bending moment $M(x)$:

$$\frac{w''}{\left(1 + w'^2\right)^{\frac{3}{2}}} = -\frac{M(x)}{EI} \tag{12}$$

In this experiment, the deflection curve of the beam is a flat curve, and $w'^2 \ll 1$, which is negligible, Equation (12) is simplified as

$$w'' = -\frac{M(x)}{EI} \tag{13}$$

Equation (14) can be obtained from the Equations (11) and (12).

$$k = -\frac{M}{EI} \tag{14}$$

The bending moment equation $M(x)$ of any cross section at a distance $x$ from the fixed end of the cantilever beam can be expressed as Equation (15).

$$M(x) = -F(l - x) \tag{15}$$

Here: $l$-length of the cantilever beam, where $l = 0.4$ m
$x$-distance between sensor center and fixed end, where $x = 0.2$ m
$EI$-cantilever beam flexural stiffness, where $EI = 5.0333$ N·m$^2$
The curvature $k$ of the cantilever beam at a distance $x$ from the fixed end can be derived from Equations (14) and (15)

$$k = -\frac{M}{EI} = \frac{F(l - x)}{EI} \tag{16}$$

The sensitivity of FBG sensors refers to the ratio of the change in the input physical quantity to the change in the output center wavelength when the sensor is in a steady state. Therefore, the sensitivity of three types of FBG sensors can be expressed as Equation (17).

$$S = \frac{\Delta\lambda}{\Delta k} \tag{17}$$

It can be judged from Figure 13 and Table 9 that for the dual FBG high-precision shape sensor loaded with vertical force, (1) the center wavelength drift of FBG2 and FBG3 is linearly related to the curvature change; (2) the sensitivity of FBG3 and FBG3 is 54~55 pm per 1 m$^{-1}$, which is lower than that of bare FBG1 and the single FBG shape sensor FBG2.

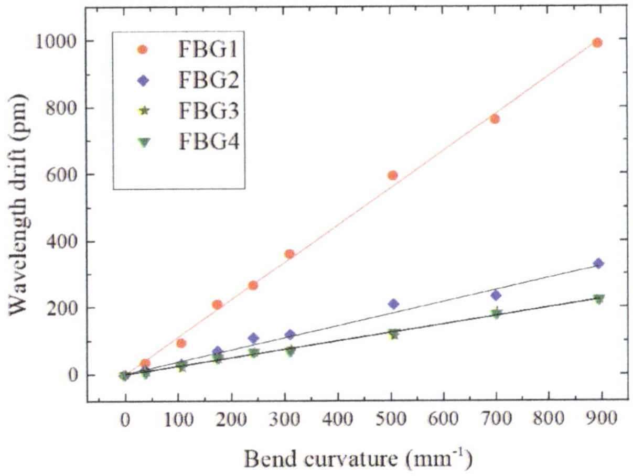


**Figure 13.** Relationship between bending and wavelength drift of three kinds of FBG sensors.

**Table 9.** Sensitivity of the three kinds of FBG sensors.

| Parameters | FBG1 | FBG2 | FBG4 | FBG3 |
|---|---|---|---|---|
| $\Delta\lambda$ (pm) | 217 | 70 | 49 | 48 |
| $\Delta k$ ($m^{-1}$) | 0.896024 | 0.896024 | 0.896024 | 0.896024 |
| $S_k$ ($pm/m^{-1}$) | 242.18 | 78.12 | 54.69 | 53.57 |

4.3.2. Calibration of the Rotation Radian Loaded with Torque

The cantilever beam will flip under the action of torque. In the test model shown in Figure 12, first, the fixture shown in the figure is welded on the cantilever beam, then torque is applied to the cantilever beam using a torque wrench, and the displacement of the edge of the beam to the position of the shape sensor is measured. The cantilever beam is in equilibrium when no torque is applied, and the cantilever beam is gradually loaded with torque to ensure that the micrometer degree increases by 0.5 mm in turn, i.e., the beam rotates by 0.021 radians with one torque loaded.

The wavelength drift of the bare grid, where the grid is at the center of the cantilever beam axis, is almost 0. The relationships between the radian and wavelength drift for the single grid shape sensor (FBG2) and the dual FBG shape sensors (FBG3 and FBG4) are shown in Figure 14.

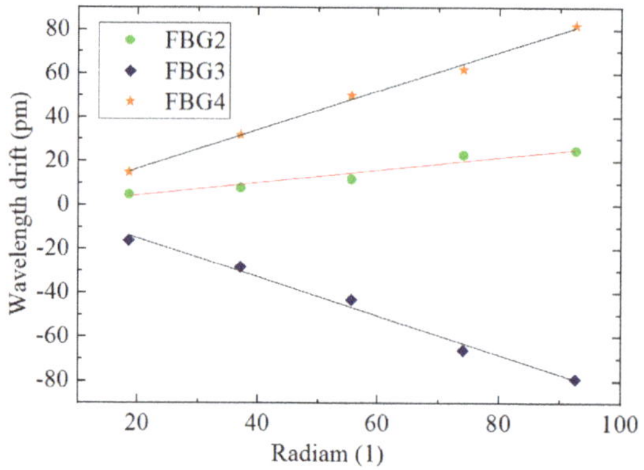


**Figure 14.** Relationship between radian and FBG wavelength drift of FBG shape sensor.

It can be judged from Figure 14 and Table 10 that for the dual FBG high-precision shape sensor loaded torque, (1) the center wavelength drift of FBG3 and FBG4 is linearly related to the radian change; (2) the sensitivity of FBG3 and FBG4 is 866.5~899 pm per 1, which is more than three times the sensitivity of the single FBG shape sensor FBG2.

**Table 10.** Radian sensitivity of the two kinds of FBG sensors.

| Parameters | FBG2 | FBG4 | FBG3 |
|---|---|---|---|
| $\Delta\lambda$ (pm) | 25 | −80 | 83 |
| $\Delta r$ (1) | 0.09233 | 0.09233 | 0.09233 |
| $Sr$ (pm/1) | 270.77 | −866.46 | 898.956 |

4.3.3. Calibration of the Dual FBG High-Precision Shape Sensor under a Combined Load

From the above analysis, it can be discovered that the sensitivity of FBG3 and FBG4 in the dual FBG high-precision shape sensor differs when different loads are applied.

When a vertical force is applied, the sensitivity to curvature is as follows

$$S_{k3} = 53.57\text{pm}/\text{m}^{-1}\ S_{k4} = 54.69\text{pm}/\text{m}^{-1}$$

The corresponding wavelength drift is shown in Equations (18) and (19), where $\Delta k$ is the curvature change.

$$\Delta\lambda_{31} = S_{k3} * \Delta k \quad (18)$$

$$\Delta\lambda_{41} = S_{k4} * \Delta k \quad (19)$$

When a torque is applied, the sensitivity to the rotation radian is as follows:

$$S_{r3} = 898.956\text{pm}/1 \; S_{r4} = -866.46\text{pm}/1$$

The corresponding wavelength drift is shown in Equations (20) and (21), where $\Delta r$ is the rotation radian change.

$$\Delta\lambda_{32} = S_{r3} * \Delta r \quad (20)$$

$$\Delta\lambda_{42} = S_{r4} * \Delta r \quad (21)$$

The forces of the two different loads mentioned above can be superimposed in the dual FBG high-precision shape sensor.

From Equations (18)–(21), we can obtain

$$\Delta\lambda_3 = S_{k3} * \Delta k + S_{r3} * \Delta r \quad (22)$$

$$\Delta\lambda_4 = S_{k4} * \Delta k + S_{r4} * \Delta r \quad (23)$$

The curvature change $\Delta k$ and the rotation radian change $\Delta r$ in the dual FBG high-precision shape sensor are expressed as Equations (24) and (25), when the combined loads are applied.

$$\Delta k = \frac{S_{r4} * \Delta\lambda_3 - S_{r3*}\Delta\lambda_4}{S_{k3} * S_{r4} - S_{k4} * S_{r3}} \quad (24)$$

$$\Delta r = \frac{S_{k4} * \Delta\lambda_3 - S_{k3*}\Delta\lambda_4}{S_{k4} * S_{r3} - S_{k3} * S_{r4}} \quad (25)$$

### 4.4. Error Analysis of Repeatability

On the experimental platform shown as Figure 12, a torque of 8 N·m is loaded on the cantilever beam at the free end of the beam, and the mass block is added in turn with weights of 101 g, 175 g, 175 g, 175 g, 175 g, and 500 g. We repeat five tests and record the relationship between the center wavelength drift of FBG3 and FBG4 and the change in curvature to obtain the curves shown in Figure 15.

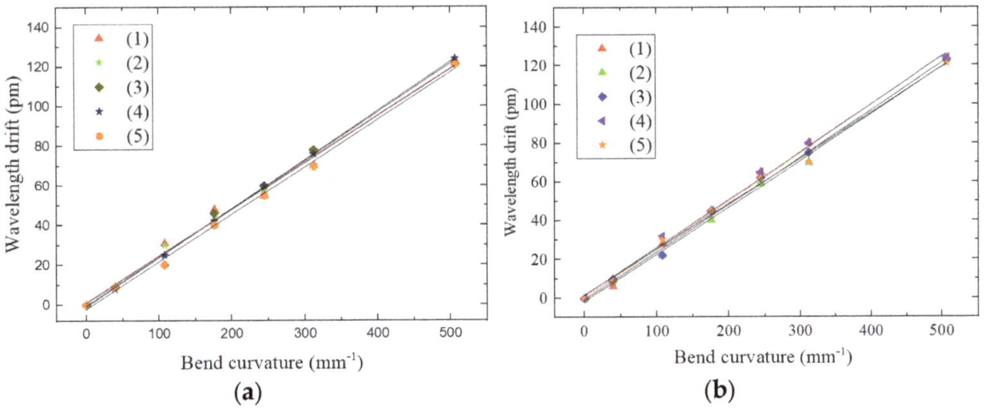


**Figure 15.** Shape sensor FBG error analysis of repeatability: (**a**) FBG3 center wavelength drift consistency, (**b**) FBG4 center wavelength drift consistency.

The repeatability errors $R$ of FBG3 and FBG4 are calculated from the sensor error formula of repeatability:

$$R3 = \frac{\Delta\lambda}{(\Delta\lambda)_{FS}} \times 100\% = \frac{2}{123} \times 100\% = 1.63\% R4 = \frac{\Delta\lambda}{(\Delta\lambda)_{FS}} \times 100\% = \frac{2}{123} \times 100\% = 1.63\%$$

Here: $\Delta\lambda_{max}$-maximum deviation value of the FBG center wavelength drift in forward-reverse stroke.

$(\Delta\lambda)_{FS}$ -the center wavelength drift of the FBG output at full scale

For the above repeatability test results, using the same operation procedure, we randomly selected a shape sensor for the test as shown in Figure 16, two of FBGs in the sensor are labeled as FBG1 and FBG2, respectively.

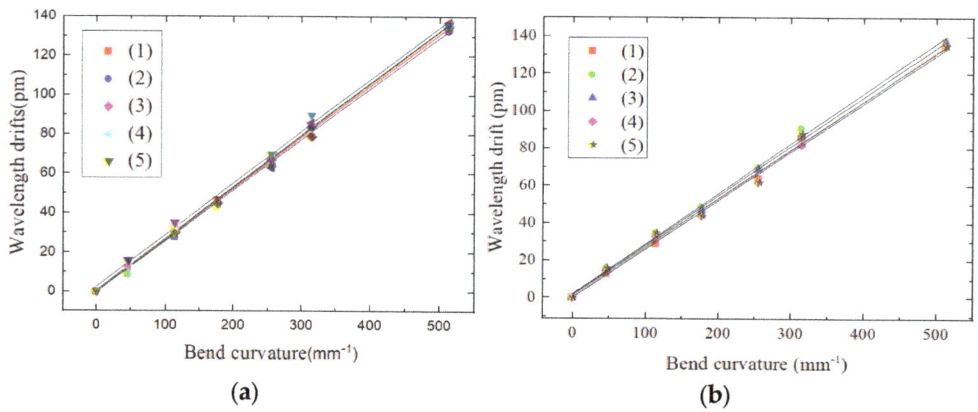


**Figure 16.** Shape sensor FBG error analysis of repeatability: (**a**) FBG1 center wavelength drift consistency, (**b**) FBG2 center wavelength drift consistency.

If the sensor has multiple output characteristic curves under the same input conditions many times, the sensor has better repeatability and smaller error, which indicates that the FBG center wavelength drift in the dual FBG high-precision shape sensor is more stable.

### 4.5. Hysteresis Error Analysis

Hysteresis error is one of the important indicators to reflect the accuracy of the sensor. The hysteresis of the sensor is easily caused by the physical properties of the sensor's sensitive component materials or design defects.

On the experimental platform shown as Figure 12, a torque of 8 N·m is loaded on the cantilever beam at the free end of the beam. First, the mass block is increased in turn with weights of 101 g, 175 g, 175 g, 175 g, 175 g and 500 g. Then, the mass block is sequentially decreased, with weights of 500 g, 175 g, 175 g, 175 g, 175 g and 101 g. We repeat five tests and record the relationship between the center wavelength drift of FBG3 and FBG4 and the change of curvature. The obtained curves are shown in Figure 17.

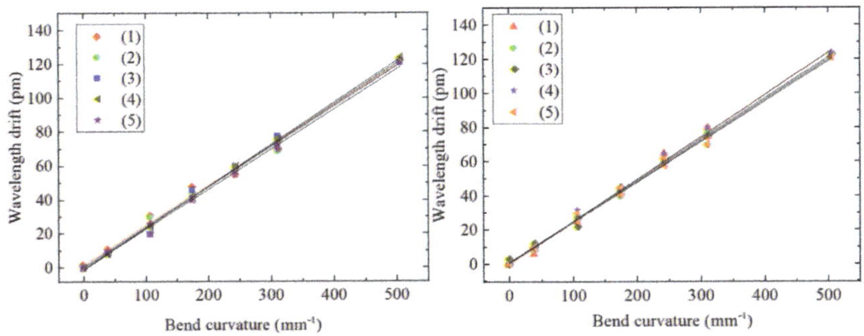


(**a**) (**b**)

**Figure 17.** Shape sensor FBG hysteresis error analysis: (**a**) FBG3 center wavelength drift in forward-reverse stroke, (**b**) FBG4 center wavelength drift in forward-reverse stroke.

The hysteresis errors of FBG3 and FBG4 are calculated from the hysteresis error formula.

$$e_{H3} = \frac{\Delta\lambda_{max}}{(\Delta\lambda)_{FS}} \times 100\% = \frac{3}{123} \times 100\% = 2.43\%$$

Here: $\Delta\lambda_{max}$-maximum deviation value of FBG center wavelength drift in forward-reverse stroke.

$(\Delta\lambda)_{FS}$-the center wavelength drift of the FBG output at full scale.

For the above hysteresis test results, using the same operation procedure, we randomly selected a shape sensor for the test as shown in Figure 18, two of FBGs in the sensor are labeled as FBG1 and FBG2, respectively.

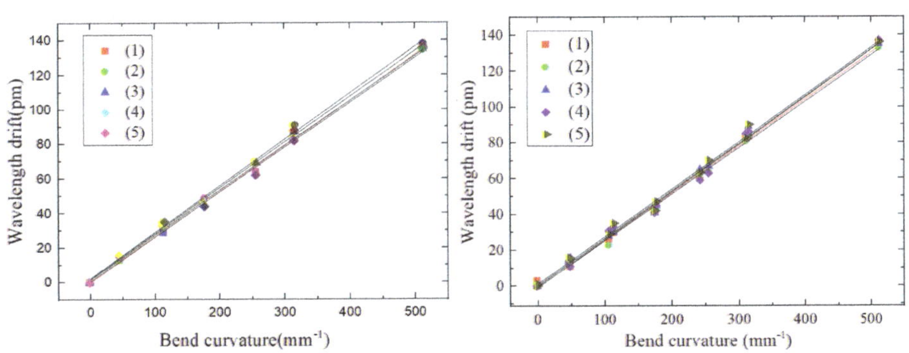


(**a**) (**b**)

**Figure 18.** Shape sensor FBG hysteresis error analysis: (**a**) FBG1 center wavelength drift in forward-reverse stroke, (**b**) FBG2 center wavelength drift in forward-reverse stroke.

From the hysteresis error in Figure 16, we can intuitively judge that the FBG center wavelength drift of the dual FBG high-precision shape sensor is stable, and the operating performance of the dual FBG high-precision shape sensor is also stable.

## 5. Shape Reconstruction Test of a Test Specimen

To verify the high accuracy and reliability of the dual FBG high-precision shape sensor in the process of shape reconstruction, which is proposed in this paper, we will carry out a reconstruction test on a test specimen installed with 19 shape sensors (38 FBGs in total), and the experimental setup is shown in Figure 19.

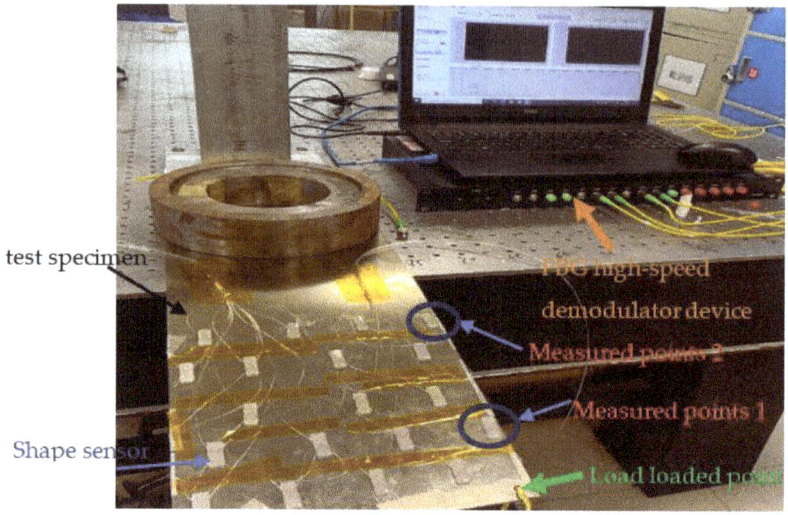


**Figure 19.** Shape sensor reconstruction experiment platform.

According to the FBG center wavelength drift in each dual FBG high-precision shape sensor, the discrete curvature information of each measurement point is obtained, and the improved Kalman filtering algorithm is used. The three-dimensional shape of the test specimen is reconstructed by the curvature, coordinate position and other information of the measured point in MATLAB software. The loading method of the test specimen is shown in Figure 18.

In Figure 20a, a weight mass is added in the middle loading point, and it is added three times in turn. In Figure 20b, a weight mass is added at the left loading point, and an upward force is loaded at the right loading point. The reconstructing shapes of the specimen are shown in Figures 20 and 21.

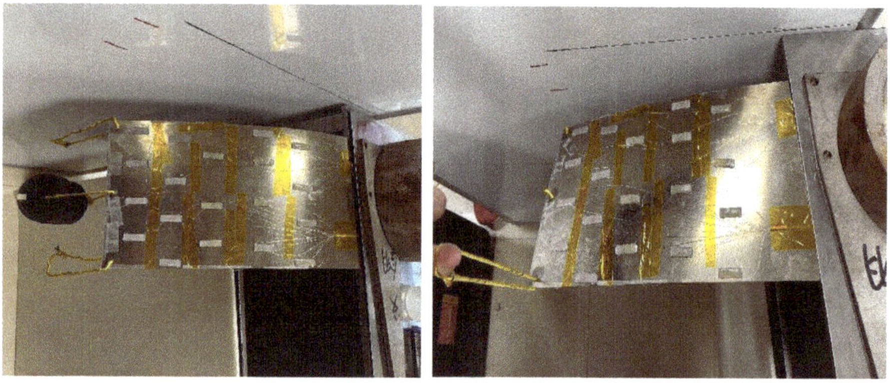

(**a**) (**b**)

**Figure 20.** Method of test piece loaded, (**a**) Load in the middle point, (**b**) Load in both side points.

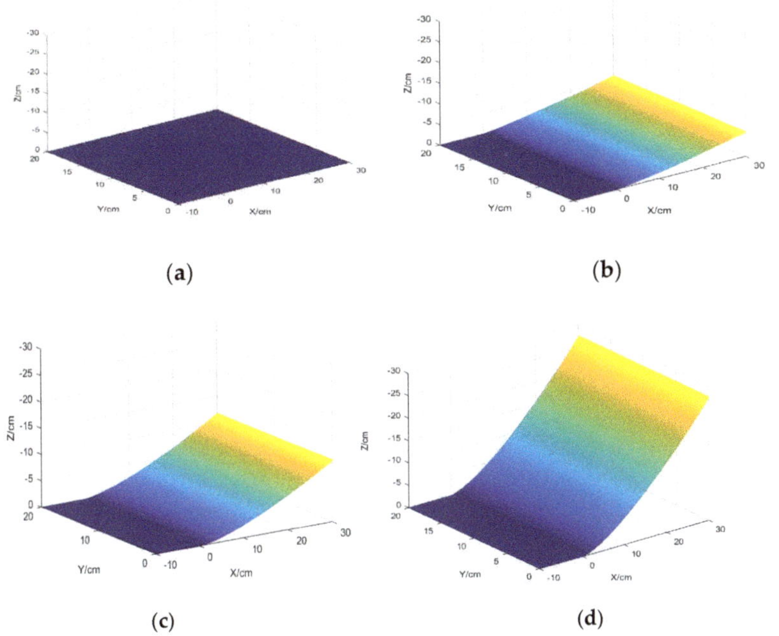


**Figure 21.** Test piece shape reconstruction loaded in the middle point: (**a**) Initial state, (**b**) state I, (**c**) state II, (**d**) state III.

In Figures 21 and 22, the change value of the measurement point in the Z-axis coordinate reflects the degree of specimen bending, which corresponds to the change in curvature of the dual FBG shape sensor, and the change in the x-axis coordinate reflects the degree of specimen rotation, which corresponds to the change in the rotation angle in the dual FBG shape sensor. To verify the accuracy of the shape reconstruction, comparing the measured and reconstructed point coordinates of the FBG shape sensor with the actual coordinates of two measurement points is shown as Tables 11 and 12. (This results comparison ignores the error caused by the reconstruction algorithm).

**Table 11.** Comparison of the results of the measurement points in Figure 21.

| Measured Points | Parameters | State I | State II | State III |
|---|---|---|---|---|
| point 1 | Reconstructed value(cm) | 1.582 | 2.497 | 5.868 |
| | Measured value(cm) | 1.671 | 2.57 | 6.13 |
| | Error (%) | 5.33 | 2.84 | 4.27 |
| point 2 | Reconstructed value(cm) | 4.65 | 8.566 | 19.52 |
| | Measured value(cm) | 4.51 | 8.151 | 20.58 |
| | Error (%) | 3.1 | 5.09 | 5.15 |

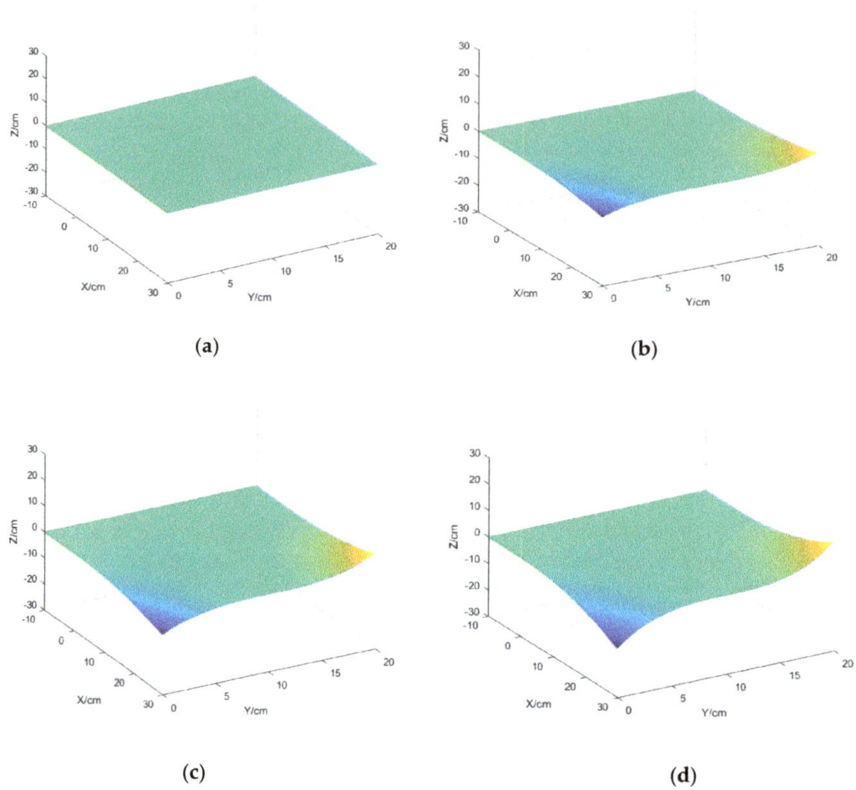


**Figure 22.** Test piece shape reconstruction loaded in both side points: (**a**) Initial state, (**b**) state IV, (**c**) state V, (**d**) state VI.

**Table 12.** Comparison of the results of the measurement points in Figure 22.

| Measured Points | Parameters | State IV | | State V | | State VI | |
|---|---|---|---|---|---|---|---|
| | | x-axi | z-axis | x-axi | z-axis | x-axi | z-axis |
| point 1 | Reconstructed values (cm)<br>Measured values (cm) | 9.917<br>9.751 | 0.497<br>0.468 | 9.899<br>9.741 | 0.6798<br>0.722 | 9.891<br>9.732 | 1.254<br>1.198 |
| | Error (%) | 1.7 | 6.13 | 1.62 | 5.84 | 1.63 | 4.67 |
| point 2 | Reconstructed values (cm)<br>Measured values (cm) | 24.51<br>24.109 | 2.848<br>2.769 | 23.99<br>23.465 | 4.48<br>4.356 | 23.08<br>22.591 | 7.074<br>7.267 |
| | Error (%) | 1.66 | 2.85 | 2.24 | 2.85 | 2.16 | 2.66 |

As shown in Tables 11 and 12, the maximum reconstruction error is 6.13% (z-axis) when the test specimen is reconstructed in shape, and the greater the curvature, the smaller the reconstruction error. The results show that the shape reconstruction of the test specimen based on the dual FBG shape sensor is highly restored, and the dual FBG shape sensor has the high feasibility and effectiveness in 3D shape reconstruction.

## 6. Conclusions

To improve the measurement accuracy of FBG shape sensors, especially in terms of structural shape reconstruction accuracy, the design idea of a high-precision shape sensor based on dual FBGs is proposed. First, a high-precision shape sensor with dual FBGs is designed and analyzed. Second, the linear relationship among the FBG wavelength drift and curvature, rotation radian loaded vertical force and torque is obtained. Finally, the verification test of shape reconstruction is completed in the laboratory. The test specimen is loaded with an arbitrary load, and the discrete curvature is collected according to the dual FBG high-precision shape sensor. Then the three-dimensional shape of the test specimen is reconstructed by using the reconstruction algorithm. Experimental results show that the maximum reconstruction error of the sensor proposed in the paper is less than 6.5%.

In summary, we can draw the following conclusion: A dual FBG high-precision shape sensor is an effective method for improving structural shape reconstruction, especially in the field of deformation monitoring such as robots, smart wearables, wings and fan blades, and has greater application potential. In the future, we will increase the sensitivity of the dual FBG high-precision shape sensor, improve the shape reconstruction algorithm and further improve the three-dimensional shape reconstruction precision of the structural deformation.

**Author Contributions:** Conceptualization and methodology, H.W. (Huifeng Wu); software, S.D.; formal analysis and data curation, H.W. (Hui Wang) and R.D.; experiment, Q.X.; writing—original draft preparation, H.W. (Huifeng Wu); writing—review and editing, L.L. All authors have read and agreed to the published version of the manuscript.

**Funding:** This research was funded by "Research on New Fiber Optic Sensing Devices and Their Industrialisation for Equipment Intelligence "(2019sdr001); "The Key Technology and Application of marine Deep Space Development and Utilization" (SKJC-KJ-2019KY02); The Basic Ability Improvement Project of Young and Middle-aged Teachers in Guangxi Universities(2018KY0657); National Natural Science Foundation of Guangxi (2020GXNSFAA297032 and 2020JJA170017).

**Institutional Review Board Statement:** Not applicable.

**Informed Consent Statement:** Not applicable.

**Data Availability Statement:** Not applicable.

**Conflicts of Interest:** The authors declare that there is no conflict of interest.

## References

1. Zhao, S.Y.; Cui, J.W.; Chen, M.M. Review on optical fiber shape sensing technology. *Opt. Precis. Eng.* **2020**, *28*, 10–29. [CrossRef]
2. Waltermann, C.; Koch, J.; Angelmahr, M.; Burgmeier, J.; Schade, W. Fiber-optical 3d shape sensing. In *Planar Waveguides and other Confined Geometries*; Springer: New York, NY, USA, 2014; pp. 227–250.
3. Shen, L.Y.; Xiao, H.; Qian, J.W.; Yang, L.J.; Zhang, Y.N. Shape reconstruction and visualization of endoscope. *Chin. J. Sci. Instrum.* **2014**, *21*, 2725–2730.
4. Wei, W.; Zhou, Z.; Xiaoping, Z.; Rui, W. Explo-ring aeroelastic stability of very flexible solar powered UAV with geometrically large deformation. *J. Northwestern Polytech. Univ.* **2015**, *33*, 1–8.
5. Pena, F.; Richards, L.; Parker, A.; Piazza, A.; Chan, P.; Hamory, P. Fiber Optic Sensing System (FOSS) Technology-A New Sensor Paradigm for Comprehensive Structural Monitoring and Model Validation Throughout the Vehicle Lifecycle. 2015. Available online: http://ntrs.nasa.gov/archive/nasa/casi.ntrs.nasa.gov/20160001157.pdf.2015 (accessed on 3 August 2021).
6. Ma, Z.; Chen, X. Fiber bragg gratings sensors for aircraft wing shape measurement: Recent applications and technical analysis. *Sensors* **2018**, *19*, 55. [CrossRef] [PubMed]
7. Sun, G.K.; Qu, D.M.; Yan, G.; Song, Y.M.; Zhu, L.Q. Bending deformation of optical fiber sensing and shape reconstruction of soft pneumatic driver. *J. Syst. Simul.* **2019**, *27*, 1052–1059.
8. Zhu, X.J.; Lu, M.Y.; Zhao, X.Y. 3D reconstruction algorithm and visualization analysis for space manipulator vibration shape. *Opt. Precis. Eng.* **2009**, *21*, 4706–4713.
9. He, Y.L.; Zhang, X.; Sun, G.K.; Song, Y.M.; Zhu, L.Q. Flexible curvature sensor based on composite substrate. *Opt. Precis. Eng.* **2019**, *27*, 1270–1276.
10. Davis, M.A.; Malinowski, E.; Chevalier, J.L.; Mcgregor, A.M.; Reaves, M. Body Shape, Position, and Posture Recognition Suit with Multi-Core Optical Shape Sensing Fiber. U.S. Patent US 9304019, 2016.
11. Monsberger, C.M.; Lienhart, W. Distributed fiber optic shape sensing of concrete structures. *Sensors* **2021**, *21*, 6098. [CrossRef] [PubMed]
12. Jang, M.; Kim, J.S.; Kang, K.; Kim, J.; Yang, S. Towards Finger Motion Capture System Using FBG Sensors. In Proceedings of the 2018 IEEE 40th Annual International Conference of the IEEE Engineering in Medicine and Biology Society (EMBC), Honolulu, HI, USA, 18–21 July 2018; pp. 3734–3737.
13. Xu, C.; Khodaei, Z.S. Shape sensing with rayleigh backscattering fibre optic sensor. *Sensors* **2020**, *20*, 4040. [CrossRef]

14. Chen, X.Y.; Zhang, Y.N.; Qian, J.W.; Shen, L.Y. Accuracy of position tracking and fabrication of thin diameter sensor. *Meas. Sci. Technol.* **2021**, *32*, 055202. [CrossRef]
15. Zhao, Z.; Soto, M.A.; Ming, T.; Thévenaz, L. Distributed shape sensing using Brillouin scattering in multi-core fibers. *Opt. Express* **2016**, *24*, 25211. [CrossRef] [PubMed]
16. Ba, D.X.; Chen, C.; Cheng, F.; Zhang, D.Y.; Lu, Z.W.; Fan, Z.G.; Dong, Y.K. A high-performance and temperature-insensitive shape sensor based on DPP-BOTDA. *IEEE Photon. J.* **2017**, *10*, 7100810. [CrossRef]
17. Szostkiewicz, U.; Soto, M.A.; Yang, Z.; Dominguez-Lopez, A.; Thevenaz, L. High-resolution distributed shape sensing using phase-sensitive optical time-domain reflectometry and multicore fibers. *Opt. Express* **2019**, *27*, 20763. [CrossRef] [PubMed]
18. Tian, C.; Wang, Z.; Sui, Q.; Jing, W.; Dong, Y.; Li, Y. Design and optimization of fbg implantable flexible morphological sensor to realize the intellisense for displacement. *Sensors* **2018**, *18*, 2342. [CrossRef]
19. Wang, C. 3D shapes detecting method of soft manipulator based on fiber Bragg grating sensor. *Control Instrum. Chem. Ind.* **2015**, *42*, 1130–1133.
20. Zhi, G.H.; Di, H.T. Wind speed monitoring system based on optical fiber curvature sensor. *Opt. Fiber Technol.* **2021**, *62*, 102467. [CrossRef]
21. Abro, A.Z.; Zhang, Y.F.; Hong, C.Y.; Ahmed, L.R.; Chen, N.L. Development of a smart garment for monitoring body postures based on fbg and flex sensing technologies. *Sens. Actuators A Phys.* **2018**, *272*, 153–160. [CrossRef]
22. Abro, Z.A.; Hong, C.; Chen, N.; Zhang, Y.; Yasin, S. A fiber bragg grating-based smart wearable belt for monitoring knee joint postures. *Text. Res. J.* **2020**, *90*, 386–394. [CrossRef]
23. He, Y.L.; Zhu, L.Q.; Sun, G.K.; Yu, M.X.; Dong, M.L. Design, Measurement and Shape Reconstruction of Soft Surgical Actuator Based on Fiber Bragg Gratings. *Appl. Sci.* **2018**, *8*, 1773. [CrossRef]
24. He, Y.; Dong, M.; Sun, G.; Meng, F.; Song, Y.; Zhu, L. Shape monitoring of morphing wing using micro-optical sensors with different embedded depth. *Opt. Fiber Technol.* **2019**, *48*, 179–185. [CrossRef]
25. Sun, G.; Wu, Y.; Li, H.; Zhu, L. 3D shape sensing of flexible morphing wing using fiber bragg grating sensing method. *Opt. Int. J. Light Electron. Opt.* **2018**, *156*, 83–92. [CrossRef]
26. Wang, Q.L.; Zhou, X.H.; Jiang, H.J.; Sun, G.K.; Zhu, L.Q. Polyimide sensing layer for bending shape measurement in soft surgical manipulators. *Optik* **2019**, *183*, 179–188. [CrossRef]
27. Leal Junior, A.G.; Theodosiou, A.; Diaz, C.; Marques, C.; Pontes, M.J.; Kalli, K. Simultaneous measurement of axial strain, bending and torsion with a single fiber bragg grating in CYTOP fiber. *J. Lightwave Technol.* **2018**, *37*, 971–980. [CrossRef]
28. Henken, K.R.; Dankelman, J.; van den Dobbelsteen, J.J.; Cheng, L.K.; van der Heiden, M.S. Error analysis of fbg-based shape sensors for medical needle tracking. *IEEE/ASME Trans. Mechatron.* **2014**, *19*, 1523–1531.
29. Sun, G.; Hu, Y.; Dong, M.; He, Y.; Yu, M.; Zhu, L. Flexible membrane curvature sensor based on multilayer polyimide substrate and optical fiber implantation. *Optik* **2018**, *176*, 559–566. [CrossRef]
30. Chen, X.; Zhang, Y.; Qian, J.; Yi, X.; Shen, L. Analysis of strain transfer characteristics of no-substrate and thin diameter sensor-sciencedirect. *Opt. Fiber Technol.* **2020**, *102*, 2367.
31. Othonos, A.; Kalli, K.; Kohnke, G.E. *Fiber Bragg Gratings: Fundamentals and Applications in Telecommunications and Sensing*; Physics Today: Boston, MA, USA, 2000.

sensors

MDPI

*Article*

# A Medium-Frequency Fiber Bragg Grating Accelerometer Based on Flexible Hinges

**Zichuang Li, Lei Liang *, Hui Wang, Shu Dai, Ke Jiang and Zhiyuan Song**

National Engineering Laboratory for Fiber Optic Sensing Technology, Wuhan University of Technology, Wuhan 430070, China; lizichuang@whut.edu.cn (Z.L.); wanghui1989@whut.edu.cn (H.W.); 290303@whut.edu.cn (S.D.); k.jiang@whut.edu.cn (K.J.); 291875@whut.edu.cn (Z.S.)
* Correspondence: lianglei@whut.edu.cn

**Abstract:** Mediumfrequency fiber Bragg grating (FBG) acceleration sensors are used in important applications in mechanical, aerospace and weapon equipment, and have strict requirements in terms of resonance frequency and sensitivity. A novel medium-frequency accelerometer, based on fiber Bragg grating and flexible hinges, is proposed in this paper. The differential structure doubles the sensitivity of the sensor while avoiding temperature effects. The structure model and principle for the sensor are introduced, the sensor's sensing characteristics are theoretically analyzed, and the structure parameters for the sensor are determined through numerical analysis. The sensing experiments show that the resonance frequency of the sensor is approximately 2800 Hz, the sensitivity is 21.8 pm/g in the flat frequency range of 50–1000 Hz, and the proposed sensor has a good temperature self-compensation function and lateral anti-interference capability.

**Keywords:** fiber Bragg grating; accelerometer; flexible hinge; vibration monitoring



**Citation:** Li, Z.; Liang, L.; Wang, H.; Dai, S.; Jiang, K.; Song, Z. A Medium-Frequency Fiber Bragg Grating Accelerometer Based on Flexible Hinges. *Sensors* **2021**, *21*, 6968. https://doi.org/10.3390/s21216968

Academic Editor: Oleg G. Morozov

Received: 6 September 2021
Accepted: 17 October 2021
Published: 20 October 2021



## 1. Introduction

Electronic acceleration sensor technology is relatively mature, and it is widely used in the field of mechanical fault diagnosis [1–3]. Compared with traditional electronic sensors, FBG sensors have many advantages, including anti-electromagnetic interference, corrosion resistance, long-distance transmission, and distributed measurement. FBG sensors are widely used in important fields, such as railways [4], bridges [5], the oil industry [6], and aerospace [7,8]. Up until the present day, FBG acceleration sensors have been widely used in vibration measurement, playing an important role in structural health monitoring [9–13]. A packaged FBG acceleration sensor is composed of a spring-mass system with one or several FBGs. Various elastic elements have been studied to form FBG acceleration sensors, including cantilever type [14–18], diaphragm type [19–21], and flexure hinge type [22–27]. In flexure hingebased acceleration sensors [23–25], the two sides of a flexure hinge are connected with the mass and the base, respectively, and the FBG is placed between the base and the inertial mass to realize the acceleration measurement. Wei et al. [26] arranged two couple FBGs to eliminate the influence of temperature and double the sensitivity of the sensor. The acceleration sensitivity reached 40 pm/g, but its resonance frequency is as low as 814.3 Hz. Wang et al. [27] realized the measurement of three-dimensional acceleration based on a composite flexible hinge. The sensor's $z$-axis sensitivity is 20.3 pm/g, and the measurement frequency range is only 0–250 Hz, which is unable to measure medium–high-frequency acceleration.

In recent years, many scholars proposed various structures for medium–high-frequency FBG acceleration sensors [28–32]. In [28], Stefani et al. proposed an FBG accelerometer based on polymer fiber and a flexible hinge structure. The sensor's resonance frequency is approximately 3 kHz, and the sensitivity is up to 19 pm/g. The study pointed out that polymer FBGs can increase the sensitivity of an accelerometer by more than four times, compared with silica FBGs. In [29], based on the flexible hinge structure, Dai et al. discussed the influence of hinge structure parameters on the sensor's sensitivity and resonance frequency, and reported an accelerometer with a resonance frequency of about 3000 Hz. Their experiments showed that the sensor sensitivity is 13.82 pm/g at an excitation frequency of 100 Hz. In [30], Guo et al. introduced a welding-packaged accelerometer based on a metal-coated FBG; the resonance frequency and sensitivity of the sensor are 3.6 kHz and 1.7 pm/g, respectively. In [31], based on the elastic structure of a steel tube and a mass block, Wang et al. designed a high-frequency FBG accelerometer. Embedded with two FBGs for axial vibration sensing, the resonance frequency and sensitivity are 3806 Hz and 4.01 pm/g, respectively. According to [32], Wu et al. reported a high-frequency FBG accelerometer with an isosceles triangular cantilever beam and a mass block. In this sensor, one FBG is pasted onto the cantilever beam for measurement and the other FBG is used as a reference. Adopting light intensity demodulation, the resonance frequency and sensitivity of the sensor are 8193 Hz and 0.46 pm/g, respectively.

For FBG acceleration sensors, the resonance frequency and sensitivity restrict each other, which is also mentioned in some related studies [21]. In existing reports about mediumhigh frequency FBG acceleration sensors, the sensitivities are often low, which leads to low resolution. A possible way to improve the sensor performance is to use FBGs in polymer optical fiber, but polymer optical fiber will bring higher transmission losses. Therefore, we try to broaden the frequency measurement range and increase the sensitivity by improving the structure of the sensor.

In this work, we propose a novel medium-frequency FBG accelerometer based on two FBGs, three flexible hinges, and two inertial mass blocks, where the three straight circular flexure hinges connect the two masses and the base into a whole. Two FBGs are then arranged symmetrically at both ends of the two masses. Such a hingemasshinge structure can improve the overall rigidity of the sensor compared with the common masshingebase structure. Meanwhile, two FBGs suspended between two masses not only doubles the sensitivity of the sensor, but also leads to a temperature self-compensation capability. In the following sections, the sensing characteristics of the sensor are analyzed theoretically, the structural parameters are determined by numerical analysis, and, finally, the sensor is experimentally studied.

## 2. Materials and Methods

### 2.1. Structure Model and Principle of the Sensor

The detailed design of the proposed medium-frequency FBG accelerometer is illustrated in Figure 1. This sensor is mainly composed of two inertial masses, three straight circular flexure hinges, two FBGs and a base. The sensor is symmetrical, the sizes of flexure hinge-2 and flexure hinge-3 are exactly the same. The two ends of the masses are provided with optical fiber grooves. The two ends of the pre-stretched FBG are fixed in the grooves by glue. The FBG is suspended above the flexible hinge, and the base and the package shell are connected by threads.

When an acceleration excitation along the measurement direction is generated externally, the two masses will rotate slightly around the center of flexible hinge-2 and flexible hinge-3, respectively, driving the FBGs to stretch or compress and converting the vibration acceleration into the strains and wavelength shifts of two FBGs. Vibration signals can be obtained from the wavelength shifts of the two FBGs. In this design, three straight round flexible hinges connect the two masses and the base into a whole. The hinge–mass–hinge structure can improve the overall rigidity of the sensor compared with the hinge-mass structure. Meanwhile, two FBGs designs double the sensor's sensitivity while realizing temperature self-compensation.

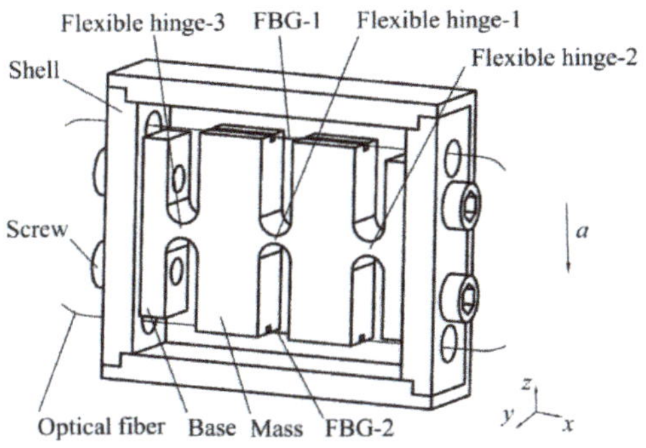


**Figure 1.** Structural diagram of the FBG accelerometer.

## 2.2. Sensitivity

When an external acceleration acts on the sensitive direction of the sensor, the two masses will rotate slightly around the center of flexible hinge-2 and flexible hinge-3, respectively, under the action of inertial force. The mechanical model of the sensor is shown in Figure 2.

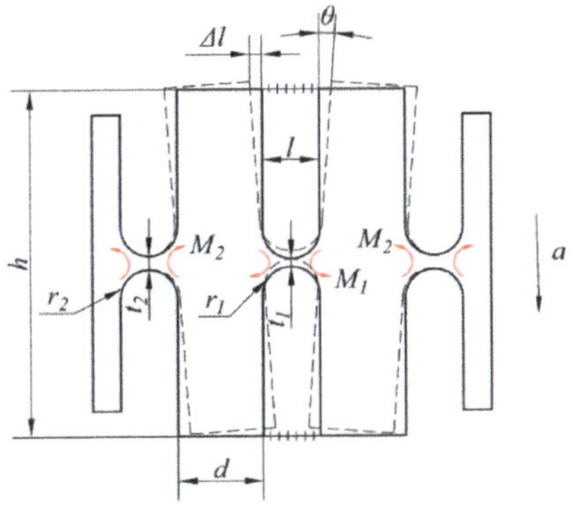


**Figure 2.** Mechanical model of the FBG accelerometer.

According to the principle of virtual displacement, the general dynamics equation for the system can be expressed as follows:

$$(-2M_1 - 2M_2)\delta\theta + 2ma\delta z - 4F_f\delta\Delta l = 0 \tag{1}$$

where $\theta$ is the rotation angle of the mass relative to flexible hinge-2, $m$ is the mass of the mass block, $a$ is the vibration acceleration in the $z$-axis of the sensor, $z$ is the displacement of the mass center of the mass in the $z$-axis, $F_f$ is the tensile force generated by the optical fiber, $l$ is the optical fiber pasting span, and $\Delta l$ is the displacement of the bonding point at one end of the optical fiber in the $x$-axis.

The magnitude of the moment $M_i$ is proportional to the stiffness $K_i$ of the straight circular flexure hinge, which can be expressed as follows:

$$M_i = K_i\theta \ (i = 1, 2) \tag{2}$$

Here, the stiffness of the straight round flexure hinge is given by [33].

$$K = \frac{EwR^2}{24} \Big/ \left[ \frac{s^3(6s^2 + 4s + 1)}{(2s+1)(4s+1)^2} + \frac{6s^4(2s+1)}{(4s+1)^{5/2}} \arctan\sqrt{4s+1} \right] \tag{3}$$

$$s = \frac{r}{t} \tag{4}$$

where $E$ is the elastic modulus of the flexible hinge, $r$ is the radius of the straight circular flexible hinge, $t$ is the thickness of the hinge waist, and $w$ is the width of the hinge.

In Equation (1), the magnitude of the pulling force can be obtained by the following:

$$F_f = 2k_f \Delta l \tag{5}$$

where the following is the elastic coefficient of the optical fiber:

$$k_f = \frac{A_f E_f}{l} \tag{6}$$

where $l$ is the bonding span of the optical fiber, and $A_f$ and $E_f$ are the cross-sectional area and elastic modulus of the optical fiber, respectively.

Since $\theta$ is very small, we can take $\sin\theta \approx \theta$. From the geometric relationship, we have the following:

$$z = (r_2 + \frac{d}{2})\theta \tag{7}$$

$$\Delta l = \frac{h}{2}\theta \tag{8}$$

When deformation occurs between spans, the strain $\varepsilon$ corresponding to the grating can be described as follows:

$$\varepsilon = \frac{2\Delta l}{l} \tag{9}$$

When the incident light passes through the FBG, part of light that meets the Bragg central wavelength condition will be reflected. External strain and temperature load will result in a shift of the Bragg central wavelength. The effects of strain and temperature on the Bragg wavelength shift can be expressed as follows:

$$\frac{\Delta\lambda}{\lambda} = (1 - P_e)\varepsilon + (\alpha_f + \xi_f)\Delta T \tag{10}$$

where $\lambda$ is the Bragg central wavelength for the FBG, $\Delta\lambda$ is the Bragg wavelength shift of the FBG, $\alpha_f$ is the thermal expansion coefficient, $\zeta_f$ is the thermo-optical coefficient, and $P_e$ is the photoelastic coefficient (theoretical value = 0.22).

For FBG-1 and FBG-2, the following apply:

$$\frac{\Delta\lambda_1}{\lambda_1} = (1 - P_e)\varepsilon_1 + (\alpha_f + \xi_f)\Delta T \tag{11}$$

$$\frac{\Delta\lambda_2}{\lambda_2} = (1 - P_e)\varepsilon_2 + (\alpha_f + \xi_f)\Delta T \tag{12}$$

where $\lambda_1$ and $\lambda_2$ represent the central wavelengths for FBG-1 and FBG-2 after prestretching, $\Delta\lambda_1$ and $\Delta\lambda_2$ represent the central wavelength shifts for FBG-1 and FBG-2 after prestretching, respectively, and $\varepsilon_1$ and $\varepsilon_2$ represent the strain amounts for FBG-1 and FBG-2 after prestretching, respectively.

FBG-1 and FBG-2 are symmetrically arranged at both ends of the mass block. When FBG-1 is stretched, FBG-2 is compressed, with $\varepsilon_1 = \varepsilon$ and $\varepsilon_2 = -\varepsilon$. The temperature sensitivity coefficients for the selected gratings are the same, and the central wavelengths are approximately equal, with $\lambda_1 = \lambda_2 = \lambda$.

By combining Equations (11) and (12) to eliminate the influence of temperature on the wavelength shift of the grating, we can obtain the following:

$$\Delta\lambda = \Delta\lambda_1 - \Delta\lambda_2 = 2(1 - P_e)\varepsilon \tag{13}$$

Combining Equations (1), (7), (8) and (13), the sensitivity of the sensor can be expressed as follows:

$$S = \frac{\Delta\lambda}{a} = \frac{4\lambda(1-P_e)\Delta l}{al} = \frac{\lambda(1-P_e)}{l}\frac{2mh(r_2+\frac{d}{2})}{k_f h^2 + K_1 + K_2} \tag{14}$$

## 2.3. Resonance Frequency

The resonance frequency is an important indicator to account for the performance of an accelerometer. Suppose $J$ is the inertia moment of the mass around the center of hinge-2 and $\theta$ is the generalized coordinate for obtaining the Lagrangian function, as follows:

$$L = T_m - V_f - V_J \tag{15}$$

The strain potential energy of the optical fiber can be expressed as follows:

$$V_f = 2 \times \frac{1}{2}k_f(h\theta)^2 \tag{16}$$

The elastic potential energy of the hinge can be obtained by the following:

$$V_J = 2 \times \frac{1}{2}(K_1 + K_2)\theta^2 \tag{17}$$

The kinetic energy of the mass block can be described as follows:

$$T_m = 2 \times \frac{1}{2}J\dot{\theta}^2 \tag{18}$$

The Lagrangian equation for the conservative force can be written as follows:

$$\frac{\partial}{\partial t}(\frac{\partial L}{\partial \dot{\theta}}) - \frac{\partial L}{\partial \theta} = 0 \tag{19}$$

Substituting Equation (15) to (18) into Equation (19), the dynamic equation for the system can be described as follows:

$$2J\ddot{\theta} + (2k_f h^2 + 2(K_1 + K_2))\theta = 0 \tag{20}$$

The resonance frequency of the system can be expressed as follows:

$$f = \frac{1}{2\pi}\sqrt{\frac{k_f h^2 + (K_1 + K_2)}{J}} \tag{21}$$

According to the moment of inertia formula and the parallel axis theorem, the moment of inertia can be obtained by the following:

$$J = m\frac{d^2 + h^2}{12} + m(r_2 + \frac{d}{2})^2 \tag{22}$$

## 2.4. Dimensional Parameter Optimization

Since the sensor is symmetrical, only the influence of flexible hinge-1 and flexible hinge-2 on the resonance frequency and sensitivity of the sensor is discussed. It can be seen from Equations (14) and (21) that the radius ($r_1$) and waist thickness ($t_1$) of hinge-1, the radius ($r_2$) and the waist thickness ($t_2$) of hinge-2, the thickness of hinge ($w$), and the width ($d$) and height ($h$) of the mass block are seven key parameters that affect the resonance frequency and sensitivity of the sensor. To obtain a higher resonance frequency and greater sensitivity, the structural parameters of the sensor need to be optimized.

Since the length of the FBG grating used is 5 mm, to increase the sensitivity, the radius $r_1$ is set to 2.5 mm. To ensure reliable fixation at both ends of the optical fiber, the width of mass block $d$ is set to 10 mm. To limit the sizes of the sensor, the height $h$ is set to 30 mm, and the thickness $w$ is 8 mm. The sensor adopts wire-cutting integrated processing. The material characteristic parameters are shown in Table 1.

**Table 1.** Parameters for the FBG accelerometer structure and material properties.

| Parameter | Parameter Description | Value (Units) |
| --- | --- | --- |
| $r_1$ | Radius of hinge-1 | 2.5 mm |
| $t_1$ | Waist thickness of hinge-1 | 1.2 mm |
| $r_2$ | Radius of hinge-2 | 2.5 mm |
| $t_2$ | Waist thickness of hinge-2 | 2.2 mm |
| $d$ | Width of mass block | 10 mm |
| $h$ | Height of mass block | 30 mm |
| $w$ | Width of hinges | 8 mm |
| $l$ | Fiber pasting span | 5 mm |
| $A_f$ | Section area of optical fiber | $1.227 \times 10^{-8}$ m$^2$ |
| $E_f$ | Young's modulus of optical fiber | 70 GPa |
| $E$ | Young's modulus of 304 steel | 210 GPa |
| $\rho$ | Density of 304 steel | 7850 kg·m$^{-3}$ |
| $\mu$ | Poisson's ratio of 304 steel | 0.3 |
| g | Gravitational acceleration | 9.8 m·s$^{-2}$ |

So, we have investigated the influence of $r_2$, $t_1$, and $t_2$ on the sensor's resonance frequency and sensitivity to determine the optimal size parameters. The dependence relationships of the resonance frequency and sensitivity on three dimensional parameters ($r_2$, $t_1$, $t_2$) are shown in Figure 3a–c.

As shown in Figure 3a, the resonance frequency decreases with increasing radius of flexible hinge-2. When 1 mm $\leq r_2 \leq$ 3 mm, the resonance frequency has a larger variation range, which can be used to adjust the working frequency range of the sensor. Compared with $t_1$ and $t_2$, $r_2$ has a greater impact on the size of the sensor, and $r_2$ should not be too large.

As shown in Figure 3b, the resonance frequency increases as the thickness of the thinnest part of flexure hinge-1 $t_1$ increases. When 0.5 mm $\leq t_1 \leq$ 2 mm, the resonance frequency changes in a smaller range.

As shown in Figure 3c, the resonance frequency increases as the thickness of the thinnest part of flexure hinge-2 $t_2$ increases. When 1 mm $\leq t_2 \leq$ 4 mm, the resonance frequency shows a large variation range, which can be used to adjust the operating frequency range. In addition, $t_2$ should not be too small, and sufficient connection strength between the mass and the base should be maintained. When 1 mm $\leq t_2 \leq$ 2.5 mm, the sensitivity range is also very large, and the mutual restriction of sensitivity and resonance frequency is obvious. To obtain a higher resonance frequency and greater sensitivity, 2 mm $\leq t_2 \leq$ 2.5 mm was selected.

This paper aims to design a medium-frequency accelerometer with a resonance frequency of about 3000 Hz. While ensuring that the sensor obtains a higher resonance frequency and greater sensitivity, we determined the structure size parameters. The structure and material characteristics of the FBG accelerometer are shown in Table 1. Based on the data in Table 1, the theoretical resonance frequency and sensitivity of the sensor are 2922.1 Hz and 23.1 pm/g, respectively.

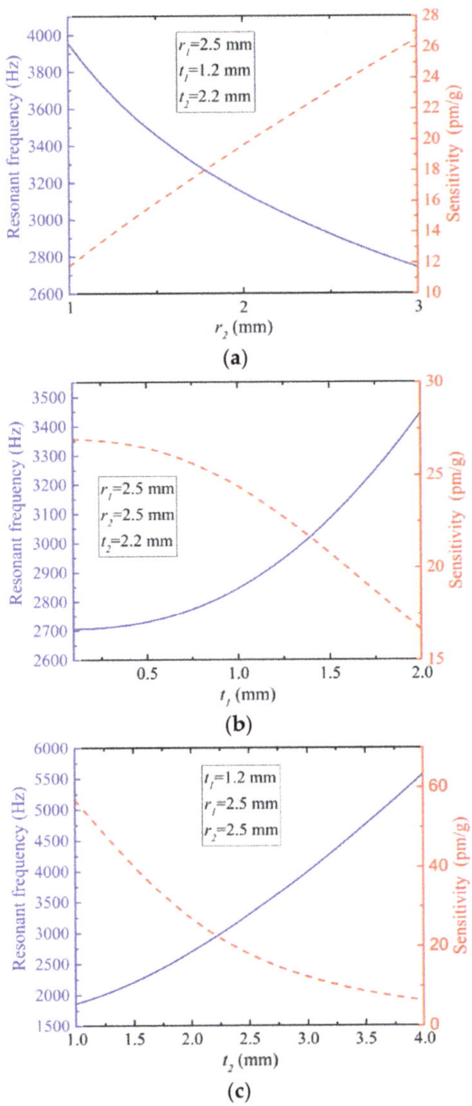


**Figure 3.** Dependence relationship of the resonance frequency and sensitivity on the (**a**) radius of hinge-2 ($r_2$), (**b**) waist thickness of hinge-1 ($t_1$), and (**c**) waist thickness of hinge-2 ($t_2$).

## 3. Simulation Analysis of the Sensor

To further study the dynamic characteristics of the sensor, ANSYS software was used to perform analysis for the sensor model. According to the size parameters given in Table 1, a three-dimensional model of the sensor was established through SOILDWORKS software and imported into ANSYS software. The material properties of the model were set according to the material characteristic parameters in Table 1. The hexdominant grid division was complete, with the grid element size being equal to 0.3 mm. Fixed support constraints were imposed on the bases at both ends of the sensor. The optical fiber was modeled by spring constraints with the same elastic coefficient, suspended above and below flexible hinge-1.

### 3.1. Model Analysis

The model was solved to obtain the sensor's deformation cloud diagram. As shown in Figure 4a, the first-order model shape involves two masses rotating slightly around the center of the flexible hinge connected to the base. The first-order resonance frequency is 2801.7 Hz, which is quite consistent with the theoretical value (2922.1 Hz). As shown in Figure 4b, the second-order model shape of the sensor involves a lateral vibration of two masses along the *y*-axis, and the second-order resonance frequency is 4472.4 Hz. These data show that the resonance frequency of the lateral vibration is 59% higher than the main resonance frequency, indicating that the sensor has a strong ability to resist lateral interference.

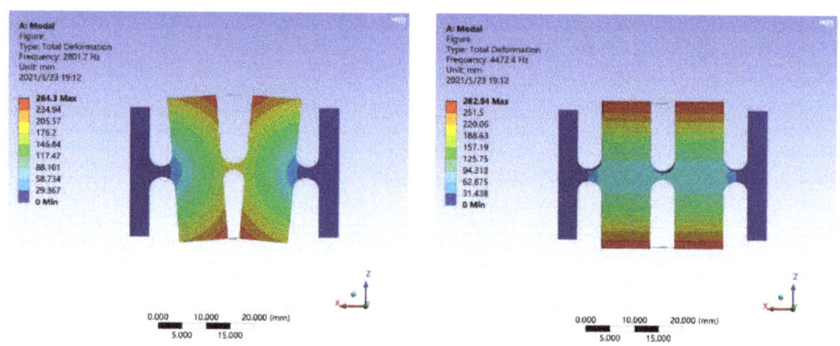


(a) (b)

**Figure 4.** Model analysis results for the FBG accelerometer: (**a**) first-order and (**b**) second-order.

### 3.2. Harmonic Response Analysis

An excitation acceleration of 1 g was applied along the *z*-axis of the sensor. The constant damping ratio was set to 0.015. The excitation frequency varied in the range from 50 Hz to 3300 Hz, with a step length of 50 Hz. The maximum displacements of the fiber's fixed points on top of the masses were calculated and converted into two FBG wavelength shifts ($\Delta\lambda_1$–$\Delta\lambda_2$). The simulated curve in Figure 5 shows the harmonic response of the sensor, and the first-order resonance frequency is approximately 2800 Hz, which is basically consistent with the theoretical value (2922.1 Hz). At an excitation signal frequency of 50 Hz, the sensor's sensitivity is 25.7 pm/g, which is basically consistent with the theoretical value (23.1 pm/g).

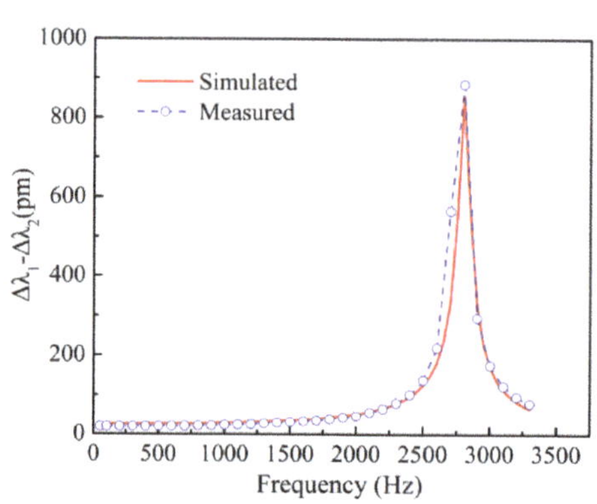


**Figure 5.** Amplitude-frequency response curves for the sensor.

## 4. Experimental Characterization of the Sensing Properties

### *4.1. Fabrication of Sensor*

According to the sensor's structure shown in Figure 1 and the parameters shown in Table 1, the sensor was directly processed into a one-piece structure by a piece of 304 stainless steel, using the slow wire electrical discharge machining (EDM) method. The machining surface of the slow wire EDM method is smooth, and the machining accuracy is high. Two FBGs (silica fiber with an outer diameter of 125 μm, Shenzhen AtGrating., Ltd., Shenzhen, China; 1545.785 nm, 1546.025 nm; 5 mm length, >70% reflectivity) were used. Adhesive 353ND (Epoxy Technology, USA) is a high-temperature-resistant epoxy designed for semiconductor, hybrid, optical fiber and medical applications. To obtain a better paste effect, firstly, about 5 mm of coating was removed at the two sides of the FBG before pasting. Then, the peeled parts were placed on the optical fiber grooves tightly. At the same time, pre-tension was applied to the FBG by hanging a weight of 300 g. When the temperature increased to 120 °C, 353ND epoxy adhesive was applied evenly to the peeled parts on both sides of the FBG, and curing was achieved by heating the epoxy at 120 °C for 15 min. After fabrication, annealing was carried out at 80 °C in an oven before the experiment.

### *4.2. Experimental System Compositions*

The experimental system for the FBG accelerometer is shown in Figure 6. The experimental system consists of the following two parts: a vibration sensing system and a demodulation system. The vibration sensing system (LAN-XI, made by the B&K Company, Copenhagen, Denmark) includes an FBG accelerometer, a vibration exciter, a power amplifier, a standard reference acceleration sensor, and a computer. The computer is responsible for controlling the frequency and amplitude of the vibration exciter. The demodulation system consists of the FBG demodulator and signal acquisition software. The FBG demodulator was self-developed, based on an FPGA-IRS demodulation module produced by BaySpec Inc. The maximum sampling frequency was 8 kHz, and the resolution was 1 pm. The demodulation system is responsible for collecting the central wavelength signal of the reflected light from the sensor grating. The proposed accelerometer was fixed onto the vibration axis of the vibration exciter, and the two FBGs in this accelerometer were connected to the demodulator through an optical fiber. The experiment was carried out at room temperature (25 °C).

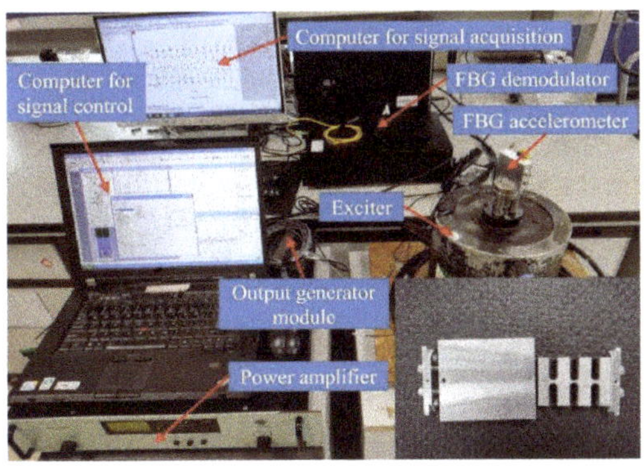


**Figure 6.** FBG accelerometer experimental system.

### *4.3. Amplitude-Frequency Response*

To study the amplitude-frequency characteristics of the sensor, the amplitude of the excitation acceleration was kept at 1 g, and the excitation frequency was varied in the range of 50 Hz to 3300 Hz, with a step of 100 Hz. The relationship between the central wavelength difference ($\Delta\lambda_1$–$\Delta\lambda_2$) of the two FBGs and the frequency under different excitation frequencies was measured, and the amplitude-frequency response curve is shown in Figure 5. The response curve shows that the resonance frequency of the accelerometer is approximately 2800 Hz, which is basically consistent with the theoretical value (2922.1 Hz) and the finite element simulation data (2801.7 Hz). The amplitude-frequency curve has better flatness below 1000 Hz, and the operating frequency range of the sensor is 50–1000 Hz.

### *4.4. Sensitivity*

In the sensitivity experiment, since the sensor adopts a dual FBG differential structure, the wavelength shift difference ($\Delta\lambda_1$–$\Delta\lambda_2$) for the two FBGs was selected as the output response. Through the vibration exciter, sinusoidal excitation signals of 100 Hz, 300 Hz, 500 Hz, 800 Hz, and 1000 Hz were given to the sensor, and the acceleration amplitude increased from 0.5 g to 3 g, with a step length of 0.5 g. Each group of experiments was repeated three times. The variations for the wavelength shift difference, with acceleration at different frequencies, are shown in Figure 7—by open square, circles, up-triangles, down-triangles, and diamonds.

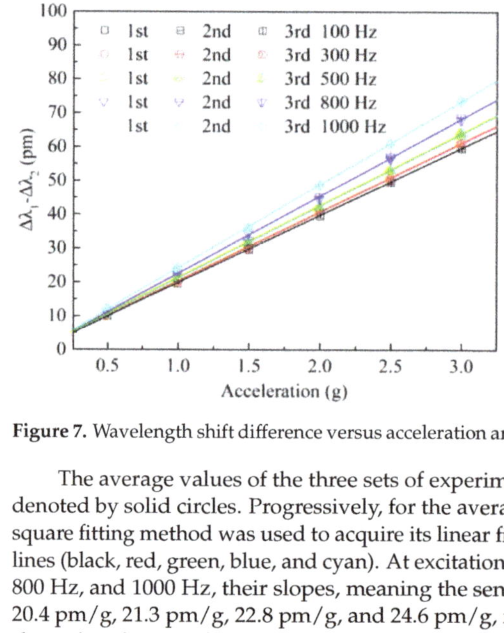


**Figure 7.** Wavelength shift difference versus acceleration amplitude at different excitation frequencies.

The average values of the three sets of experimental data at different frequencies are denoted by solid circles. Progressively, for the average values at each frequency, the least-square fitting method was used to acquire its linear fitting line, which was denoted by solid lines (black, red, green, blue, and cyan). At excitation frequencies of 100 Hz, 300 Hz, 500 Hz, 800 Hz, and 1000 Hz, their slopes, meaning the sensitivities of the sensor, are 19.9 pm/g, 20.4 pm/g, 21.3 pm/g, 22.8 pm/g, and 24.6 pm/g, respectively. These experimental data show that the wavelength shift difference for the sensor has a good linear relationship with the acceleration amplitude. Within the working frequency of 50–1000 Hz, the sensor obtained an average sensitivity of 21.8 pm/g. The main reasons for the difference between theoretical sensitivity and experimental sensitivity are dimensional deviations in the sensor, in its slow wire EDM and packaging.

The Bessel formula was used to calculate the standard deviation. First, all the sub-samples' standard deviations $\sigma_i$ were calculated, where $i$ = 1, 2 . . . $N$, and $N$ = 6 is the number of calibration points. Then, the standard deviation of the sensor was calculated using $\sigma = \sqrt{\frac{1}{N}\sum_{i=1}^{N}\sigma_i^2}$. Finally, the repeatability error was calculated as $e_R = \eta\sigma/\lambda_{FS}$, where $\eta$ = 3, which is the coverage factor, and $\lambda_{FS}$ is the maximum wavelength shift. As shown in Figure 7, when the frequency of the excitation signal is 100 Hz, 300 Hz, 500 Hz, 800 Hz, and 1000 Hz, the repeatability errors for the sensor are 2.7%, 2.4%, 2.7%, 3.3%, and 2.2%, respectively.

### 4.5. Temperature Self-Compensation

Since FBGs are sensitive to both strain and temperature, to study the temperature self-compensation performance of the sensor, a temperature control box (SDEI, SDJ402F, adjustment range of 10–200 °C, resolution of 0.01 °C, measurement accuracy of 0.05 °C) was used to test the sensor. The sensor was placed in the temperature control box, and the temperature was adjusted from 20 °C to 80 °C, in steps of 10 °C. The central wavelength shift was measured for the two FBGs at different temperatures, and the temperature response curves obtained for $\Delta\lambda_1$, $\Delta\lambda_2$, and $\Delta\lambda_1$–$\Delta\lambda_2$ are shown in Figure 8. The temperature sensitivity coefficients for FBG-1 and FBG-2 are $K_{T1}$ = 22.01 pm/°C and $K_{T2}$ = 21.67 pm/°C, respectively, and their linear correlation coefficients are 0.999. The experimental results show that the sensitivity coefficients for the two FBGs are basically the same, and the sensor has good temperature self-compensation capability.

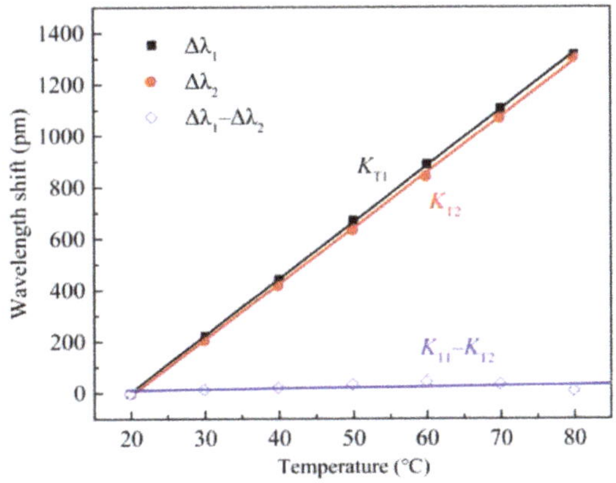


**Figure 8.** Temperature responses and self-compensation for the FBG accelerometer.

### 4.6. Cross-Interference Characteristic

Anti-interference ability is also an important indicator for accelerometers, so an anti-interference experiment was also finished, where the sensor was fixed onto the vibration exciter along the main axis of the vibration exciter and in its vertical direction, respectively. A sinusoidal excitation signal, with frequency of 300 Hz and amplitude of 1 g, was applied to the sensor. The wavelength shift differences of the sensor were obtained in the main and lateral directions, as shown in Figure 9. The peak–peak value of the central wavelength difference for the sensing in the main direction is about 41 pm, but the peak–peak value in the lateral direction does not exceed 2 pm; therefore, the lateral interference degree for the sensor is less than 5%, which shows that the designed sensor has a good anti-interference ability.

Table 2 gives a characteristics comparison of the proposed medium–high-frequency FBG accelerometer with other FBG accelerometers, based on different structures, including resonance frequency, sensitivity, FBG type, and temperature self-compensation capabilities.

**Table 2.** Summary of the characteristics of reported medium–highfrequency FBG accelerometers.

| Ref | Resonance Frequency | Sensitivity | Fiber Type | Temperature Self-Compensation |
|---|---|---|---|---|
| Stefani [28] | 3000 Hz | 19 pm/g | Polymer FBG | No |
| Dai [29] | 2918 Hz | 13.82 pm/g | Silica FBG | No |
| Guo [30] | 3600 Hz | 1.7 pm/g | Metalized FBG | No |
| Wang [31] | 3806 Hz | 4.01 pm/g | Silica FBG | Yes |
| Wu [32] | 8356 Hz | 0.46 pm/g | Silica FBG | Yes |
| This article | 2800 Hz | 21.8 pm/g | Silica FBG | Yes |

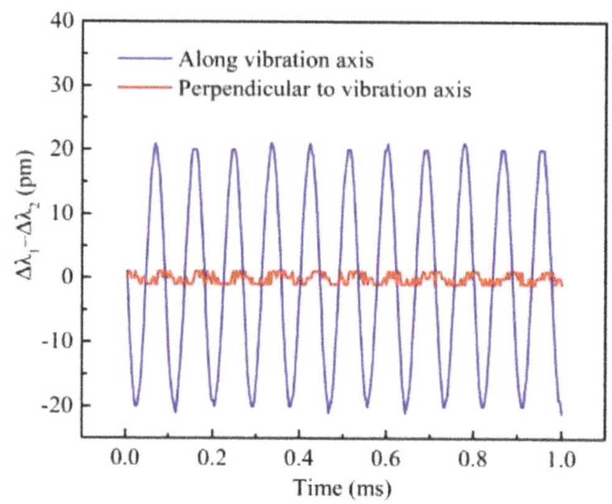


**Figure 9.** Wavelength shift difference of the sensor when the vibration is along or perpendicular to the vibration axis of exciter.

## 5. Conclusions

A novel medium-frequency FBG accelerometer, based on two FBGs, three flexible hinges, and two inertial mass blocks, was designed and prepared. Theoretical analysis and finite element simulation were used to determine the optimal parameters of the sensor. The sensing experiments show that the proposed sensor had a resonance frequency of about 2800 Hz and a sensitivity of 21.8 pm/g in the flat frequency range of 50–1000 Hz. The sensor's resonance frequency and sensitivity, given by theoretical analysis, finite element simulation, and sensing experiments, are basically consistent. In addition, the experiments also show that the sensor has a good temperature self-compensation function (temperature sensitivity $K_{T1}$–$K_{T2}$ $\approx$ 0.34 pm/°C) and lateral anti-interference capability (lateral interference degree is less than 5%). These performances provide a novel and reliable method for the engineering application of medium-frequency FBG acceleration sensors.

**Author Contributions:** Conceptualization and methodology, Z.L.; validation and formal analysis, Z.L.; data curation, Z.L. and Z.S.; writing—original draft preparation, Z.L.; writing—review and editing, L.L., H.W., S.D. and K.J. All authors have read and agreed to the published version of the manuscript.

**Funding:** This work was supported in part by the Provincial Science and Technology Innovation Strategy Project of Zhongshan under Grant 2019sdr001 and in part by the Optical Fiber Sensing Submarine Cable System for Deep-Sea Riser Safety Monitoring under Grant SKJC-2020-01-016.

**Institutional Review Board Statement:** Not applicable.

**Informed Consent Statement:** Not applicable.

**Data Availability Statement:** Not applicable.

**Conflicts of Interest:** The authors declare no conflict of interest.

## References

1. Baccarini, L.M.R.; Silva, V.V.R.; Menezes, B.R.; Caminhas, W.M. SVM practical industrial application for mechanical faults diagnostic. *Expert Syst. Appl.* **2011**, *38*, 6980–6984. [CrossRef]
2. Li, P.; Kong, F.; He, Q.; Liu, Y. Multiscale slope feature extraction for rotating machinery fault diagnosis using wavelet analysis. *Measurement* **2013**, *46*, 497–505. [CrossRef]
3. Safizadeh, M.S.; Latifi, S.K. Using multi-sensor data fusion for vibration fault diagnosis of rolling element bearings by accelerometer and load cell. *Inf. Fusion.* **2014**, *18*, 1–8. [CrossRef]
4. Catalano, A.; Bruno, F.A.; Pisco, M.; Cutolo, A.; Cusano, A. An Intrusion Detection System for the Protection of Railway Assets Using Fiber Bragg Grating Sensors. *Sensors* **2014**, *14*, 18268–18285. [CrossRef] [PubMed]

5. Ye, X.W.; Su, Y.H.; Xi, P.S. Statistical Analysis of Stress Signals from Bridge Monitoring by FBG System. *Sensors* **2018**, *18*, 491. [CrossRef]
6. Qiao, X.; Shao, Z.; Bao, W.; Rong, Q. Fiber Bragg Grating Sensors for the Oil Industry. *Sensors* **2017**, *17*, 429. [CrossRef] [PubMed]
7. Shen, J.; Zeng, X.; Luo, Y.; Cao, C.; Wang, T. Research on Strain Measurements of Core Positions for the Chinese Space Station. *Sensors* **2018**, *18*, 1834. [CrossRef] [PubMed]
8. Goossens, S.; Pauw, B.D.; Geernaert, T.; Salmanpour, M.S.; Khodaei, Z.S.; Karachalios, E.; Castillo, D.S.; Thienpont, H.; Berghmans, F. Aerospace-grade surface mounted optical fibre strain sensor for structural health monitoring on composite structures evaluated against in-flight conditions. *Smart Mater. Struct.* **2019**, *28*, 065008. [CrossRef]
9. Antunes, P.; Lima, H.; Varum, H.; André, P. Optical fiber sensors for static and dynamic health monitoring of civil engineering infrastructures: Abode wall case study. *Measurement* **2012**, *45*, 1695–1705. [CrossRef]
10. Linessio, R.P.; Sousa, K.D.; Silva, T.D.; Bavastri, C.A.; Antunes, P.F.D.; Silva, J.C.C. Induction Motors Vibration Monitoring Using a Biaxial Optical Fiber Accelerometer. *IEEE Sens. J.* **2016**, *16*, 8075–8082. [CrossRef]
11. Zhang, Z.; Liu, C.; Li, H.; He, Z.; Zhao, X. Optical fiber grating vibration sensor for vibration monitoring of hydraulic pump. *Photonic Sens.* **2018**, *7*, 140–147. [CrossRef]
12. Yüksel, K.; Kinet, D.; Moeyaert, V.; Kouroussis, G.; Caucheteur, C. Railway monitoring system using optical fiber grating accelerometers. *Smart Mater. Struct.* **2018**, *27*, 105033. [CrossRef]
13. Lim, K.S.; Zaini, M.K.A.; Ong, Z.C.; Abas, F.Z.M.; Salim, M.A.B.M.; Ahmad, H. Vibration Mode Analysis for a Suspension Bridge by Using Low-Frequency Cantilever-Based FBG Accelerometer Array. *IEEE Trans. Instrum. Meas.* **2021**, *70*, 1–8. [CrossRef]
14. Liu, Q.; Jia, Z.; Fu, H.; Yu, D.; Gao, H.; Qiao, X. Double Cantilever Beams Accelerometer Using Short Fiber Bragg Grating for Eliminating Chirp. *IEEE Sens. J.* **2016**, *16*, 6611–6616. [CrossRef]
15. Parida, O.P.; Thomas, J.; Nayak, J.; Asokan, S. Double-L Cantilever-Based Fiber Bragg Grating Accelerometer. *IEEE Sens. J.* **2019**, *19*, 11247–11254. [CrossRef]
16. Udos, W.; Lee, Y.-S.; Lim, K.-S.; Ong, Z.-C.; Zaini, M.K.A.; Ahmad, H. Signal enhancement of FBG-based cantilever accelerometer by resonance suppression using magnetic damper. *Sens. Actuators A* **2020**, *304*, 111895. [CrossRef]
17. Li, S.; Feng, Z.; Ma, Q.; Wang, R.; Zhou, R.; Qiao, X. Fiber Bragg grating accelerometer based on symmetrical tilting cantilever beams and solder glass packaging for harsh environment. *Opt. Fiber Technol.* **2021**, *65*, 102579. [CrossRef]
18. Zhao, X.; Jia, Z.; Fan, W.; Liu, W.; Gao, H.; Yang, K.; Yu, D. A Fiber Bragg Grating acceleration sensor with temperature compensation. *Optik* **2021**, *241*, 166993. [CrossRef]
19. Liu, Q.; He, X.; Qiao, X.; Sun, T.; Grattan, K.T.V. Design and Modeling of a High Sensitivity Fiber Bragg Grating-Based Accelerometer. *IEEE Sens. J.* **2019**, *19*, 5439–5445. [CrossRef]
20. Li, T.; Tan, Y.; Han, X.; Zheng, K.; Zhou, Z. Diaphragm Based Fiber Bragg Grating Acceleration Sensor with Temperature Compensation. *Sensors* **2017**, *17*, 218. [CrossRef]
21. Zhang, F.; Jiang, S.; Wang, C.; Ni, J.; Zhao, Q. Broadband and High Sensitivity FBG Accelerometer Based on Double Diaphragms and h-Shaped Hinges. *IEEE Sens. J.* **2021**, *21*, 353–359. [CrossRef]
22. Qiu, L.; Liang, L.; Li, D.; Xu, G. Theoretical and experimental study on FBG accelerometer based on multi-flexible hinge mechanism. *Opt. Fiber Technol.* **2017**, *38*, 142–146. [CrossRef]
23. Zhang, Y.; Zhang, W.; Zhang, Y.; Chen, L.; Yan, T.; Wang, S.; Yu, L.; Li, Y.P. 2-D Medium–High Frequency Fiber Bragg Gratings Accelerometer. *IEEE Sens. J.* **2017**, *17*, 614–618. [CrossRef]
24. Song, H.; Wang, Q.; Liu, M.; Cai, Q. A Novel Fiber Bragg Grating Vibration Sensor Based on Orthogonal Flexure Hinge Structure. *IEEE Sens. J.* **2020**, *20*, 5277–5285. [CrossRef]
25. Yan, B.; Liang, L. A Novel Fiber Bragg Grating Accelerometer Based on Parallel Double Flexible Hinges. *IEEE Sens. J.* **2020**, *20*, 4713–4718. [CrossRef]
26. Wei, L.; Yu, L.; Wang, J.; Jiang, D.; Liu, Q.; Liu, Z. An FBG-Sensing Two-Dimensional Vibration Sensor Based on Multi-Axis Flexure Hinge. *IEEE Sens. J.* **2019**, *19*, 3698–3710. [CrossRef]
27. Wang, H.; Liang, L.; Zhou, X.; Tu, B. New Fiber Bragg Grating Three-Dimensional Accelerometer Based on Composite Flexure Hinges. *Sensors* **2021**, *21*, 4715. [CrossRef]
28. Stefani, A.; Andresen, S.; Yuan, W.; Herholdt-Rasmussen, N.; Bang, O. High sensitivity polymer optical fiber-Bragg-grating-based accelerometer. *IEEE Sens. J.* **2012**, *24*, 763–765. [CrossRef]
29. Dai, Y.; Yin, G.; Liu, B.; Xu, G.; Karanja, J.M. Medium-high frequency FBG accelerometer with integrative matrix structure. *Appl. Opt.* **2015**, *54*, 3115–3121. [CrossRef]
30. Guo, Y.; Zhang, D.; Zhou, Z.; Xiong, L.; Deng, X. Welding-packaged accelerometer based on metal-coated FBG. *Chin. Opt. Lett.* **2013**, *11*, 070604.
31. Wang, X.; Guo, Y.; Xiong, L.; Wu, H. High-Frequency Optical Fiber Bragg Grating Accelerometer. *IEEE Sens. J.* **2018**, *18*, 4954–4960. [CrossRef]
32. Wu, H.; Lin, Q.; Han, F.; Zhao, L.; Jiang, Z. Design and analysis of high-frequency fiber Bragg grating vibration sensor. *Meas. Sci. Technol.* **2021**, *32*, 025108. [CrossRef]
33. Wu, Y.F.; Zhou, Z.Y. Design of flexure hinges. *Eng. Mech.* **2002**, *19*, 136–140.

sensors

*Article*

# Design and Modeling of a Fully Integrated Microring-Based Photonic Sensing System for Liquid Refractometry

**Grigory Voronkov [1], Aida Zakoyan [1], Vladislav Ivanov [1], Dmitry Iraev [1], Ivan Stepanov [1], Roman Yuldashev [1], Elizaveta Grakhova [1], Vladimir Lyubopytov [1], Oleg Morozov [2,3,*] and Ruslan Kutluyarov [1]**

1 Ufa University of Science and Technology, 32, Z. Validi St., Ufa 450076, Russia
2 Kazan National Research Technical University named after A. N. Tupolev-KAI (KNRTU-KAI), 10, Karl Marx Street, Kazan 420111, Russia
3 Kazan Federal University, 18, Kremlyovskaya Str., Kazan 420008, Russia
* Correspondence: microoil@mail.ru

**Abstract:** The design of a refractometric sensing system for liquids analysis with a sensor and the scheme for its intensity interrogation combined on a single photonic integrated circuit (PIC) is proposed. A racetrack microring resonator with a channel for the analyzed liquid formed on the top is used as a sensor, and another microring resonator with a lower Q-factor is utilized to detect the change in the resonant wavelength of the sensor. As a measurement result, the optical power at its drop port is detected in comparison with the sum of the powers at the through and drop ports. Simulations showed the possibility of registering a change in the analyte refractive index with a sensitivity of 110 nm per refractive index unit. The proposed scheme was analyzed with a broadband source, as well as a source based on an optoelectronic oscillator using an optical phase modulator. The second case showed the fundamental possibility of implementing an intensity interrogator on a PIC using an external typical single-mode laser as a source. Meanwhile, additional simulations demonstrated an increased system sensitivity compared to the conventional interrogation scheme with a broadband or tunable light source. The proposed approach provides the opportunity to increase the integration level of a sensing device, significantly reducing its cost, power consumption, and dimensions.

**Keywords:** integrated photonics; silicon photonics; refractometry; optoelectronic oscillator; interrogation



**Citation:** Voronkov, G.; Zakoyan, A.; Ivanov, V.; Iraev, D.; Stepanov, I.; Yuldashev, R.; Grakhova, E.; Lyubopytov, V.; Morozov, O.; Kutluyarov, R. Design and Modeling of a Fully Integrated Microring-Based Photonic Sensing System for Liquid Refractometry. *Sensors* **2022**, *22*, 9553. https://doi.org/10.3390/s22239553

Academic Editor: Francesco De Leonardis

Received: 19 October 2022
Accepted: 2 December 2022
Published: 6 December 2022



## 1. Introduction

Nowadays, sensing systems for label-free determination of the substance concentration in a liquid play a significant role in modern medicine [1], experimental biology [2], and applied chemistry [3]. In the general case, the structure of a photonic sensing system includes an optical sensor circuit, a light source, a detector, and an interrogation and control circuit. The technological maturity of integrated photonics makes it possible to effectively implement the individual elements of such a system in an integrated form to create high-performance lab-on-a-chip devices. At the same time, increasing the integration level with the required stability of the characteristics is an obvious way to implement refractometric sensing systems, which are in demand in a wide range of applications.

Among the various approaches to implementing refractometry systems, phase-sensitive circuits are the most attractive in terms of technological viability and efficiency. Widely used schemes for detecting changes in the physical properties of an analyte include fiber (FBG) and waveguide (WBG) Bragg gratings [4–7], waveguide Mach–Zehnder interferometers (MZI) [8–10], and microring (MRR) or microdisk (MDR) resonators [11–13].

FBG sensors have proven to be very simple to implement yet highly efficient structures for environmental measurements [14,15]. A shift in the resonant wavelength can be caused by a grating deformation or various types of interactions between the analyte and the grating structure. These interactions are achieved by forming a phase shift in the lattice [5] or integrating a Fabry–Perot interferometer, which allows for obtaining a sensor sensitivity of up to 1210.49 nm per refractive index unit (RIU) [7].

In addition to fiber sensors, high performance can be obtained using integrated sensors due to their ability to fabricate MZI, MRR, and WBG on photonic integrated circuits (PICs) using technological platforms such as silicon nitride ($Si_3N_4$) or silicon-on-insulator (SOI). SOI structures are most often used for operation in the near-infrared and mid-infrared ranges due to their relatively low propagation losses and small allowed bending radii of the Si waveguides. These structures also have a large set of components that are supported for monolithic integration, including fast diode phase modulators and photodetectors [16].

Among all existing types of sensors on PICs, MRR is the most used [17]. MRRs have great potential to implement label-free detectors because they are fast, energy-efficient, and have a significant sensitivity to detect biomolecules in liquids and gases [12,13]. MRR fabrication technology is widely applied, and silicon MRRs can be integrated into a microfluidic system [11]. The principle of operation is based on a shift in the resonant wavelengths of the ring caused by a change in the refractive index (RI) near the sensor, which occurs when the studied parameter of the substance varies.

The sensitivity of MRR sensors currently varies from 70 to 490 nm/RIU depending on the type of waveguide and the selected operating mode [11]. Specifically, for the MRR sensors on SOI, sensitivity does not exceed 100 nm/RIU [18,19]. MRR cascading [13,20], subwavelength grating microring [21,22], or phase-shifted Bragg grating microring [23] have been proposed to increase the sensitivity.

The main problem in utilizing MRRs as sensors is a need for a tunable mode-hop-free laser or a broadband source to excite the resonances within the structure. Hybrid integration of a semiconductor laser directly onto a chip is a technologically feasible but rather expensive solution. In this context, the generation of broadband radiation directly on a PIC could be a solution capable of circumventing these technological limitations.

The output characteristics of a sensing system are greatly determined by the efficiency of the interrogation scheme. The typical solution for assessing a change in the optical spectrum is using commercial interrogators based on tunable lasers [24]. However, the pricey equipment and the system bulkiness are among the main disadvantages of this approach. For the practical application of optical sensors, indirect methods have been developed to analyze the optical emission spectrum for high-speed interrogation with high resolution [25]. Depending on the analyzed parameters of the output electrical signal, three methods of interrogation are distinguished: by intensity, when the photocurrent power is measured [26]; by frequency, when the frequency of the signal at the output of the photodetector is estimated [27,28]; and by time, when the shape of the output signal envelope is analyzed [29].

Intensity interrogation can provide fast sensor scanning, but the radiation source dispersion or the low stability of the optical source affects the measurement accuracy [29]. Frequency interrogation can be implemented by an optoelectronic oscillator (OEO), where the generated frequency depends on the resonant characteristics of the sensor. High-precision measurements can be achieved by the ultra-high wavelength resolution of the OEO (up to 360 fm [28]). However, this method requires additional devices for measuring the microwave signal frequency. Time interrogation enables a high scanning rate (up to 48.6 MHz [29]), but the resolution of such systems is inferior to those described above. In addition, the accuracy of the signal shape estimation is related to the capacity of the analog-to-digital converter (ADC). Thus, for integrated implementation, it is preferable to use intensity interrogation because its limitations are determined mainly by the shortcomings of the radiation source.

PIC-based interrogation systems are extremely interesting in sensing applications [30], especially in the case of wearable devices and medical diagnostics [24]. In addition, monolithic integration of the interrogation system and the sensor into a single miniature chip will potentially reduce losses and the footprint, consequently creating ultra-compact sensing systems with high performance. However, to our knowledge, PIC-based sensing systems with such an integration level have not been previously discussed.

In this paper, we propose for the first time, to our knowledge, the design of a sensing system for liquids refractometry that is suitable for the monolithic integration of both the sensor and interrogator parts (except only the light source) on a single SOI PIC. Two different design options are considered, including an external broadband light source and an OEO. The latter allows for an off-the-shelf, single-mode laser in the intensity interrogation scheme.

The remainder of this paper is organized as follows: in the second section, the proposed schemes of an integrated sensing system and the main design problems are discussed; the third section describes the MRR-based sensor and interrogator design features and characteristics; the fourth section presents the simulation results for the intensity interrogation scheme based on an MRR and a broadband radiation source, as well as for the scheme based on an MRR and an OEO source; in the fifth section, the application aspects of the designed interrogation system together with an FBG and a phase-shifting FBG (PS-FBG) are considered; and in the Discussion section, the possible options for PIC-based implementation of the proposed system on different technological platforms are considered, as well as the modeling results for the influence of the waveguide width deviation on the system operability.

## 2. PIC-Based Sensing System Concept and Operation Principle

### 2.1. Microring-Based Sensing System with a Broadband Optical Source

The structure of the proposed sensing system is shown in Figure 1. The input port of the microring racetrack resonator, which acts as a sensor, receives a broadband signal from an optical source. The effective RI of the MRR's waveguide depends on the RI of the medium surrounding the MRR, which determines the position of the resonant peaks in the transmission spectrum at the MRR's drop port. The proposed system is designed to detect and measure changes in the properties of aqueous solutions. To do this, a microfluidic channel can be connected to the MRR sensor by analogy with [11].

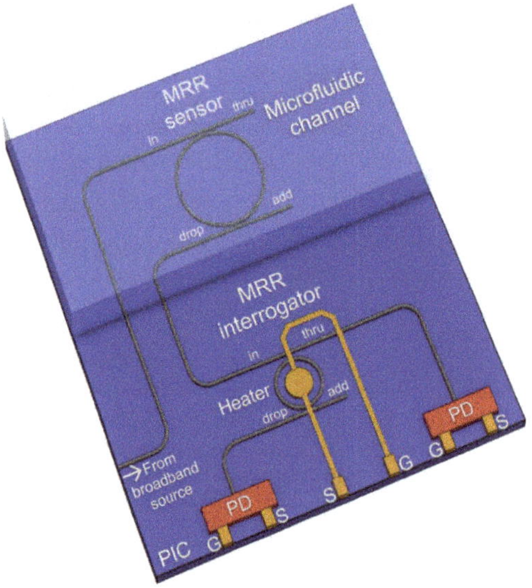


**Figure 1.** 3D design draft of the proposed integrated MRR-based sensing system with a broadband source (not to scale).

To trace the sensor resonant wavelength without using a mode-hop-free tunable laser or an optical spectrum analyzer, which are non-compliant with planar integration, we propose an interrogation scheme by intensity, where another MRR is utilized for the sensor interrogation.

In this case, the interrogation process is a transformation of a change in the resonant wavelength at the input port of the interrogator into the intensity of optical power at the through and drop ports of the interrogator (Figure 2).

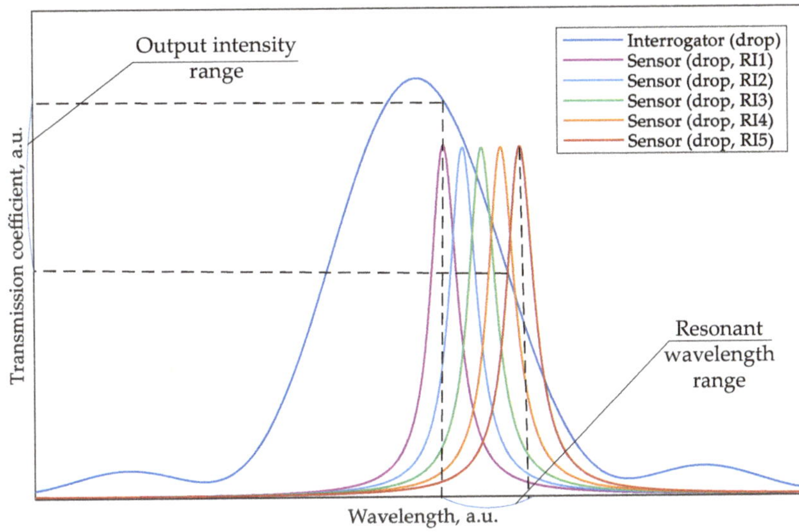


**Figure 2.** Principle of MRR-based interrogation by intensity.

The signal at the output of the sensor's drop port is determined by its resonant characteristic. The MRR resonant peak width at the half power level (FWHM) is given by:

$$FWHM = \frac{(1 - r_1 r_2 a)\lambda_{res}^2}{\pi n_g L \sqrt{r_1 r_2 a}} \tag{1}$$

where $a = \exp(-\alpha L/2)$ is the ring amplitude transmission coefficient; $\alpha$ is the ring waveguide power attenuation coefficient; $r_1$ and $r_2$ are the amplitude transmission coefficients of the directional couplers to the in-through and add-drop waveguides, respectively; $\lambda_{res}$ is the MRR resonant wavelength ($\lambda_{res} = Ln_{eff}/m$, where positive integer $m$ is the resonance number); $L$ is the MRR circumference; and $n_g$ and $n_{eff}$ are the values of the waveguide group and effective refractive indices, respectively. The optical power transmission coefficients at the through and drop ports of the MRR are determined, respectively, by the equations:

$$T_p = \frac{I_{pass}}{I_{input}} = \frac{r_2^2 a^2 - 2 r_1 r_2 a \cos\varphi + r_1^2}{1 - 2 r_1 r_2 a \cos\varphi + (r_1 r_2 a)^2} \tag{2a}$$

$$T_d = \frac{I_{drop}}{I_{input}} = \frac{(1 - r_1^2)(1 - r_2^2) a}{1 - 2 r_1 r_2 a \cos\varphi + (r_1 r_2 a)^2} \tag{2b}$$

As mentioned in the Introduction, one of the main drawbacks of the intensity interrogation system is its dependence on the instability of the optical source output power. To address this shortcoming, we propose using the relative value defined as:

$$P = 10\lg\left(\frac{p_{drop}}{p_{drop} + p_{thru}}\right) \quad (3)$$

where $p_{drop}$ and $p_{thru}$ are the output powers at the sensor MRR's drop and through ports, respectively. Thus, the proposed interrogation scheme estimates the power level, $p_{drop}$, at the drop port of the interrogator MRR relative to its total output power ($p_{drop}$ + $p_{thru}$). These powers may be calculated using the MRR transmission spectrum (2a, 2b). In this case, considering that $p_{thru} = P_{in}T_p$ and $P_{in} = P_{in0} + \Delta P_{in}$, the relative power will be determined as follows:

$$P = 10\lg\left(\frac{T_d}{T_d + T_p}\right) \quad (4)$$

Thus, the measured relative power does not depend on fluctuations in the optical source output power.

### 2.2. Matching the Frequency Characteristics of the Sensor, Interrogator, and Broadband Source

An analysis of Figure 2 allows us to formulate the essential requirements for the sensing system consisting of the two MRRs. First, the FWHM of the sensor MRR should be substantially smaller than the FWHM of the interrogator MRR to provide a greater sensitivity and measurement range of the interrogator. Practically, the FWHM of the sensor MRR can be reduced by increasing its circumference without changing the coupling coefficient in directional couplers, e.g., by inserting straight waveguide sections between the MRR half rings (using racetrack-MRR). Therefore, the quality factor, $Q$, of the interrogator MRR, defined as:

$$Q = \frac{\lambda_{res}}{FWHM} \quad (5)$$

should be lower than that of the sensor. The actual relationship between the FWHMs of the sensor and interrogator MRRs should be determined for specific application scenarios and the possible range of the sensor's resonant wavelength change. Ideally, this range should be completely located within the monotonic section of the utilized resonant peak in the interrogator transmission spectrum.

Second, it should be considered that the MRR resonances occur periodically with a free spectral range (FSR) determined by:

$$FSR = \frac{\lambda^2}{n_g L} \quad (6)$$

and thus, the resonant peaks in the sensor and the interrogator transmission spectra must intersect only in the wavelength range corresponding to the operating range of the sensing system (see Figure 2). In practice, this condition can be fulfilled in two ways. The first option is to select the bandwidth of the optical source to be equal to the half FWHM of the interrogator's resonance and its central wavelength coinciding with the middle of the monotonic section of that resonance. The second option is to choose the geometry of the sensor and interrogator MRRs so that the resonant peaks, except for the analyzed sensor peak and the corresponding operating interrogator peak, would not overlap in a sufficiently wide wavelength range exceeding the bandwidth of the optical source. The second option is universal and practically achievable in terms of selecting a source, so let us consider it in more detail. On the one hand, the large radius of the sensor ring ensures its high-quality factor and, consequently, its sensitivity. On the other hand, increasing the radius of the sensor ring leads to a decrease in its FSR, which can lead to the response of the interrogator to be closely spaced sensor resonant peaks. An example of such a negative overlap is shown in Figure 3.

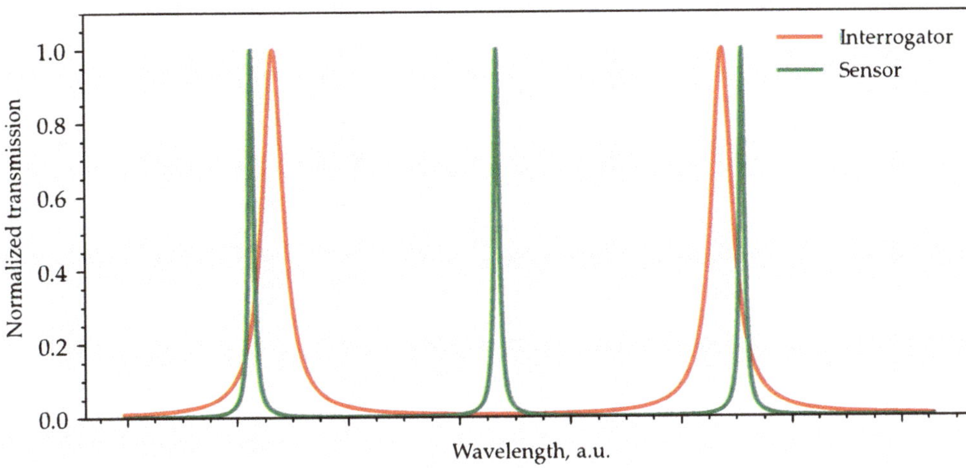


**Figure 3.** An example of improper overlapping of the interrogator and sensor transmission spectra, which can lead to destructive crosstalk between signals from two sensors' resonant peaks.

The left peak of the interrogator transmission spectrum is situated on the decline region of the sensor's resonant characteristic, whereas the right peak is on the rise region. Then, with an increase in the sensor resonant wavelength in response to the analyte RI change, the signal from the left peak at the interrogator's drop port will increase, and the signal from the right peak will decrease, which will prevent correct measurements. To reduce this probability, the FSR of the interrogator must be suitably large, which also provides the required lower quality factor, thus increasing the RI measuring range. It should be considered that commercially available optical sources have either a very narrow band (CW lasers) or, in the case of SLD sources, provide a bandwidth of 30–40 nm at a 3 dB level [31]. A narrower band (up to 20 nm) can be obtained by installing an additional filter based on AWG (arrayed waveguide grating) [32] or FBG at the source output. Such a filter can be either external, combined with the source, or integrated directly into the sensing system PIC (in the case of using AWG). A larger source bandwidth increases the probability that several interrogator's resonant peaks fall into the source band and the likelihood of a destructive overlap between the sensor and the interrogator characteristics. Thus, the choice of MRR dimensions is determined not only by a trade-off between the system sensitivity and the range of measured RI values but also by the characteristics of the used optical source.

### 2.3. *Microring-Based Sensing System with an Optoelectronic Oscillator*

The general problem with intensity interrogation is the need for a broadband optical source [26] or a cascade of narrow-band sources [6], which complicates the integrated implementation of the interrogator and increases its cost. Thus, it is crucial to consider how to obtain a feeding optical signal on a PIC and how to reduce its influence on the sensing system characteristics. Let us consider a Mach–Zehnder phase modulator covered by a positive feedback loop as a source. Such a system performs the function of an optoelectronic oscillator (OEO) [33]. Figure 4 shows the structure of the proposed PIC-based interrogator, using a frequency self-adjustable OEO as an optical radiation source. It consists of two key components: a phase MZM and MRR, where the latter performs as a notch filter at the drop port and a bandpass filter at the through port. It should be noted that the photodetector can be made either as a discrete external device or implemented on a PIC, as shown in Figure 4.

The microwave elements are shown tentatively. They may be mounted on the printed circuit board (PCB) as discrete planar devices or implemented as a monolithic microwave integrated circuit. The electrical circuit design is out of the scope of this paper.

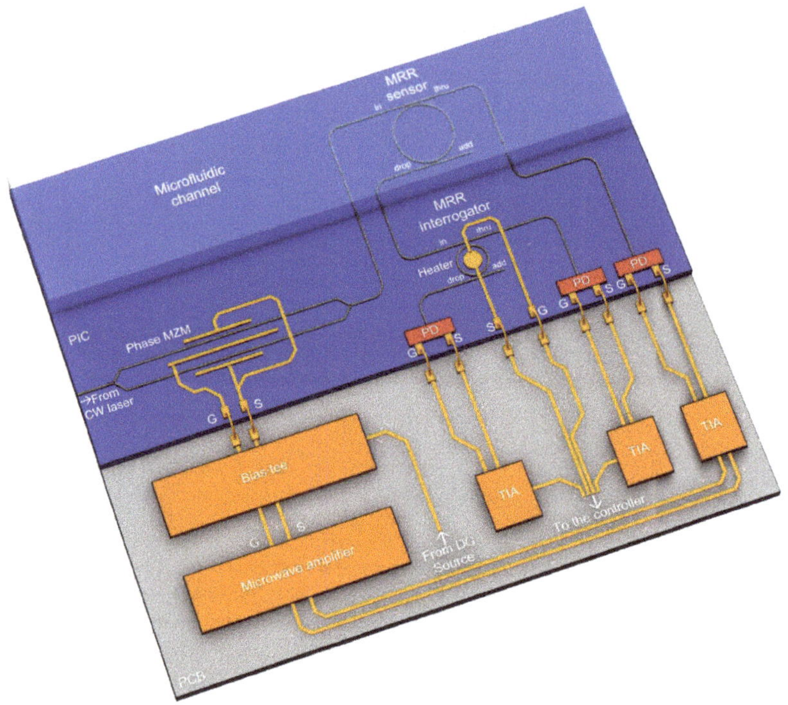


**Figure 4.** 3D design draft of the proposed integrated sensing system with an OEO (not to scale; microwave elements are shown tentatively).

The operating principle of an OEO was described in [33]. To start the generation, the narrow-band continuous-wave (CW) optical signal from an external laser diode is used as the carrier signal for the phase Mach–Zehnder modulator (MZM). At the output of the phase MZM, a phase-modulated (PM) optical signal is generated and sent to the MRR sensor, which serves as an optical notch filter. A feature of the PM signal spectrum is the phase opposition of the spectral components, which are symmetrically relative to the carrier. Therefore, when a PM signal is applied to the photodiode (PD), the antiphase components compensate each other, and the electrical signal at the output of the PD is determined only by the intensity of the optical carrier. However, when the notch filter suppresses one of the spectral components of the PM optical signal at the PD input, one of the spectral components becomes uncompensated. This causes a microwave signal to appear at the PD output. The frequency of this signal is determined by the difference between the CW wavelength and the resonant wavelength of the notch filter (i.e., the sensor MRR):

$$f = \left| \frac{c}{\lambda_{CW}} - \frac{c}{\lambda_{MRR}} \right| \tag{7}$$

where $c$ is the velocity of light in vacuum; and $\lambda_{CW}$ and $\lambda_{MRR}$ are the laser source wavelength and the resonant wavelength of the sensor, respectively. The microwave signal through the amplifier is fed to the input of the phase MZM, thereby closing the positive feedback loop.

Following the described operation principle, the OEO source bandwidth at the sensor drop port will be very narrow, as it is defined by the width of one spectral component in the signal, phase-modulated by a narrow-band microwave signal (the confirming simulation results are given in Section 4.2). In this case, the main wavelength of this optical signal is unambiguously related to the sensor transmission spectrum and changes along with it. Thus, the scheme shown in Figure 4 allows the generation of the input signal for the intensity interrogator without using a broadband optical source.

## 3. Sensing System Elements

### 3.1. Microring-Based Sensor

The layout of a label-free MRR sensor for liquid analysis is shown in Figure 1. It represents a silicon racetrack MRR on a $SiO_2$ buried-oxide layer, operating in a real-time regime. Hereinafter, the material dispersion in the waveguides is modeled by using the RI wavelength dependencies of the materials from [34]. The following MRR sensor geometric parameters were set: an outer radius set at 18 μm, the straight waveguide length of the racetrack set at 3 μm, the gap in the directional coupler of the in-through ports set at 0.18 μm, the gap in the directional coupler of the add-drop ports set at 0.25 μm, and the Si waveguide height/width set at 0.22/0.4 μm. The frequency response of the MRR sensor was calculated using the FDTD (Finite Difference Time Domain) method in the Ansys Lumerical software as follows: a simulation time of 7000 fs, a mesh accuracy of 3, and a minimum mesh step of 0.25 nm. The calculation was carried out for the TE-mode. The dispersion, loss, effective RI, and group RI were defined using the Finite Difference Eigenmode Solver the in Lumerical MODE. The MRR calculation method was employed as presented in [35,36].

The gap value for the in-through waveguide was selected based on the critical coupling criterion [37]. Using FDTD, the attenuation in the MRR waveguide and the coupling coefficients of the directional couplers were calculated. The resulting gap value for the in-through waveguide equals 0.18 μm. The presence of an add-drop waveguide introduces additional losses into the MRR system caused by the transfer of energy from the ring to the drop port. Therefore, the add-drop port coupler's gap value was chosen to have a high Q-factor in the under-coupling regime, providing enhanced sensitivity [17]. The resonant properties of the MRR were evaluated using FWHM (Equation (1)) and Q-factor (Equation (5)) using the following values: the ring transmission was measured using FDTD to be $a = 0.971$; the amplitude coupling coefficients to the in-through and add-drop waveguides are $r_1 = 0.935$ and $r_2 = 0.961$, respectively; the MRR circumference is $L = 2\pi R + 2L_s$ ($L_s = 3$ μm is the length of the straight waveguide in the racetrack, $R = 18$ μm is the ring radius); and $n_g = 4.55$ and $n_{eff} = 2.21$ are the values of the waveguide group and effective RIs, respectively, determined numerically using the Ansys Lumerical MODE software R1.4 2021.

Figure 5 shows a fragment of the transmission spectrum at the through and drop ports of the sensor for the minimum value of the analyte RI (1.311). According to Figure 5, with the gap values set as described above, the FWHM equals 190 pm (whereas in the case of equal gap values for both couplers of 0.18 μm, the FWHM would be 270 pm). Thus, an increase in the gap with the add-drop waveguide to 0.25 μm reduced the coupling coefficient and FWHM, providing the calculated quality factor $Q \approx 8000$. However, a further gap expansion leads to a decrease in the power at the drop port, making it more difficult to register.

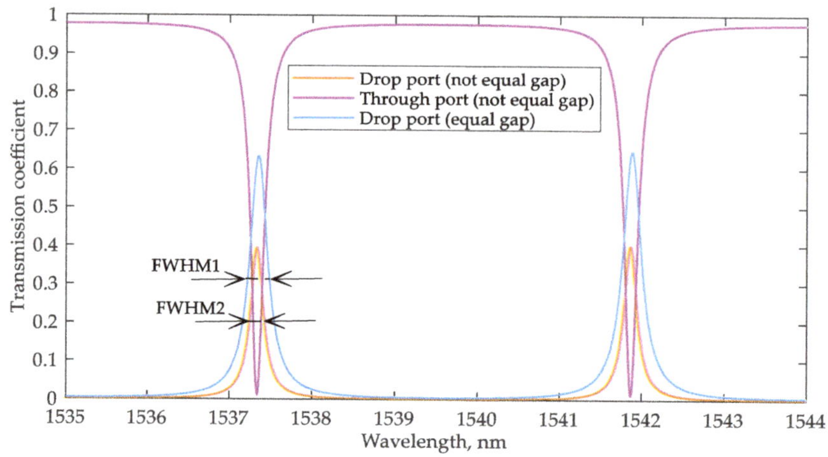


**Figure 5.** Transmission spectra of the through and drop ports of the sensor MRR.

As mentioned in the Introduction, a variation in any parameter of a liquid flowing over the sensor surface (e.g., composition, concentration, or temperature) shifts the resonant wavelengths of the MRR. Figure 6 demonstrates this process depending on the analyte RI sweeping from 1.311 to 1.315 with $\Delta n$ = 0.001 (the analyzed resonant peak around 1537.4 nm is shown). To obtain the sensor MRR transmission spectrum depending on the analyte RI, the FDTD calculation of the sensor MRR was redone for each analyte RI value above the MRR waveguide within the channel area (see Figure 1). The resonant wavelength shift, $\Delta\lambda$, shows almost linear behavior, with $\Delta\lambda$ = 0.11 nm at $\Delta n$ = 0.001. From these results, the sensor sensitivity, $S$, can be evaluated as follows:

$$S = \frac{\Delta\lambda}{\Delta n_{eff}} = 110\,\frac{\text{nm}}{\text{RIU}} \tag{8}$$

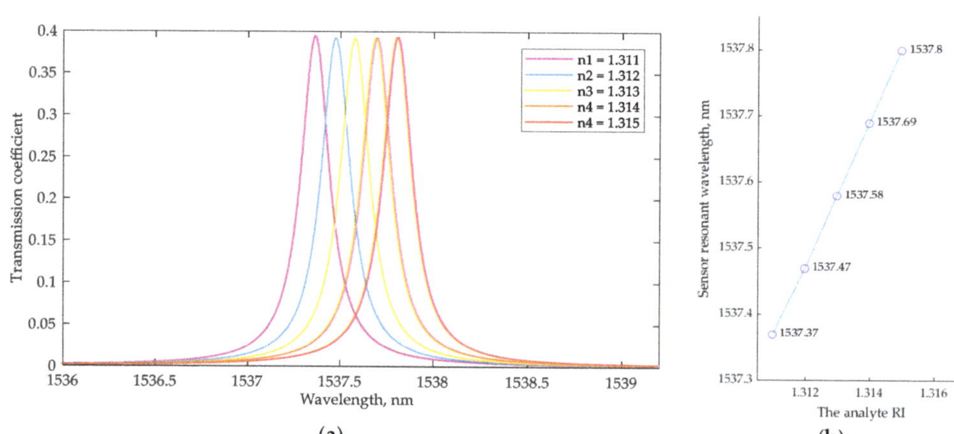


**Figure 6.** (**a**) The dependence of the sensor MRR transmission spectrum at the drop port on the analyte RI; (**b**) the dependence of the analyzed resonant wavelength of the sensor on the analyte RI.

The resulting sensitivity value slightly exceeds the results presented in [18], where a design with different gaps in the straight and add-drop couplers was also used. This improvement can be explained by the method applied for calculating the resonator's design parameters. At the same time, because we used a single MRR sensor in the simulation, the sensitivity is lower than in [21–23].

### *3.2. Microring-Based Interrogator*

To fulfill the requirements of the interrogator transmission spectrum discussed in Section 2, the parameters of the interrogator's MRR were defined as follows: the outer radius was set at 10 μm, the gap in both directional couplers was set at 0.1 μm, and the Si waveguide height/width were set at 0.22/0.4 μm. The interrogator MRR frequency response and corresponding Q-factor and FWHM values were calculated using the FDTD method in the Ansys Lumerical software, similar to the sensor MRR. The corresponding Q-factor of the interrogator's MRR read ~1400, and the FWHM equaled 1.1 nm.

Figure 7 shows a fragment of the interrogator MRR transmission spectra at the through and drop ports.

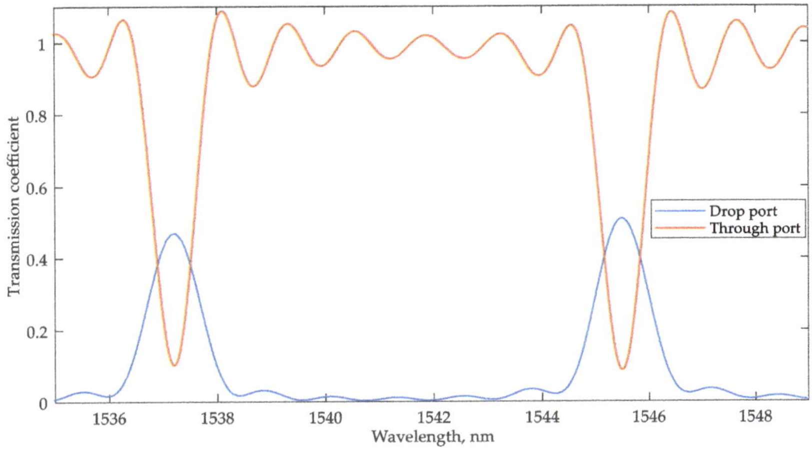


**Figure 7.** Transmission spectra at the through and drop ports of the interrogator MRR.

## 4. Sensing System Simulation

### *4.1. System Simulation with a Broadband Source*

A simulation of the whole sensing system was carried out following the scheme shown in Figure 1 in the Lumerical INTERCONNECT environment. The transmission spectra of the sensor and interrogator (for the minimum considered analyte RI of 1.311) are shown in Figure 8a. The resonant peaks of the sensor and interrogator MRRs overlap at the wavelength of 1537.37 nm (this resonant peak of the sensor is selected as the analyzed one); the next resonant peak of the sensor is at 1541.91 nm (so, the FSR equals 4.54 nm), and the previous resonant peak is at 1532.83 nm. There is a partial overlapping of the resonant peaks of the sensor and interrogator at 1546.52 nm, and the next resonant peak of the interrogator is at 1553.4 nm (the FSR of the interrogator equals 8.07 nm). Therefore, the broadband source may have a spectrum width of up to 20 nm, but the wavelengths from 1533 nm to 1535 nm and from 1550 nm to 1553 nm are useless because the transmission of the interrogator is very close to zero within those ranges. Therefore, to accelerate the simulation, the broadband source was emulated with an optical network analyzer (ONA) unit with a wavelength range from 1535 nm to 1550 nm, and the output powers at the interrogator PDs were calculated according to the equation:

$$P_{pd} = Z_{out} \left( \int_{\lambda_{\min}}^{\lambda_{\max}} P_l(\lambda) R(\lambda)^2 d\lambda \right)^2 \quad (9)$$

where $P_l(\lambda)$ is the optical power at the PD input; $R(\lambda)$ is the PD responsivity; $Z_{out}$ is the PD load resistance; and $\lambda_{\min}$ and $\lambda_{\max}$ are the minimum and the maximum wavelengths of the broadband source, respectively (1535 nm and 1550 nm, respectively, in our case).

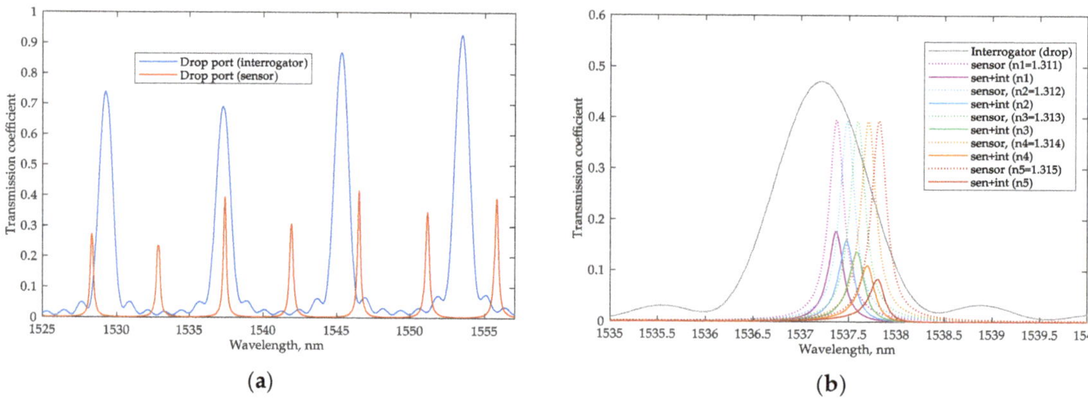


**Figure 8.** (**a**) Transmission spectra at the drop ports of the sensor and interrogator MRRs; (**b**) Wavelength dependence of the system transmission coefficient for the different analyte RIs.

The source signal enters the in-port of the sensor, and when the resonance condition is met, the signal appears at the sensor drop port and enters the in-port of the interrogator. Figure 8b shows how the optical power intensity at the drop port of the interrogator changes depending on the analyte RI. Suppose the analyzed resonant wavelength of the sensor and the utilized resonant wavelength of the interrogator coincide. In that case, there will be a maximum transmission coefficient and maximum intensity of the optical signal at the drop port of the interrogator. If the analyzed resonance of the sensor hits within the FSR of the interrogator, then the transmission coefficient will be minimal. Almost all the optical power will pass to the through port of the interrogator.

To plot the dependence of the relative power at the system output on the analyte RI, we first analyzed the optical intensity at the through and drop ports of the sensor. The results for each port were separately converted into units of electrical power according to Equation (9), considering the wavelength-dependent responsivity of the PD (Figure 9a). The typical SOI-based structure described in [38] was used for the PD numerical model. Then, for each value of the RI, the relative power at the interrogator drop port was calculated using Equation (3).

The results are shown in Figure 9b. As can be seen from the graph, a change in the RI of 0.001 leads to a change in the relative power of at least 1.35 dB, which makes it possible to estimate the minimum aggregate sensitivity of the entire system by analogy with Equation (8):

$$S_{\min} = \frac{\Delta P_{\min}}{\Delta n} = 1350 \frac{\text{dB}}{\text{RIU}} \tag{10}$$

The averaged system sensitivity over the entire measurement range equals $\overline{S} = 1570\ \text{dB/RIU}$.

The simulation results confirm the validity of the requirements outlined in Section 2. The choice of the MRR geometry provided a higher Q-factor for the sensor and a lower Q-factor for the interrogator. The ratio of the sensor and interrogator FSRs made it possible to use a broadband source (with spectrum width up to 20 nm) without the negative resonances overlapping effect. Of course, it is possible to expand the bandwidth of a broadband source with a decrease in the interrogator MRR radius, but this would lead to a further decrease in its quality factor and the slope of the resonant characteristic, which would reduce the system sensitivity.

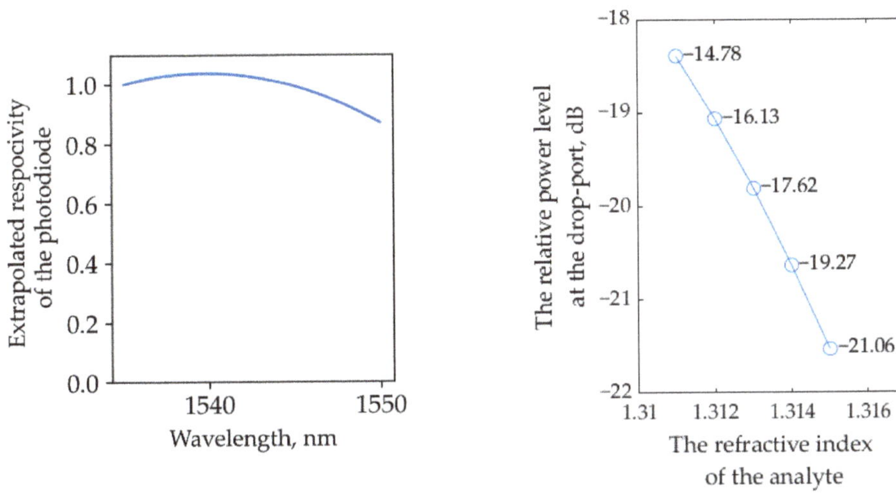


(a) (b)

**Figure 9.** (**a**) Modeled wavelength dependence of the photodiode responsivity; (**b**) Modeled dependence of the relative power at the system output on the analyte RI in the scheme with a broadband source.

*4.2. Simulating the System with Optoelectronic Oscillator as a Source*

Let us consider the sensing system scheme with the OEO utilized as a radiation source, according to Figure 4. The possibility of an integrated OEO design on the SOI platform was demonstrated in [39]. The simulation was performed in the Lumerical INTERCONNECT environment with the following circuit parameters: a CW laser carrier wavelength of 1538 nm; a CW laser output power of 2 mW; a microwave amplifier gain of 65 dB; the PD model is the same as in Section 4.1; the PD dark current of 60 nA; the PD, amplifier, and MZM RF bandwidth $f_{max}$ = 50 GHz; and the MZM $V_\pi$ = 4 V. At the current stage of research, the idealized phase MZM transfer function is used without parasitic amplitude modulation because the real phase modulator characteristics will be strongly dependent on the parameters of the photonic integration platform used for the system implementation. The parameters of the MRR sensor and MRR interrogator are given above in Section 3. The choice of the CW laser wavelength is determined by the range of the analyzed sensor resonance wavelengths and the allowable microwave signal frequency, which is determined by the frequency characteristics of the PD, modulator, and amplifier. To set the value of the laser carrier wavelength for analyzing a certain sensor resonance peak, it is necessary to define the maximum value of the microwave signal frequency, $f_{max}$. Then, it is easy to convert Equation (7) into an expression for determining the required optical carrier wavelength. For definiteness, we assume that the wavelength of the CW laser is greater than the analyzed resonance wavelength. In this case, the highest value of the microwave signal frequency will correspond to the lowest value of the MRR resonant wavelength, $\lambda_{MRR\mathrm{min}}$. Then:

$$f_{\mathrm{max}} = \frac{c}{\lambda_{MRR\mathrm{min}}} - \frac{c}{\lambda_{CW}} \tag{11}$$

$$\lambda_{CW} = \frac{c \cdot \lambda_{MRR\mathrm{min}}}{c - f_{\mathrm{max}}\lambda_{MRR\mathrm{min}}} \tag{12}$$

Figure 10 shows the simulated optical spectra at the through and drop ports of the sensor MRR. The dependence of the relative power at the interrogator drop port on the analyte RI, obtained from the simulation, is shown in Figure 11.

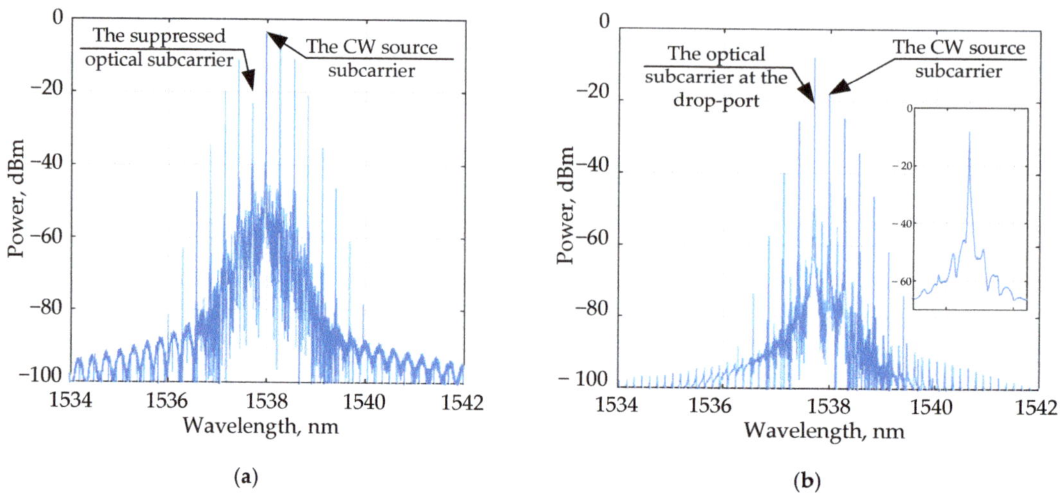


**Figure 10.** Optical spectrum at the through port (**a**) and the drop port (**b**) of the sensor in the scheme with an OEO. In inset: the magnified spectrum of the optical subcarrier at the drop-port.

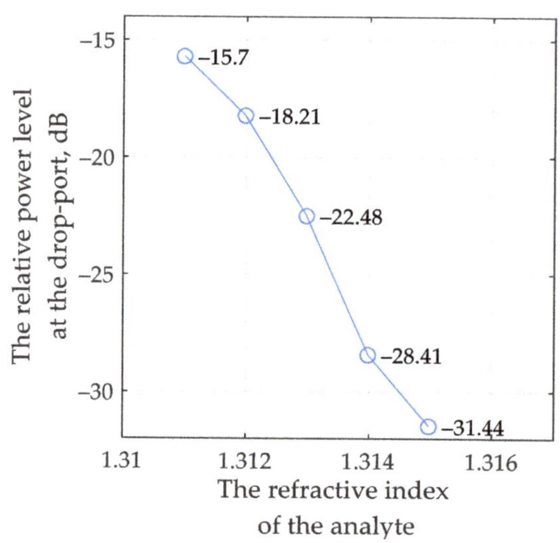


**Figure 11.** Modeled dependence of the relative power at the system output on the analyte RI in the scheme using an OEO as a source.

It can be seen that a change in the RI of 0.001 leads to a change in the output relative power of at least 2.51 dB, which makes it possible to evaluate the minimum aggregate sensitivity of the entire sensing system by analogy with Section 3.1:

$$S_{\min} = \frac{\Delta P_{\min}}{\Delta n} = 2510 \frac{\text{dB}}{\text{RIU}} \tag{13}$$

The averaged system sensitivity, in this case, is $\overline{S} = 3935$ dB/RIU.

Thus, when using an OEO, the system's sensitivity is much higher than when using an optical broadband source. However, the use of an OEO in the described intensity interrogation system has some specifics. First, the interrogation wavelength range is limited by the RF bandwidth of the phase modulator, RF amplifier, and PD (see Equation (7)). For example, at the maximum OEO generation frequency of 30 GHz, the interrogation range will be 235 pm, and when the frequency range is extended to 50 GHz, the interrogation wavelength range will increase to 395 pm. At the same time, this limitation in the analyzed spectrum width removes the problem mentioned above regarding the unintended mixing of the useful signal from the analyzed resonant peak of the sensor MRR and the parasitic signals from its other resonant peaks.

Second, the use of an OEO causes additional delays. Figure 12 shows that it takes about 5 ns to start generating in a steady-state operational mode. During this time, the output of the interrogator will not be reliable.

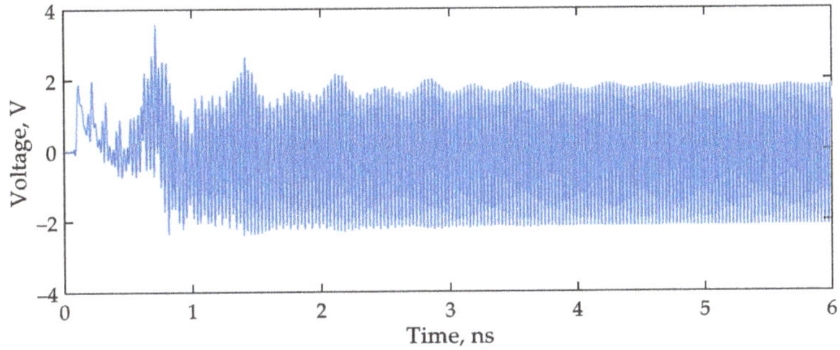


**Figure 12.** The waveform of the microwave signal at the MZM input.

Moreover, the scheme with an OEO is anticipated to be more power-consuming due to the need for optoelectronic and electro-optic conversions and microwave signal amplification in the feedback loop.

Despite the shortcomings described, the interrogation scheme using an OEO also has obvious advantages. First, there will be only one resonant peak in the signal spectrum at the drop port of the MRR sensor (Figure 10b); second, the linewidth of the optical spectrum generated by the OEO is significantly narrower than the resonance FWHM of the MRR used as a sensor. Thus, the optical spectrum width, according to the simulation, was less than 10 pm, even though the FWHM of the sensor is 190 pm. These advantages avoid the negative overlap of adjacent peaks of the sensor transmission spectrum described in Section 2 and, as can be seen from the simulation results, increase the sensitivity of the sensor system.

## 5. Bragg-Grating Interrogation Using an MRR-Based Interrogator

Fully integrated sensing systems are of interest, as described in the Introduction. However, in monitoring extended objects (i.e., pipelines and bridges), it is advisable to create quasi-distributed sensor systems in which the sensors and the interrogators are geographically spaced. Today, in such systems, FBG sensors are widely used, so it makes sense to also evaluate the compatibility of the proposed PIC-based interrogator with FBG. The simulation was carried out for a typical grating using the parameters presented in Table 1. The transmission and reflection spectra of the FBG are shown in Figure 13. For representativeness, an ONA was used as an input optical signal source. The interrogator output signals were calculated by integrating the ONA output signals over the spectrum width of the broadband source. The spectral band of the broadband source was assumed to be the same as in the model with the MRR sensor (1535–1550 nm).

**Table 1.** FBG sensor parameters.

| Name | Value |
| --- | --- |
| Outer cladding diameter | 150 μm |
| Inner core diameter | 50 μm |
| Grating period | 0.5 μm |
| Number of periods | 20,000 |
| Effective refractive index | 1.5 |
| Periodic variation in the refractive index of the core | $10^{-3}$ |
| Material | $SiO_2$ |

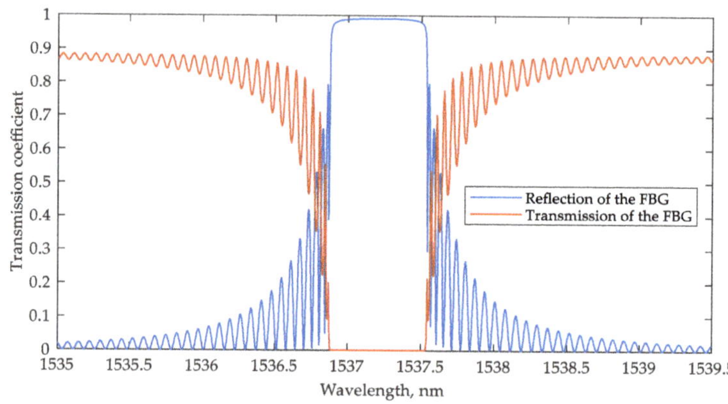


**Figure 13.** FBG transmission and reflection spectra at the resonant wavelength $\lambda_B$ = 1537.2 nm.

The modeling has shown that the scheme using an OEO as a source cannot work properly with standard FBG due to the relatively wide FBG's output spectrum. However, an OEO can be applied with appropriately designed fiber grating sensors, e.g., a narrow-band phase-shifting Bragg grating (PS-FBG).

The MRR-based interrogation scheme was modeled with FBG replacing the MRR sensor. Figure 14a shows the first operation option when the FBG transmission signal is processed, whereas Figure 13b demonstrates the second option when FBG is connected through an optical circulator to operate with the reflected signal.

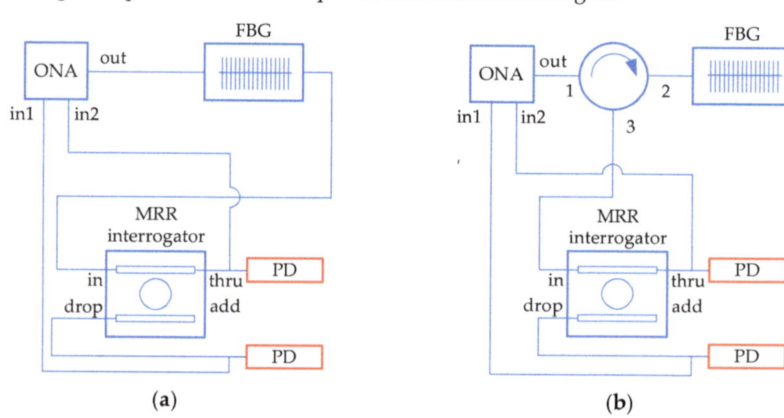


**Figure 14.** The simulation schemes of FBG interrogation: (**a**) by analyzing the transmitted light; (**b**) by analyzing the reflected light. Numbers 1–3 denote the order of the circulator's ports.

Figure 15a,b show the interrogator drop port transmission spectra and demonstrate how the transmission changes depending on the shift of the Bragg wavelength. The simulated relative power at the drop port of the interrogator versus the resonant wavelength of the FBG is provided in Figure 15.

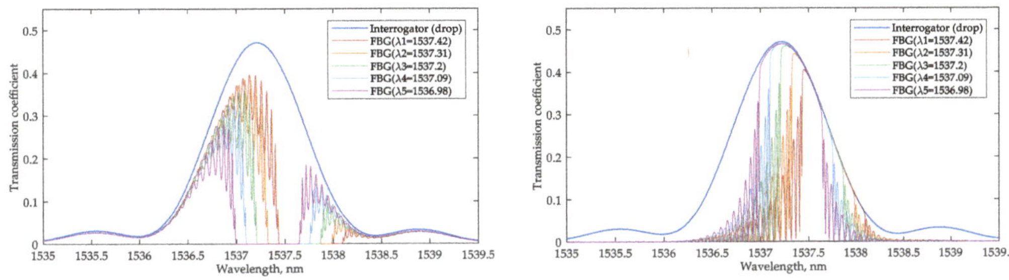


(**a**) (**b**)

**Figure 15.** The combined transmission spectra at the interrogator drop port depending on the Bragg wavelength of the FBG: (**a**) for the scheme that analyzes the transmitted light; (**b**) for the scheme that analyzes the reflected light.

The sensitivity $S_{BG}$ can be calculated as follows:

$$S_{BG} = \frac{\Delta P}{\Delta \lambda} \tag{14}$$

where $\Delta P$ is the difference between the neighboring values of the relative power levels at the interrogator drop port and $\Delta\lambda$ is the corresponding change in the resonant wavelength. According to Equation (14), the minimum system sensitivity for analyzing the transmitted and reflected light are 7.2 dB/nm and 2.27 dB/nm, respectively. The average sensitivity may be calculated as the ratio between the relative power change and the resonant wavelength variation; the average sensitivity equals 13.7 dB/nm and 2.8 dB/nm for analyzing the transmitted and reflected light, respectively.

By comparing these two interrogation options for an FBG sensor (Figure 16), it can be noticed that the sensitivity turns out to be higher when operating with the reflected light. At the same time, for the option that analyzes the transmitted light, the sensitivity is flatter over a wider range of the FBG's resonant wavelength shifts. This result can be explained by the shape of the interrogator's transmission spectrum. When operating in reflected light, the FBG works as a bandpass filter, passing the signal to the interrogator near the high-steepness region of its transmission spectrum. When operating in the transmitted light, the FBG works as a notch filter, and the wavelengths at the in-port of the interrogator fall into the low slope region of the filter characteristic.

Due to the relatively wide spectrum width of the reflected signal, it is not possible to use a classical Bragg grating in an interrogation scheme with an OEO. However, the PS-FBG has a very narrow stopband spectrum when analyzing the reflected light. We simulated the interrogation system with the PS-FBG (parameters are given in Table 2) following the scheme shown in Figure 17a. The grating spectrum is shown in Figure 17b. The simulated relative power at the drop port of the interrogator versus the resonant wavelength of the PS-FBG is provided in Figure 18. The average sensitivity of the sensing system in the model reads about 37 dB/nm.

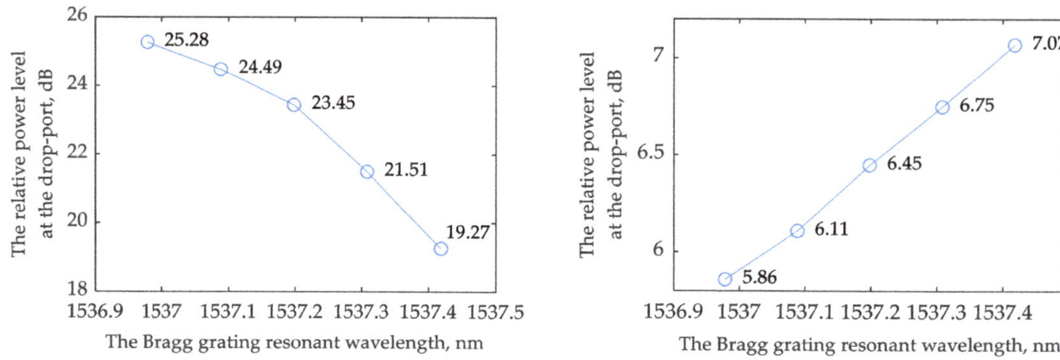


(a) (b)

**Figure 16.** Relative power at the interrogator drop port versus the resonant wavelength of the FBG: (**a**) for the scheme that analyzes the transmitted light; (**b**) for the scheme that analyzes the reflected light.

**Table 2.** PS-FBG sensor parameters.

| Name | Value |
|---|---|
| Outer cladding diameter | 150 μm |
| Inner core diameter | 50 μm |
| Grating period | 0.51 μm |
| Phase-shifted aria | 0.51 μm |
| Number of periods | 4000 |
| Effective refractive index | 1.5 |
| Periodic variation in the refractive index of the core | $10^{-3}$ |
| Material | $SiO_2$ |

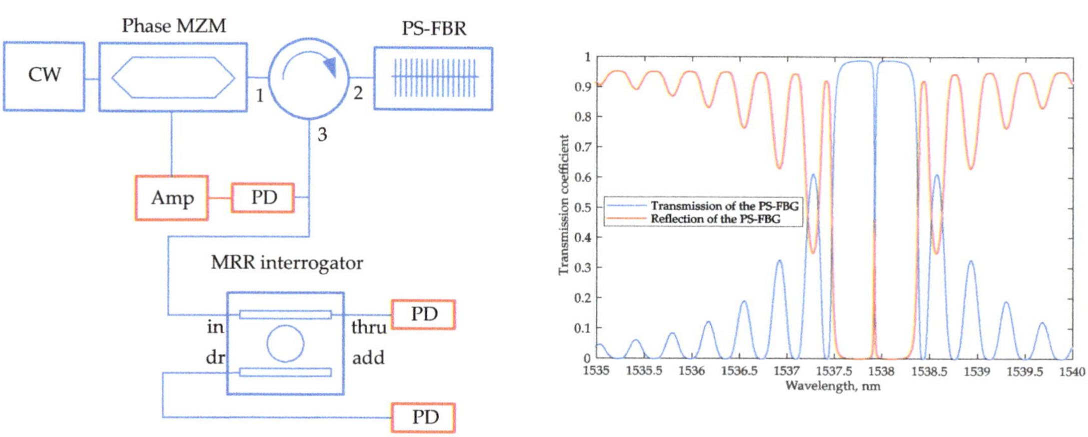


(a) (b)

**Figure 17.** (**a**) PS-FBG interrogation scheme with an OEO; (**b**) PS-FBG transmission and reflection signals spectra at the resonant wavelength $\lambda_B$ = 1537.2 nm. Numbers 1–3 denote the order of the circulator's ports.

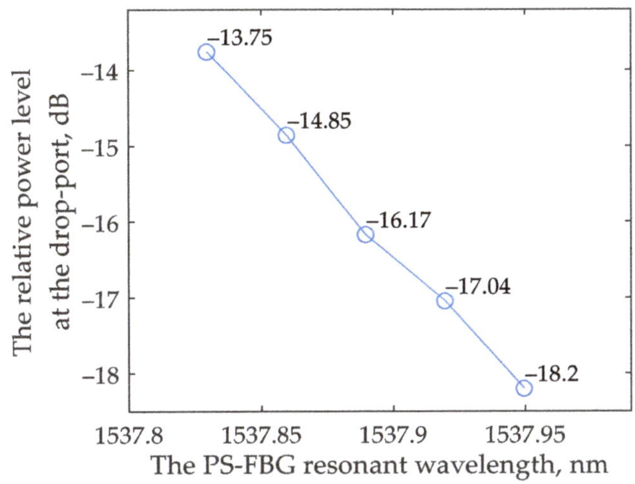


**Figure 18.** Relative power at the interrogator drop port versus the resonant wavelength of the PS-FBG in the scheme with OEO.

In turn, the PS-FBG cannot be used in the intensity interrogation scheme when operating in transmitted light due to the wide output of the optical spectrum. However, its use when analyzing reflected light (by analogy with Figure 14b) is possible.

The simulated relative power at the drop port of the interrogator versus the resonant wavelength of the PS-FBG in the scheme broadband source is provided in Figure 19. The corresponding average sensitivity of the sensing system is about 55 dB/nm.

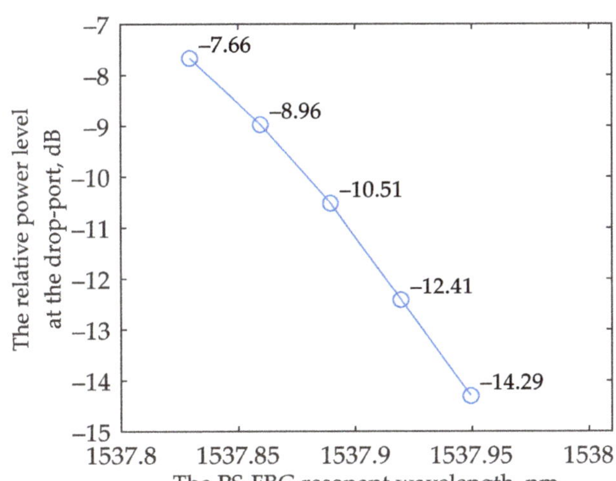


**Figure 19.** Relative power at the interrogator drop port versus the resonant wavelength of the PS-FBG in the scheme with broadband source.

Although the scheme with the broadband source provides higher system sensitivity, it should be underscored that it requires a more complicated source with a certain bandwidth, which may be hard to realize on a PIC. System sensitivity for the interrogation scheme with an OEO is lower but remains sufficient to distinguish the output electrical signal variations using modern electronic components.

## 6. Discussion

The proposed integrated sensing system can be fabricated on the SOI platform. Other commonly used technological platforms such as silicon nitride, indium phosphide (InP), and polymer-based have several limitations [40], making them less suitable for the fabrication of the considered scheme.

Thus, on the $Si_3N_4$ platform, modulators can be implemented with heaters only, but the thermo-optical effect provides operation times down to 30 μs [41], which limits the performance of the interrogation scheme with an OEO. Meanwhile, for polymer-based platforms, modulation speeds of about 250 ns can be achieved with a stress-optical effect [42], and in commercially available InP platforms, the phase modulators based on p-i-n hetero-junction can provide a modulation bandwidth >20 GHz [43]. However, all aforementioned platforms are of lower contrast [44] compared to SOI, which leads to an increased radius of the MRRs. Owing to this, the FSR values become too small, which significantly complicates the system design required to avoid unintended interference from the side resonant peaks.

For a fully integrated system, laser source implementation on a chip is required. Only the InP platform can provide the full range of photonic devices. However, in the standard substrate-based InP platform, there are no small bend radii of the waveguides. The waveguide radii similar to those of the SOI platform can be achieved on the novel InP membrane platform [45], which is currently under development and appears promising for full integration of the considered sensing system. Alternatively, the hybrid integration of InP elements could be used to overcome this problem. For instance, passive elements can be implemented using an SOI or $Si_3N_4$ process and then connected to the InP chip with active elements via 45° mirrors or butt-joint coupling [40]. In addition, flip-chip integration is available for the silicon nitride platform [46]. These integration approaches provide insertion losses from 1 dB to 2 dB [40,46] that are lower than when using discrete elements connected via optical fiber.

A promising solution is using monolithic hybrid platforms [47,48], which provide a smaller footprint and better reliability than conventional hybrid integration methods. It is worth noting that both platforms have optical sources with a bandwidth higher than 20 nm [48,49].

One of the problems in implementing PIC-based devices is the inaccuracies in the feature size of the fabricated chips. For the scheme under consideration, the deviation in the width of the waveguides appears to be the primary type of fabrication inaccuracy that can affect the system performance, as it changes the effective and group indices of the waveguides and the coupling coefficients of the directional couplers in MRRs. According to the experimental data presented in [35], the intra-wafer standard deviation of the waveguide width for the typical SOI fabrication process based on 248 nm photolithography is about 3.9 nm. In the commercially available state-of-the-art 193 nm immersion lithography and dry etch fabrication process, the intra-wafer waveguide width standard deviation can be as low as 2.5 nm [50]. However, according to [51], for different wafers, the standard deviation of waveguide width can reach 6.4 nm. Based on the wafer maps of the waveguide geometry variations, their influence on the performance of certain integrated components can be analyzed, and the corresponding fabrication yield can be predicted and optimized, as presented in [52].

We have analyzed the possibility of maintaining the system operability with errors in the waveguide width up to 8 nm, which is guaranteed to consider the characteristics of available platforms. The performed analysis showed (Figure 20) that the sensitivity of 110 nm/RIU is retained; although, changing the waveguide width significantly affects the resonant frequencies of the MRR sensor and the transfer function at the drop port. The change in the position of the resonant frequencies of the sensor MRR relative to the frequency characteristics of the interrogator can be compensated using a built-in heater for the interrogator MRR (Figures 1 and 4). The same method can be applied in the case of FBG sensors interrogation to compensate for their possible Bragg wavelength deviation, which practically does not exceed tenths of a nanometer. The systematic amplitude and phase inaccuracies during the FBGs inscription and their mitigation have been described in the literature [53,54]. In a lower-quality interrogator MRR, the waveguide width deviation only leads to a shift in the resonant frequencies (Figure 21), which can also be corrected using a heater. Changes in the FWHM and Q-factor are negligible. Therefore, the sensitivity of the entire system will not change. Thus, the proposed schemes will remain operable within a practically relevant range of fabrication errors in the waveguide width.

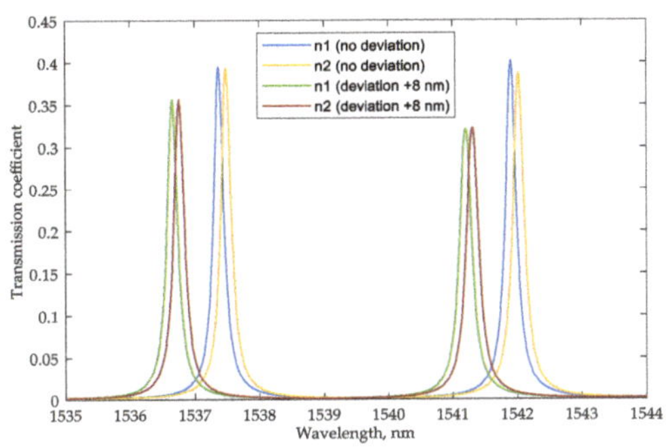


(a)

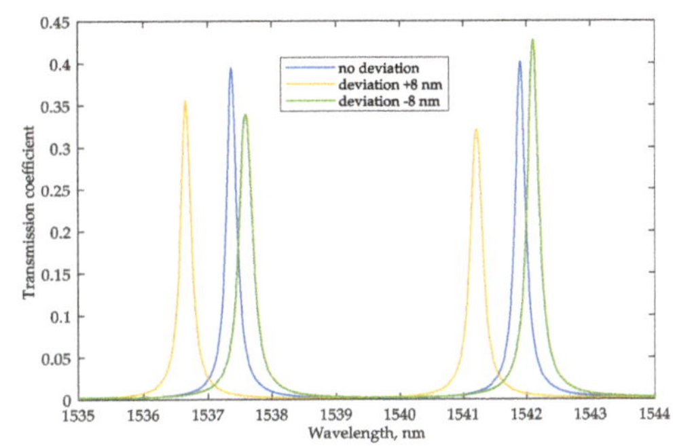


(b)

**Figure 20.** (**a**) Sensor's transmission spectra for sensing an RI increment of 0.001 (n1 = 1.311, n2 = 1.312) in the case of fabrication deviation of the waveguide width of +8 nm; (**b**) sensor's transmission spectra for the same analyte RI (n = 1.311) for the fabrication deviation of the waveguide width of ±8 nm.

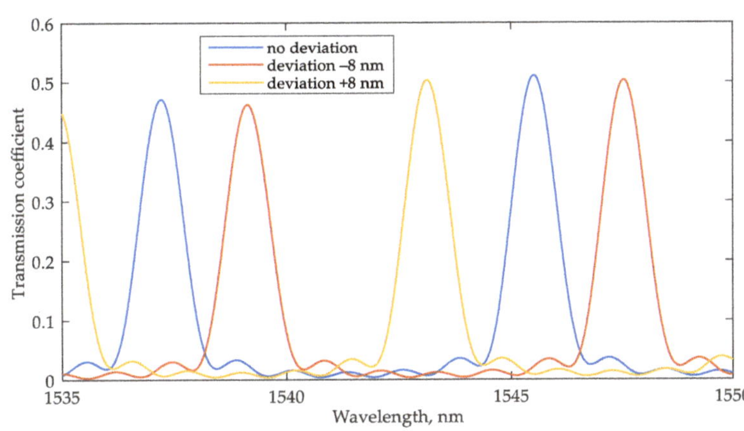


**Figure 21.** Interrogator's transmission spectra for the fabrication deviation in the waveguide width of ±8 nm.

## 7. Conclusions

The concept of a sensing system based on microring add-drop resonators, suitable for integration on the SOI platform, is proposed. It includes a sensor based on the racetrack MRR, which, according to simulation results, has a sensitivity of 110 nm/RIU and the interrogator based on the MRR with a lower Q-factor.

A novel scheme for generating an input optical signal for feeding the sensor using the OEO is also proposed. The key feature of this scheme is that it does not require either an external broadband source or a tunable laser for its operation. At the same time, the average sensitivity of the system with an OEO is 3935 dB/RIU, which is even higher than that for the configuration with a broadband or tunable source (1570 dB/RIU). Despite the limitation of the sensor's transmission spectrum width that can be analyzed, this scheme allows for interrogating the sensor targeted exclusively on the analyzed resonant peak and avoiding crosstalk from the other resonant peaks. Noteworthy, the transient process of the OEO causes an additional delay in measurements. Yet, the simulation showed that the introduced delay is lower than 10 ns, which is relevant only at sampling rates exceeding 100 MHz.

As modeling revealed, the interrogation scheme with an OEO cannot be applied with typical FBG-based sensors due to their relatively wide stop band. Therefore, system performance with the FBG sensor was evaluated using a broadband source. The scheme provides classical FBG interrogation in reflected and transmitted light. The system sensitivity in the former case is considerably higher; although, it is less monotonic because the FBG output spectrum is wide and may cover the interrogator transmission spectrum regions with different slopes. At the same time, interrogation in transmitted light does not require an optical circulator, providing a less expensive and more compact scheme suitable for full integration. That fact makes it a good candidate for applications with undemanding high sensitivity. Nevertheless, the OEO could be a promising integrable solution as a source operating with narrow-band gratings, such as PS-FBG. For such gratings, simulation was performed with a broadband source and an OEO. The results have shown that the former provides a higher system sensitivity but requires a more complicated source that may be difficult for PIC-based realization; the latter provides a sufficient system sensitivity when using it with a control circuit realized on modern electronic components.

**Author Contributions:** Conceptualization, G.V. and R.K.; methodology, G.V., V.L. and I.S.; software, G.V., A.Z., V.I., D.I., R.Y. and I.S.; validation, R.K., V.L., and O.M.; formal analysis, R.K. and E.G.; investigation, G.V., A.Z., V.I. and D.I.; resources, R.K. and E.G.; data curation, R.K. and V.L.; writing—original draft preparation, G.V., A.Z. and V.I.; writing—review and editing, G.V., R.K., E.G., V.L. and O.M.; visualization, G.V., A.Z. and V.I.; supervision, R.K.; project administration, E.G.; funding acquisition, E.G. and R.K. All authors have read and agreed to the published version of the manuscript.

**Funding:** The research is supported by the Ministry of Science and Higher Education of the Russian Federation within the state assignment for the FSBEI HE "USATU" (agreement No. 075-03-2021-014) and conducted in the research laboratory "Sensor systems based on integrated photonics devices" of the Eurasian Scientific and Educational Center.

**Institutional Review Board Statement:** Not applicable.

**Informed Consent Statement:** Not applicable.

**Data Availability Statement:** Not applicable.

**Conflicts of Interest:** The authors declare no conflict of interest.

## References

1. Luan, E.; Shoman, H.; Ratner, D.M.; Cheung, K.C.; Chrostowski, L. Silicon photonic biosensors using label-free detection. *Sensors* **2018**, *18*, 3519. [CrossRef] [PubMed]
2. Ciminelli, C.; Dell'Olio, F.; Conteduca, D.; Armenise, M.N. Silicon photonic biosensors. *IET Optoelectron.* **2019**, *13*, 48–54. [CrossRef]
3. Chandrasekar, R.; Lapin, Z.J.; Nichols, A.S.; Braun, R.M.; Fountain, A.W. Photonic integrated circuits for department of defense-relevant chemical and biological sensing applications: State-of-the-art and future outlooks. *Opt. Eng.* **2019**, *58*, 1. [CrossRef]
4. Morais, E.; Pontes, M.J.; Marques, C.; Leal-Junior, A. Liquid level sensor with two FBGs embedded in a PDMS diaphragm: Analysis of the linearity and sensitivity. *Sensors* **2022**, *22*, 1268. [CrossRef] [PubMed]
5. Yang, D.; Liu, Y.; Wang, Y.; Zhang, T.; Shao, M.; Yu, D.; Fu, H.; Jia, Z. Integrated optic-fiber sensor based on enclosed EFPI and structural phase-shift for discriminating measurement of temperature, pressure and RI. *Opt. Laser Technol.* **2020**, *126*, 106112. [CrossRef]
6. Li, H.; An, Z.; Mao, Q.; Zuo, S.; Zhu, W.; Zhang, S.; Zhang, C.; Li, E.; García, J.D.P. SOI waveguide bragg grating photonic sensor for human body temperature measurement based on photonic integrated interrogator. *Nanomaterials* **2021**, *12*, 29. [CrossRef] [PubMed]

7. Liu, Y.; Liu, X.; Zhang, T.; Zhang, W. Integrated FPI-FBG composite all-fiber sensor for simultaneous measurement of liquid refractive index and temperature. *Opt. Lasers Eng.* **2018**, *111*, 167–171. [CrossRef]
8. Vogelbacher, F.; Kothe, T.; Muellner, P.; Melnik, E.; Sagmeister, M.; Kraft, J.; Hainberger, R. Waveguide Mach-Zehnder biosensor with laser diode pumped integrated single-mode silicon nitride organic hybrid solid-state laser. *Biosens. Bioelectron.* **2022**, *197*, 113816. [CrossRef]
9. Sun, X.; Dai, D.; Thylén, L.; Wosinski, L. High-sensitivity liquid refractive-index sensor based on a Mach-Zehnder interferometer with a double-slot hybrid plasmonic waveguide. *Opt. Express* **2015**, *23*, 25688. [CrossRef]
10. Crespi, A.; Gu, Y.; Ngamsom, B.; Hoekstra, H.J.W.M.; Dongre, C.; Pollnau, M.; Ramponi, R.; van den Vlekkert, H.H.; Watts, P.; Cerullo, G.; et al. Three-dimensional Mach-Zehnder interferometer in a microfluidic chip for spatially-resolved label-free detection. *Lab Chip* **2010**, *10*, 1167. [CrossRef]
11. Wu, S.; Guo, Y.; Wang, W.; Zhou, J.; Zhang, Q. Label-free biosensing using a microring resonator integrated with poly-(dimethylsiloxane) microfluidic channels. *Rev. Sci. Instrum.* **2019**, *90*, 035004. [CrossRef] [PubMed]
12. Gangwar, R.K.; Qin, J.; Wang, X. Porous silicon–based microring resonator for temperature and cancer cell detection. *Front. Phys.* **2022**, *10*, 929033. [CrossRef]
13. Bahadoran, M.; Seyfari, A.K.; Sanati, P.; Chua, L.S. Label free identification of the different status of anemia disease using optimized double-slot cascaded microring resonator. *Sci. Rep.* **2022**, *12*, 5548. [CrossRef]
14. Sun, X.; Zeng, L.; Hu, Y.; Duan, J. Fabrication and sensing application of phase shifted bragg grating sensors. *Materials* **2022**, *15*, 3720. [CrossRef] [PubMed]
15. Fernández Gavela, A.; Grajales García, D.; Ramirez, J.; Lechuga, L. Last advances in silicon-based optical biosensors. *Sensors* **2016**, *16*, 285. [CrossRef]
16. Kazanskiy, N.L.; Khonina, S.N.; Butt, M.A. Advancement in silicon integrated photonics technologies for sensing applications in near-infrared and mid-infrared region: A review. *Photonics* **2022**, *9*, 331. [CrossRef]
17. Cardenosa-Rubio, M.C.; Robison, H.M.; Bailey, R.C. Recent advances in environmental and clinical analysis using microring resonator-based sensors. *Curr. Opin. Environ. Sci. Health* **2019**, *10*, 38–46. [CrossRef] [PubMed]
18. Zhang, P.; Ding, Y.; Wang, Y. Asymmetrical microring resonator based on whispering gallery modes for the detection of glucose concentration. *Optik* **2018**, *171*, 642–647. [CrossRef]
19. Huang, W.; Luo, Y.; Zhang, W.; Li, C.; Li, L.; Yang, Z.; Xu, P. High-sensitivity refractive index sensor based on Ge–Sb–Se chalcogenide microring resonator. *Infrared Phys. Technol.* **2021**, *116*, 103792. [CrossRef]
20. Bogaerts, W.; De Heyn, P.; Van Vaerenbergh, T.; De Vos, K.; Kumar Selvaraja, S.; Claes, T.; Dumon, P.; Bienstman, P.; Van Thourhout, D.; Baets, R. Silicon microring resonators. *Laser Photonics Rev.* **2012**, *6*, 47–73. [CrossRef]
21. Huang, L.; Yan, H.; Xu, X.; Chakravarty, S.; Tang, N.; Tian, H.; Chen, R.T. Improving the detection limit for on-chip photonic sensors based on subwavelength grating racetrack resonators. *Opt. Express* **2017**, *25*, 10527. [CrossRef] [PubMed]
22. Wu, N.; Xia, L. Side-mode suppressed filter based on anangular grating-subwavelength grating microring resonator with high flexibility in wavelength design. *Appl. Opt.* **2019**, *58*, 7174. [CrossRef] [PubMed]
23. Zhao, C.Y.; Zhang, L.; Zhang, C.M. Compact SOI optimized slot microring coupled phase-shifted bragg grating resonator for sensing. *Opt. Commun.* **2018**, *414*, 212–216. [CrossRef]
24. Słowikowski, M.; Kaźmierczak, A.; Stopiński, S.; Bieniek, M.; Szostak, S.; Matuk, K.; Augustin, L.; Piramidowicz, R. Photonic integrated interrogator for monitoring the patient condition during MRI diagnosis. *Sensors* **2021**, *21*, 4238. [CrossRef]
25. Yao, J. Microwave photonic sensors. *J. Light. Technol.* **2021**, *39*, 3626–3637. [CrossRef]
26. Tozzetti, L.; Bontempi, F.; Giacobbe, A.; Pasquale, F.D.; Faralli, S. Fast FBG interrogator on chip based on silicon on insulator ring resonator add/drop filters. *J. Light. Technol.* **2022**, *40*, 5328–5336. [CrossRef]
27. Chew, S.X.; Yi, X.; Yang, W.; Wu, C.; Li, L.; Nguyen, L.; Minasian, R. Optoelectronic oscillator based sensor using an on-chip sensing probe. *IEEE Photonics J.* **2017**, *9*, 5500809. [CrossRef]
28. Yao, J. Optoelectronic oscillators for high speed and high resolution optical sensing. *J. Light. Technol.* **2017**, *35*, 3489–3497. [CrossRef]
29. Yao, J. Photonic generation of microwave arbitrary waveforms. *Opt. Commun.* **2011**, *284*, 3723–3736. [CrossRef]
30. Marin, Y.E.; Nannipieri, T.; Oton, C.J.; Pasquale, F.D. Current status and future trends of photonic-integrated FBG interrogators. *J. Light. Technol.* **2018**, *36*, 946–953. [CrossRef]
31. Shidlovski, V.R. Superluminescent diode light sources for OCT. In *Optical Coherence Tomography*; Biological and Medical Physics, Biomedical Engineering; Drexler, W., Fujimoto, J.G., Eds.; Springer: Berlin/Heidelberg, Germany, 2008; pp. 281–299. ISBN 978-3-540-77549-2.
32. Rank, E.A.; Agneter, A.; Schmoll, T.; Leitgeb, R.A.; Drexler, W. Miniaturizing optical coherence tomography. *Transl. Biophoton.* **2022**, *4*, e202100007. [CrossRef]
33. Yao, X.S.; Maleki, L. Optoelectronic microwave oscillator. *J. Opt. Soc. Am. B* **1996**, *13*, 1725. [CrossRef]
34. Palik, E.D.; Ghosh, G. *Handbook of Optical Constants of Solids*; Academic Press: San Diego, CA, USA, 1998; ISBN 978-0-12-544415-6.
35. Lu, Z.; Jhoja, J.; Klein, J.; Wang, X.; Liu, A.; Flueckiger, J.; Pond, J.; Chrostowski, L. Performance prediction for silicon photonics integrated circuits with layout-dependent correlated manufacturing variability. *Opt. Express* **2017**, *25*, 9712. [CrossRef]
36. Chrostowski, L.; Hochberg, M. *Silicon Photonics Design: From Devices to Systems*, 1st ed.; Cambridge University Press: Cambridge, UK, 2015; ISBN 978-1-107-08545-9.

37. Milvich, J.; Kohler, D.; Freude, W.; Koos, C. Integrated phase-sensitive photonic sensors: A system design tutorial. *Adv. Opt. Photonics* **2021**, *13*, 584. [CrossRef]
38. Liow, T.Y.; Ang, K.W.; Fang, Q.; Song, J.F.; Xiong, Y.Z.; Yu, M.B.; Lo, G.-Q.; Kwong, D.L. Silicon modulators and germanium photodetectors on SOI: Monolithic integration, compatibility, and performance optimization. *IEEE J. Sel. Top. Quantum Electron.* **2010**, *16*, 307–315. [CrossRef]
39. Zhang, W.; Yao, J. Silicon photonic integrated optoelectronic oscillator for frequency-tunable microwave generation. *J. Light. Technol.* **2018**, *36*, 4655–4663. [CrossRef]
40. Venghaus, H.; Grote, N. (Eds.) *Fibre Optic Communication: Key Devices*, 2nd ed.; Springer Series in Optical Sciences; Springer: Berlin/Heidelberg, Germany; New York, NY, USA, 2017; ISBN 978-3-319-42365-4.
41. Soref, R. Tutorial: Integrated-photonic switching structures. *APL Photonics* **2018**, *3*, 021101. [CrossRef]
42. Hosseini, N.; Dekker, R.; Hoekman, M.; Dekkers, M.; Bos, J.; Leinse, A.; Heideman, R. Stress-optic modulator in TriPleX platform using a piezoelectric lead zirconate titanate (PZT) thin film. *Opt. Express* **2015**, *23*, 14018. [CrossRef]
43. Augustin, L.M.; Santos, R.; den Haan, E.; Kleijn, S.; Thijs, P.J.A.; Latkowski, S.; Zhao, D.; Yao, W.; Bolk, J.; Ambrosius, H.; et al. InP-based generic foundry platform for photonic integrated circuits. *IEEE J. Sel. Top. Quantum Electron.* **2018**, *24*, 6100210. [CrossRef]
44. Bowers, J.E.; Liu, A.Y. A comparison of four approaches to photonic integration. In Proceedings of the Optical Fiber Communication Conference, OSA. Los Angeles, CA, USA, 19–23 March 2017; p. 16930096.
45. Jiao, Y.; Nishiyama, N.; van der Tol, J.; van Engelen, J.; Pogoretskiy, V.; Reniers, S.; Kashi, A.A.; Wang, Y.; Calzadilla, V.D.; Spiegelberg, M.; et al. InP membrane integrated photonics research. *Semicond. Sci. Technol.* **2020**, *36*, 013001. [CrossRef]
46. Theurer, M.; Moehrle, M.; Sigmund, A.; Velthaus, K.-O.; Oldenbeuving, R.M.; Wevers, L.; Postma, F.M.; Mateman, R.; Schreuder, F.; Geskus, D.; et al. Flip-chip integration of InP to SiN photonic integrated circuits. *J. Light. Technol.* **2020**, *38*, 2630–2636. [CrossRef]
47. Hu, Y.; Liang, D.; Beausoleil, R.G. An advanced III-V-on-silicon photonic integration platform. *Opto-Electron. Adv.* **2021**, *4*, 200094. [CrossRef]
48. Yan, Z.; Han, Y.; Lin, L.; Xue, Y.; Ma, C.; Ng, W.K.; Wong, K.S.; Lau, K.M. A monolithic InP/SOI platform for integrated photonics. *Light Sci. Appl.* **2021**, *10*, 200. [CrossRef]
49. Besancon, C.; Fanneau, P.; Neel, D.; Cerulo, G.; Vaissiere, N.; Make, D.; Pommereau, F.; Fournel, F.; Dupre, C.; Baron, T.; et al. Laser array covering 155 Nm wide spectral band achieved by selective area growth on silicon wafer. In Proceedings of the 2020 European Conference on Optical Communications (ECOC), IEEE. Brussels, Belgium, 6–10 December 2020; pp. 1–4.
50. Xing, Y.; Dong, J.; Khan, U.; Bogaerts, W. Capturing the effects of spatial process variations in silicon photonic circuits. *ACS Photonics* **2022**. [CrossRef]
51. Siew, S.Y.; Li, B.; Gao, F.; Zheng, H.Y.; Zhang, W.; Guo, P.; Xie, S.W.; Song, A.; Dong, B.; Luo, L.W.; et al. Review of silicon photonics technology and platform development. *J. Light. Technol.* **2021**, *39*, 4374–4389. [CrossRef]
52. Bogaerts, W.; Xing, Y.; Khan, U. Layout-aware variability analysis, yield prediction, and optimization in photonic integrated circuits. *IEEE J. Sel. Top. Quantum Electron.* **2019**, *25*, 6100413. [CrossRef]
53. Buryak, A.V.; Stepanov, D.Y. Correction of systematic errors in the fabrication of fiber bragg gratings. *Opt. Lett.* **2002**, *27*, 1099. [CrossRef]
54. Miller, G.A.; Flockhart, G.M.H.; Cranch, G.A. Technique for correcting systematic phase errors during fibre bragg grating inscription. *Electron. Lett.* **2008**, *44*, 1399. [CrossRef]

sensors

MDPI

*Article*

# Wavelength-Tunable Vortex Beam Emitter Based on Silicon Micro-Ring with PN Depletion Diode

Ivan V. Stepanov [1], Denis M. Fatkhiev [1,*], Vladimir S. Lyubopytov [2,1], Ruslan V. Kutluyarov [1], Elizaveta P. Grakhova [1], Niels Neumann [3], Svetlana N. Khonina [4,5] and Albert K. Sultanov [1]

1 School of Photonics Engineering and Research Advances (SPhERA), Ufa State Aviation Technical University, 450008 Ufa, Russia; stepanov.iv@ugatu.su (I.V.S.); kutluyarov.rv@ugatu.su (R.V.K.); grakhova.ep@ugatu.su (E.P.G.); sultanov.ah@ugatu.su (A.K.S.)
2 Center for Photonic Science and Engineering, Skolkovo Institute of Science and Technology, 121205 Moscow, Russia; v.lyubopytov@skoltech.ru
3 Chair of Radio Frequency and Photonics Engineering, TU Dresden, 01062 Dresden, Germany; niels.neumann@tu-dresden.de
4 Department of Technical Cybernetics, Samara National Research University, 443086 Samara, Russia; khonina@ipsiras.ru
5 Image Processing Systems Institute Branch of the Federal Scientific Research Center "Crystallography and Photonics" of Russian Academy of Sciences, 443001 Samara, Russia
* Correspondence: fatkhiev.dm@ugatu.su

**Abstract:** Herein we propose a design of a wavelength-tunable integrated vortex beam emitter based on the silicon-on-insulator platform. The emitter is implemented using a PN-depletion diode inside a microring resonator with the emitting hole grating that was used to produce a vortex beam. The resonance wavelengths can be shifted due to the refractive index change associated with the free plasma dispersion effect. Obtained numerical modeling results confirm the efficiency of the proposed approach, providing a resonance wavelength shift while maintaining the required topological charge of the emitted vortex beam. It is known that optical vortices got a lot of attention due to extensive telecommunication and biochemical applications, but also, they have revealed some beneficial use cases in sensors. Flexibility in spectral tuning demonstrated by the proposed device can significantly improve the accuracy of sensors based on fiber Bragg gratings. Moreover, we demonstrate that the proposed device can provide a displacement of the resonance by the value of the free spectral range of the ring resonator, which means the possibility to implement an ultra-fast orbital angular momentum (de)multiplexing or modulation.

**Keywords:** orbital angular momentum; vortex beam; microring resonator; pn-depletion diode



**Citation:** Stepanov, I.V.; Fatkhiev, D.M.; Lyubopytov, V.S.; Kutluyarov, R.V.; Grakhova, E.P.; Neumann, N.; Khonina, S.N.; Sultanov, A.K. Wavelength-Tunable Vortex Beam Emitter Based on Silicon Micro-Ring with PN Depletion Diode. *Sensors* **2022**, *22*, 929. https://doi.org/10.3390/s22030929

Academic Editor: Francesco De Leonardis

Received: 11 December 2021
Accepted: 24 January 2022
Published: 25 January 2022



## 1. Introduction

Since the unique properties of optical beams carrying orbital angular momentum (OAM), also referred to as optical vortices, have been discovered in [1], the request for further research and development in this field has been growing steadily. This was not unreasonable, as applications of the vortex beams turned out to be interesting in a wide variety of areas. One of their most known use cases is trapping and moving particles with the optical tweezers and spanners [2,3]. Optical tweezers empowered with OAM have demonstrated the manipulation of particles with multiple degrees of freedom, as well as the simultaneous trapping of multiple particles [4,5]. Optical beams carrying OAM, e.g., Bessel beams [6], along with other kinds of structured light [7] have also found their application in such a remarkable topic as quantum communications [8], specifically in higher-dimensional quantum key distribution [9], entanglement swapping [10], and multidimensional entanglement [11].

Another major field for vortex beams is optical communications where OAM is usually considered as an additional degree of freedom for multiplexing. The exponentially growing demand for network traffic [12,13] resulted in the fiber-optic lines utilizing time, wavelength, and polarization division multiplexing, which have almost reached the Shannon limit [13,14]. Therefore, the next step on the way to increase the throughput of fiber transmission lines was the space division multiplexing (SDM) [15–17]. SDM technology is based on the use of a degree of freedom determined by the transverse distribution of the electromagnetic (EM) field, that corresponds to multiplexing of spatially separated optical fields in multicore fibers (MCF) or using several linear polarized (LP) modes in few-mode fibers (FMF) [18]. SDM concept can be applied in both fiber-optic [17–20] and atmospheric [21–23] optical communication lines.

A common property of optical modes carrying OAM is the presence of a multiplier $e^{i\ell\varphi}$, where $\ell$ is the azimuthal mode index. These modes are the eigenfunctions of the angular momentum operator and carry the OAM proportional to $\ell$ [1]. As the OAM modes represent a basis of orthogonal functions which can be divided spatially by its order $\ell$, it is convenient to use them in the SDM approach [19,24]. Moreover, OAM multiplexing can be combined with other multiplexing technologies and multilevel modulation formats for increasing throughput to the Tbit level [20,25].

Finally yet importantly, beams carrying OAM are widely used in sensing. In biochemistry, optical vortices were used to detect the molecules of amino acids, nucleotides, and sugars [26]. Diffraction limit [27] and super resolution [28] imaging was reached with focusing of vortex beams. Another example is a temperature sensor consisting of a fiber Bragg grating (FBG), an optical fiber path used to eliminate errors, and a Gaussian beam, interfering with the OAM beam transmitted through the Bragg grating [29]. The principle of operation of this temperature sensor lies in a combination of the thermo-optical effect and the effect of thermal expansion, which appear in the Bragg grating when the temperature changes and leads to a shift in the central wavelength of the reflected spectrum. In turn, the phase difference between the Gaussian beam and the vortex beam leads to the rotation of their interference diagram. The temperature measurement step corresponds to the rotation of the radiation pattern. A similar method can be used to make highly accurate measurements of microstrains caused by pressure and displacement. It is also important to note the recent successes in plasmonic vortex studies: nanometrology approaches [30], generation of the high order plasmonic vortices [31], and OAM-SPR (surface plasmon resonance) based refractive index sensing [32] are making a breakthrough.

In this paper, we propose and numerically verify a novel scheme of real-time OAM order switch for radiated optical vortex beam using a pn-depletion diode integrated into the ring waveguide. The most common solution that can be used to excite optical beams with a helical phase front, and which we employed in the proposed design, is a micro-ring resonator [33]. Usually, this is a ring-shaped waveguide with grating elements for the light beam emission, and a bus waveguide located at a small gap, which couples an input beam into the resonator. Such μm-scale structure was first demonstrated in [34]. Such devices are capable to emit vector optical vortices with definite and quantized OAM states. It is possible due to the ring (or disk) resonators supporting whispering gallery modes (WGM), which carry high orders of OAM. The grating provides a periodic modulation of an effective refractive index, and its working principle is analogous to the operation principle of grating couplers in straight waveguides. The light wave is scattered by the grating elements, and as a result, a part of the radiated power is deflected in the direction of the constructive interference. Because the waveguide has a ring shape and supports WGM, according to the Huygens principle the wavefront of the emitted light should point to the azimuthal direction $\varphi$ and be helical.

To our knowledge, there has been one system demonstrated based on a single ring resonator that realizes real-time OAM switching. In [35], authors developed an approach to a fast electrically-controlled vortex order switch and demonstrated a scheme with the switching time of down to 20 μs. In this scheme, heaters are used to change the refractive index of the waveguide, and as a result, the effective WGM index changes, which causes the restructuring of the emitted OAM mode. In our case, an inversely-biased pn-junction offers more energy-efficient and faster switching [36] compared to the existing schemes. Therefore, we believe, that this scheme can be used also as a fast (GHz-scale) electro-optical OAM modulator.

The paper is organized as follows: in Section 2 we describe the working principle of the proposed scheme; in Section 3 the device modeling results are presented; and in Section 4 the methodology and the results of analysis of the emitted beams propagation are described.

## 2. Principle of Operation

A schematic view of the proposed device is depicted in Figure 1. The device consists of a straight input waveguide and a ring resonator with a light-emitting grating (etched holes on top of the ring waveguide) and a pn-depletion diode, imprinted over the part of the ring. The diode cross-section in detail is shown in Figure 2 and its dimensional quantities listed in the Table 1.

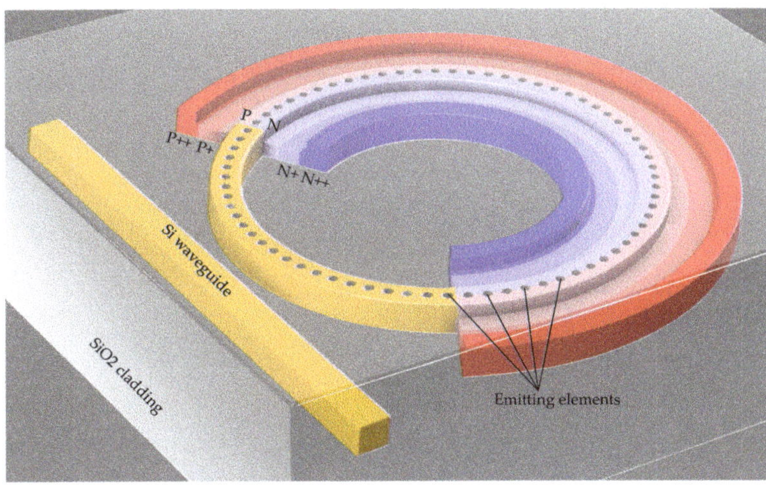


**Figure 1.** 3D view of the proposed device. The pn-depletion diode follows the Si rib bent waveguide. The different shades of blue and red represent the different electron (blue) and hole (red) concentrations where darker shades correspond to higher concentrations.

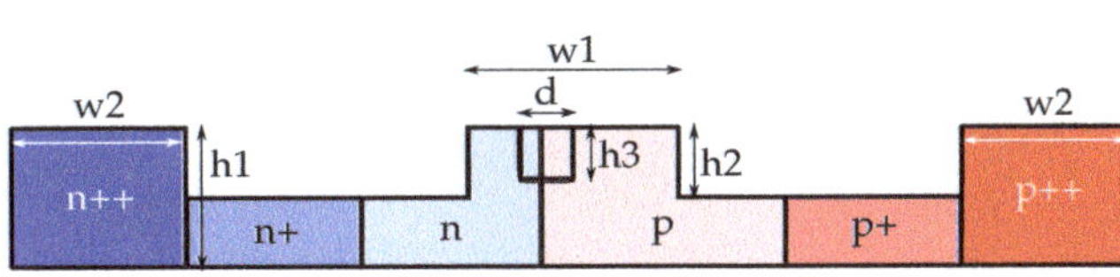


**Figure 2.** Cross-section of the doped part of the ring. N and P are for the electrons and holes concentrations, respectively: ++ is $10^{20}$ cm$^{-3}$, + is $10^{19}$ cm$^{-3}$, and without sign is $5 \times 10^{18}$ cm$^{-3}$. d and h3 denote the hole diameter and height, respectively.

The main principle of vortex beam generation is similar to the presented in [37], where the order of the radiated OAM carrying beam $\ell$ satisfies the following condition:

$$\ell = p - gq, \tag{1}$$

where $p$ is the WGM order in the ring, $q$ is the number of grating elements in the ring resonator, and $g$ is the diffraction order, which is an integer and can be calculated as [37]:

$$\left(n_{eff} - 1\right)\frac{2\pi R}{q\lambda} < g < \left(n_{eff} + 1\right)\frac{2\pi R}{q\lambda}, \tag{2}$$

where $R$ is the ring resonator radius, $n_{eff}$ is the effective index of the ring waveguide, and $\lambda$ is the operating wavelength.

To switch the emitted vortex order and the emitter resonance wavelength, it is necessary to change the effective index of the ring resonator. There are two main methods to modify the effective index: to use thermo-optical effect and electro-optical effect (mainly plasma dispersion effect). In the context of optical switches and modulators, the first effect is usually used in cases when GHz response frequencies are not required [36]. The main disadvantages of the thermo-optical effect for modulation and switching are milli- to microsecond response and mW order power consumption [38]. In contrast, the electro-optical effect is characterized by nanosecond to sub-nanosecond response and in most cases much lower, or at least comparable to the thermo-optical case, power consumption [36]. On the other hand, realizations of electro-optical effects impose higher losses and crosstalk [36], however, the advantage of quick response usually overrides these problems.

In our device, we propose to use an inversely-biased pn-depletion diode, integrated into the ring resonator, providing to change the refractive index of the ring waveguide due to the plasma dispersion effect [39].

To model the device, we used modified coefficients for the Soref and Bennet model, where a change in absorption coefficient and refractive index for wavelength 1.55 μm (C-band) can be expressed as [40]:

$$\Delta\alpha = \Delta\alpha_e + \Delta\alpha_h = 8.88 \times 10^{-21} \Delta N_e^{1.167} + 5.84 \times 10^{-20} \Delta N_h^{1.109}, \quad (3)$$

$$-\Delta n = \Delta n_e + \Delta n_h = 5.4 \times 10^{-22} N_e^{1.011} + 1.53 \times 10^{-18} \Delta N_h^{0.838}, \quad (4)$$

where $\Delta N_e$ and $\Delta N_h$ are the changes in number of electrons and holes in the active region, respectively.

The resonant wavelength $\lambda_{res}$ of the ring can be calculated as [41]:

$$\lambda_{res} = \frac{n_{eff} L}{m}, m = 1, 2, 3, \ldots, \quad (5)$$

where $L$ is the circumference of the ring resonator.

**Table 1.** List of geometric values in Figure 2.

| Dimension | w1 | w2 | h1 | h2 | h3 | d |
|---|---|---|---|---|---|---|
| Value | 0.545 | 0.5 | 0.22 | 0.11 | 0.07 | 0.15 |

The proposed device can be fabricated using a standard silicon-on-insulator (SOI) platform, or other platforms supporting doping, where the waveguides with pn-junction can be implemented. For our simulations, we considered a generic SOI platform with the Si layer thickness of 220 nm. The designed waveguide structures in this platform can be typically fabricated using 193 nm deep ultraviolet photolithography, or electron-beam lithography (EBL).

## 3. Simulation Results

For the numerical modeling and simulation of the device, we used the Ansys Lumerical software. The first step was calculating the carrier number in the cross-section of the ring waveguide. This distribution was simulated in Lumerical Device and further exported into a mat-file for application in the following modeling steps. The obtained carrier number distributions are presented in Figure 3.

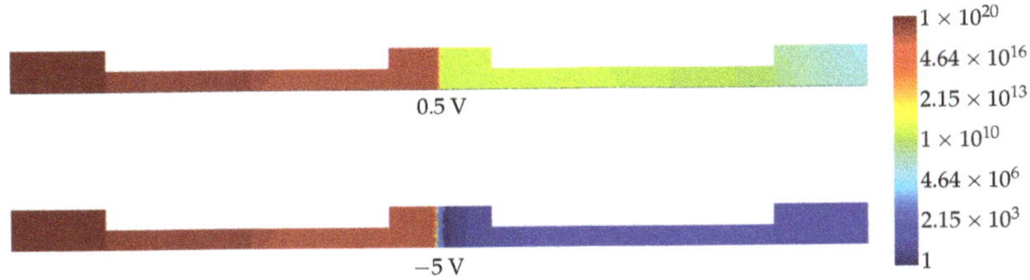


**Figure 3.** Distributions of electrons in the cross-section of the rib waveguide for different applied voltage values.

In the next step, we used the obtained carrier distributions to account for the effective index change due to the free plasma dispersion effect in Lumerical FDTD. To implement this, Lumerical's silicon material model with the modified coefficients from Equation (4) have been applied. Next, we calculated the resonance characteristics of the microring emitter when voltages of 0.5 V and −5 V are applied between the inner and outer parts of the ring waveguide. Finally, we investigated the emitted field distributions (after spreading 4 µm from the device) at the resonance wavelengths of interest (near 1550 nm).

As our simulations show, for the ring resonators with radii of less than 25 µm there is no possibility to obtain the resonance shift equal or greater than the device free spectral range (FSR). The main reason is that the FSR value of the small rings is biggish, and the length of the doped region is too short to obtain the corresponding phase shift. This limitation does not allow us to realize the OAM order modulation for such small rings, but as shown in Figure 4 we can utilize pn-diode to adjust the resonance characteristics of the OAM emitter.

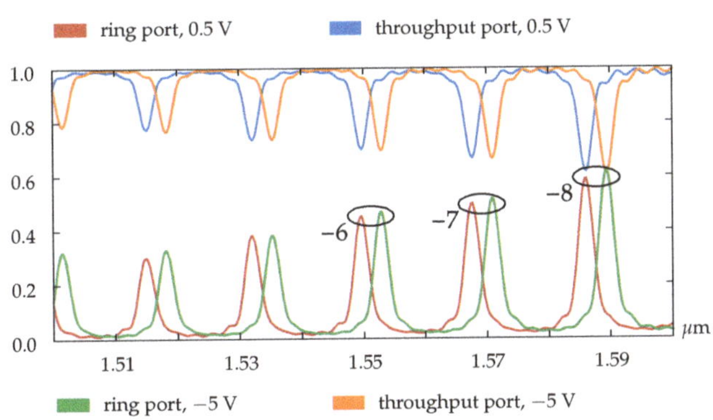


**Figure 4.** Normalized transmission spectra for the ring with radius 5.5 µm.

Figure 5 shows that the vortex order does not change within one FSR. This effect confirms that we have only functionality of the resonance adjustment for small rings.

In the case of larger rings (especially for the rings with radii of greater than 25 µm), the FSR value becomes small enough to realize the change in the OAM state. As can be seen from Figure 6, the resonances shift over the one FSR value. Through this effect, the OAM order modulation of the optical signal can be obtained, as can be seen in Figure 7.

In addition, note that the resonant curves have smaller peaks due to the lower coupling in the case of the larger ring. The weakening of coupling can be explained by the increased complexity and hence the heterogeneity of the emitter, containing about 300 grating elements, which becomes difficult for optimizing due to doping. Nevertheless, we have shown that the proposed scheme generally is capable to implement the OAM order switching.

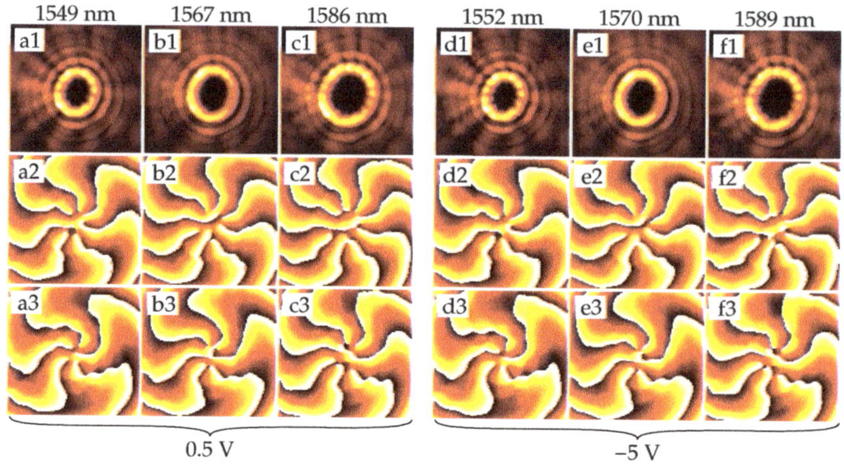


**Figure 5.** Intensity and phase distributions of the emitted fields from the ring with radius 5.5 µm. (**a1–a3**) is an intensity, $arg(E_x)$ and $arg(E_y)$ at a given resonant wavelength, respectively. Phase distributions are obtained after passing the field through the quarter-wave plate, so its $x$-component's azimuthal order is above by one, and its $y$-component's azimuthal order is below by one than the actual order of the generated vector vortex beam. Distribution patterns (**b1–b3**)–(**f1–f3**) are obtained similarly. (**a1–a3**), (**b1–b3**), and (**c1–c3**) refer to the resonances at the voltage of 0.5 V, and (**d1–d3**), (**e1–e3**), and (**f1–f3**) to the resonances at −5 V.

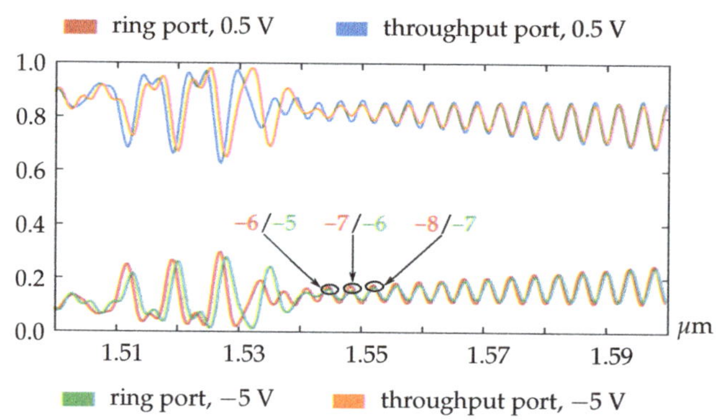


**Figure 6.** Normalized transmission spectra for the ring with radius 26.5 µm.

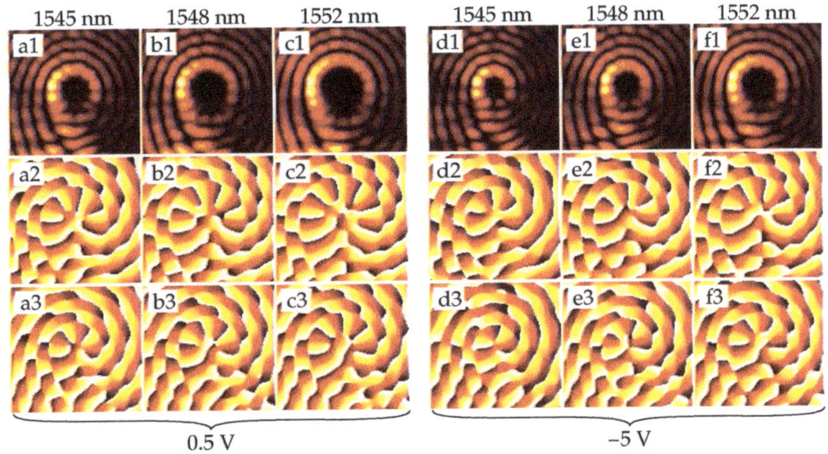


**Figure 7.** Intensity and phase distributions of the emitted fields from the ring with radius 26.5 μm. **(a1–a3)** is an intensity, $arg(E_x)$ and $arg(E_y)$ at a given resonant wavelength, respectively. Phase distributions are obtained after passing the field through the quarter-wave plate, so its $x$-component's azimuthal order is above by one, and its $y$-component's azimuthal order is below by one than the actual order of the generated vector vortex beam. Distribution patterns **(b1–b3)**–**(f1–f3)** are obtained similarly. **(a1–a3)**, **(b1–b3)**, and **(c1–c3)** refer to the resonances at the voltage of 0.5 V, and **(d1–d3)**, **(e1–e3)**, and **(f1–f3)** to the resonances at −5 V.

## 4. Analysis of the Emitted Field Propagation

To ensure that the resulting beams retain their vortex structure as they propagate through free space (for example, before injecting them into the fiber), we performed calculations using some obtained field distributions from the emitter. The near-field distributions for the vortices of 3rd (field 1) and 7th (field 2) orders are shown in Figure 8.

As it can be seen, field 1 is a radially polarized field with the 3rd order vortex phase component. Field 2 is a hybrid-polarized (superposition of radial and azimuthal polarization) field with a 7th order vortex phase component. There is comparably high intensity in the $E$-field component $|E_z|^2$ due to the diffraction in the near field [42–44].

Primarily, we calculated the electromagnetic field in the lens focal plane (focal length $f = 1.3$ mm, numerical aperture of the lens $NA = 0.01$) by using the transverse electric field components of the incident beam and the vector propagation operator [45–47]:

$$\begin{bmatrix} \mathbf{E}(r,\varphi,z) \\ \mathbf{H}(r,\varphi,z) \end{bmatrix} = -\frac{if}{\lambda} \int_0^{\theta_{\max}} \int_0^{2\pi} \begin{bmatrix} \mathbf{V}_E(\theta,\phi) \\ \mathbf{V}_H(\theta,\phi) \end{bmatrix} \exp\{ik[r\sin\theta\cos(\phi-\varphi) + z\cos\theta]\}\sqrt{\cos\theta}\sin\theta d\theta d\phi \quad (6)$$

where $\sin(\theta_{\max})$ corresponds to the lens numerical aperture, polarization vectors are defined on the transverse electric field components of the incident beam $E_{0x}(\theta,\phi)$ and $E_{0y}(\theta,\phi)$ applying the following equations:

$$\mathbf{V}_E(\theta,\phi) = \begin{bmatrix} A(\theta,\phi) & C(\theta,\phi) \\ C(\theta,\phi) & B(\theta,\phi) \\ -D(\theta,\phi) & -E(\theta,\phi) \end{bmatrix} \begin{pmatrix} E_{0x}(\theta,\phi) \\ E_{0y}(\theta,\phi) \end{pmatrix}, \quad (7)$$

$$\mathbf{V}_H(\theta,\phi) = \begin{bmatrix} C(\theta,\phi) & -A(\theta,\phi) \\ B(\theta,\phi) & -C(\theta,\phi) \\ -E(\theta,\phi) & D(\theta,\phi) \end{bmatrix} \begin{pmatrix} E_{0x}(\theta,\phi) \\ E_{0y}(\theta,\phi) \end{pmatrix}, \quad (8)$$

$$\begin{cases} A(\theta,\phi) = 1 - \cos^2\phi(1-\cos\theta), \\ B(\theta,\phi) = 1 - \sin^2\phi(1-\cos\theta), \\ C(\theta,\phi) = -\sin\phi\cos\phi(1-\cos\theta), \\ D(\theta,\phi) = \cos\phi\sin\theta, \\ E(\theta,\phi) = \sin\phi\sin\theta. \end{cases} \quad (9)$$

The calculation results, corresponding to the field in the far zone, are shown in Figure 9. It can be seen that the beam structure has changed, becoming close to the radially polarized Bessel beam of the third order, but the phase and polarization states of the beam are preserved. Similarly, a hybrid-polarized seventh-order Bessel beam was formed in the far-field.

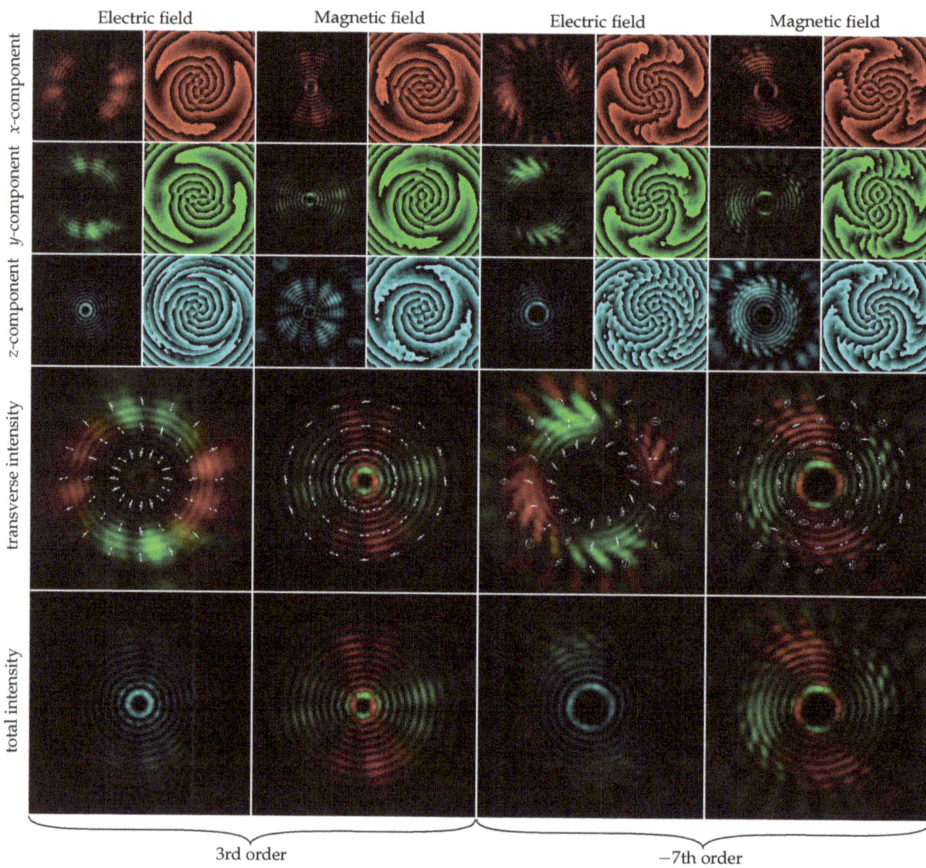


**Figure 8.** Distribution patterns of the emitted field in the near field. Size of the field plots is 26 µm × 26 µm.

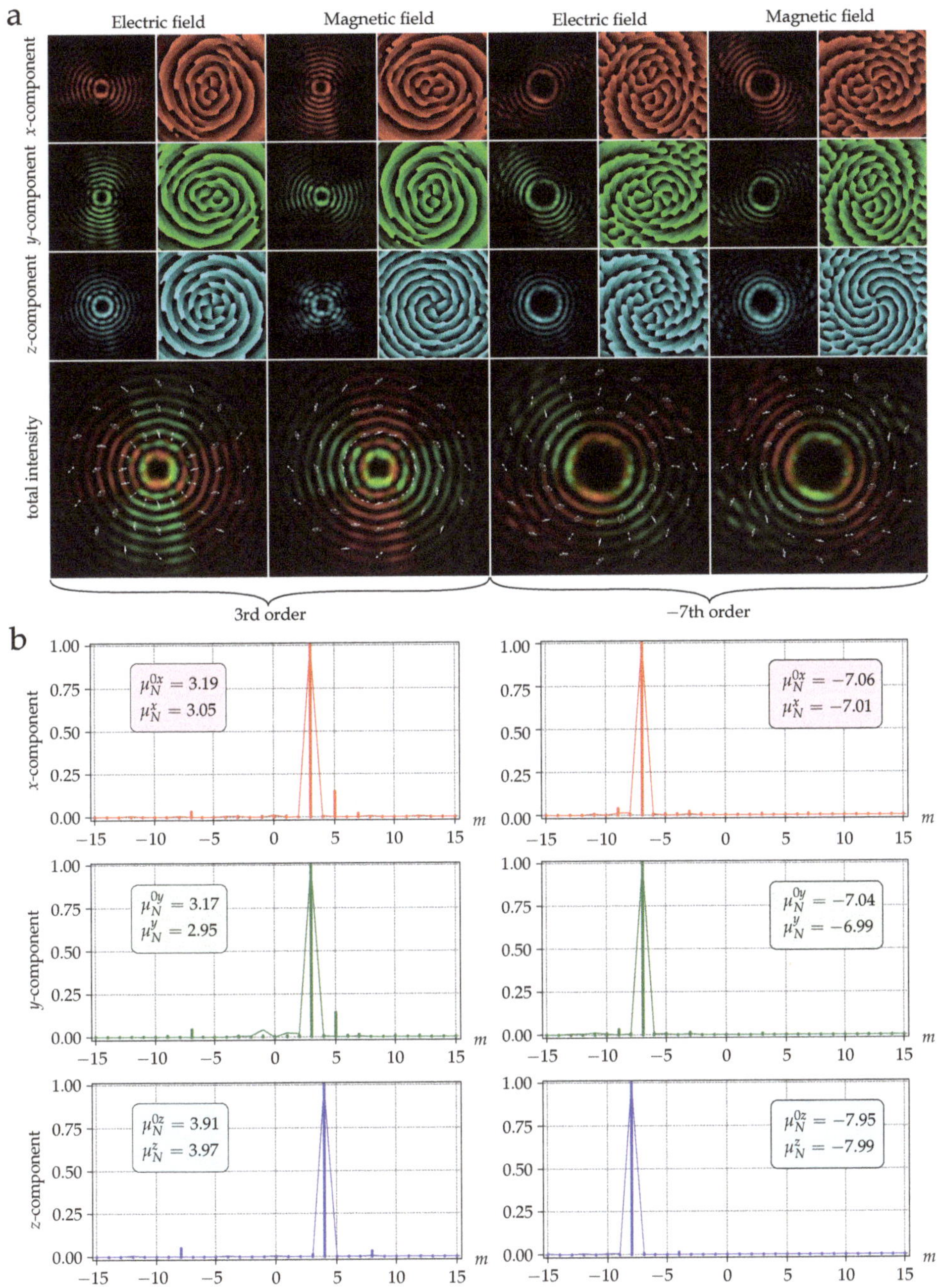


**Figure 9.** (**a**) Distribution patterns of the emitted fields in far-field. The size of the field plots is 2 mm × 2 mm. (**b**) Normalized intensities of the expansion coefficients in the basis of angular harmonics (OAM spectra) of the components of the electric field 1 (left column) and field 2 (right column): for the incident field $|c_m^{0t}|^2$ (vertical lines) and for far-field $|c_m^t|^2$ (envelopes).

Figure 9 also shows the results of the decomposition of each component of the electric fields 1 and 2, based on angular harmonics for the incident field [48]:

$$c_m^{0t} = \frac{1}{2\pi} \int_0^{\theta_{\max}} \int_0^{2\pi} E_{0t}(\theta, \phi) \exp(-im\phi) d\theta d\phi, \tag{10}$$

and in far field:

$$c_m^t = \frac{1}{2\pi} \int_0^{R} \int_0^{2\pi} E_t\left(r, \varphi, z_f\right) \exp(-im\varphi) r dr d\varphi, \tag{11}$$

where $t = \{x, y, z\}$ and $m$ is integer.

Using the coefficients from Equations (10) and (11), the OAM value for each field component can be calculated by the following formula [49]:

$$\mu_N^t = \left(\sum_{m=-N}^{N} m\left|c_m^t\right|^2\right)\left(\sum_{m=-N}^{N} \left|c_m^t\right|^2\right)^{-1} \tag{12}$$

where $2N + 1$ is the number of calculated decomposition coefficients (we used $N = 15$).

As follows from the results shown in Figure 9, the value of OAM (12) is practically preserved in the far-field for all the field components of the vector vortex beams. In this case, the OAM value for the longitudinal component differs by one from the OAM values of the transverse components, which is in full agreement with the theory [44–46].

## 5. Discussion

In this paper, we proposed the novel design of the microring-based vortex beam emitter. First, the proposed scheme provides a possibility to realize the adjustment of the spectral properties of optical vortex emitter and, correspondingly, reduce the dependence of the emitter resonances on the fabrication errors. Also, for the structures with larger radii (starting from approximately 25 μm), it becomes possible to realize the OAM switching in the wavelength domain, which can be used for electro-optical OAM modulation of the signal.

As mentioned above, the proposed scheme is expected to provide much faster resonance adjustment or OAM order switching with much lower power consumption compared to the schemes based on thermo-optical effect following from the operating characteristics of the inversely biased pn-junction. Moreover, the proposed integrated scheme is much smaller than the conventional discrete optics for generating optical vortices in free space. Generally, our device can be useful for future data transmission systems with spatial division (de)multiplexing, for OAM encoding, or in sensing systems.

It is worth noting, that tunable vortex mode emitter can be especially useful in applications where the beam's OAM order and fine-tuning of its spectral characteristics are of decisive importance. For example, FBG temperature sensors, which are based on the interaction of OAM radiation with the transmission medium, are influenced by the measured external parameters. Theoretically, our proposed device can be used for spin [50] and lateral motion [51] detecting schemes. These methods are based on the light-matter interaction, which couples OAM and mechanical momentum. Also, it would be interesting to use such an emitter in an OAM-controlled hybrid plasmonic circuit for optical logic operations [52] and in other photonic circuits for information processing.

For further development of the proposed device, the metal mirror placed in the buried oxide [53] can be applied to suppress the cylindrical vector Bessel modes of higher orders, which are formed sideways, and therefore improve the emission efficiency. There is also a possibility to further optimize the coupling between the bus waveguide and the ring resonator to increase the depth of the resonances and raise the emitted vortex beam power.

## 6. Conclusions

In summary, we showed that by integrating a pn-diode in the ring waveguide of the OAM emitter, the carrier number can be controlled enabling a refractive index modulation, that allows rapidly changing the transmission spectrum of the emitter. This effect allows adjusting the emitter resonance to the desired wavelength while maintaining the required OAM in the case of rings with radii smaller than 25 μm. For the larger ring radii, it is possible to implement the OAM (de)multiplexing or modulation.

Modeling results have shown that our device in case of the small ring radius (5.5 μm) emits an optical vortex with the topological charge $\ell = -6$ at the wavelength of 1549 nm and the voltage level of 0.5 V, and with the same azimuthal order at 1552 nm and $-5$ V. This provides a resonance wavelength shift while maintaining the required topological charge of the emitted vortex beam. In the case of the larger ring (with the radius of 26.5 μm) the voltage change provides the change in the topological charge of the emitted vortex beam ($-7$ and $-6$ at the wavelength of 1548 nm for the voltage levels of 0.5 V and $-5$ V, respectively). In a detailed study of the available beam distributions, we also ensured that the radiated beams maintain their topological charges while propagating in free space.

We believe that the presented results will be useful both for the further development of OAM-powered devices and for moving towards full-fledged photonic integration.

**Author Contributions:** Conceptualization, V.S.L., S.N.K. and A.K.S.; methodology, V.S.L., D.M.F. and I.V.S.; software, I.V.S. and D.M.F.; validation, V.S.L. and R.V.K.; formal analysis, S.N.K., N.N. and A.K.S.; investigation, V.S.L., D.M.F., and I.V.S.; resources, R.V.K. and A.K.S.; data curation, V.S.L., I.V.S. and D.M.F.; writing—original draft preparation, I.V.S. and D.M.F.; writing—review and editing, V.S.L., R.V.K. and D.M.F.; visualization, I.V.S., D.M.F. and S.N.K.; supervision, E.P.G. and A.K.S.; project administration, E.P.G., R.V.K. and A.K.S.; funding acquisition, E.P.G., A.K.S. and N.N. All authors have read and agreed to the published version of the manuscript.

**Funding:** This work was partially funded under the grant of the Russian Science Foundation Project № 19-49-04112 (sections on the device concept and operation principle, simulations of the device parameters), and partially funded by the Ministry of Science and Higher Education of the Russian Federation within the state assignment for the Ufa State Aviation Technical University (agreement № 075-03-2021-014 dated 29.09.2021 [FEUE-2021-0013]) and conducted in the research laboratory "Sensor systems based on integrated photonics devices" of the Eurasian Scientific and Educational Center (section on the analysis of the emitted field propagation).

**Institutional Review Board Statement:** Not applicable.

**Informed Consent Statement:** Not applicable.

**Data Availability Statement:** Not applicable.

**Acknowledgments:** Authors thank the Institute of Computer Studies of the Ufa State Aviation Technical University for the technical support of the performed numerical computations on the HPC cluster [https://www.ugatu.su/supercomputer/] (accessed on 1 September 2021).

**Conflicts of Interest:** The authors declare no conflict of interest.

## References

1. Allen, L.; Beijersbergen, M.W.; Spreeuw, R.J.C.; Woerdman, J.P. Orbital angular momentum of light and the transformation of Laguerre-Gaussian laser modes. *Phys. Rev. A* **1992**, *45*, 8185. [CrossRef] [PubMed]
2. He, H.; Friese, M.E.J.; Heckenberg, N.R.; Rubinsztein-Dunlop, H. Direct Observation of Transfer of Angular Momentum to Absorptive Particles from a Laser Beam with a Phase Singularity. *Phys. Rev. Lett.* **1995**, *75*, 826–829. [CrossRef]
3. Grier, D.G. A revolution in optical manipulation. *Nature* **2003**, *424*, 810–816. [CrossRef] [PubMed]
4. Chapin, S.C.; Germain, V.; Dufresne, E.R. Automated trapping, assembly, and sorting with holographic optical tweezers. *Opt. Express* **2006**, *14*, 13095–13100. [CrossRef]
5. Padgett, M.; Bowman, R. Tweezers with a twist. *Nat. Photonics* **2011**, *5*, 343–348. [CrossRef]
6. Vicente, O.C.; Caloz, C. Bessel beams: A unified and extended perspective. *Optica* **2021**, *8*, 451–457. [CrossRef]
7. Forbes, A.; Nape, I. Quantum mechanics with patterns of light: Progress in high dimensional and multidimensional entanglement with structured light. *AVS Quantum Sci.* **2019**, *1*, 011701. [CrossRef]
8. Ndagano, B.; Nape, I.; Cox, M.A.; Rosales-Guzman, C.; Forbes, A. Creation and Detection of Vector Vortex Modes for Classical and Quantum Communication. *J. Light. Technol.* **2018**, *36*, 292–301. [CrossRef]

9. Mafu, M.; Dudley, A.; Goyal, S.; Giovannini, D.; McLaren, M.; Padgett, M.J.; Konrad, T.; Petruccione, F.; Lütkenhaus, N.; Forbes, A. Higher-dimensional orbital-angular-momentum-based quantum key distribution with mutually unbiased bases. *Phys. Rev. A* **2013**, *88*, 032305. [CrossRef]
10. Zhang, Y.; Agnew, M.; Roger, T.; Roux, F.S.; Konrad, T.; Faccio, D.; Leach, J.; Forbes, A. Simultaneous entanglement swapping of multiple orbital angular momentum states of light. *Nat. Commun.* **2017**, *8*, 632. [CrossRef] [PubMed]
11. Liu, J.; Nape, I.; Wang, Q.; Vallés, A.; Wang, J.; Forbes, A. Multidimensional entanglement transport through single-mode fiber. *Sci. Adv.* **2020**, *6*, eaay0837. [CrossRef] [PubMed]
12. Richardson, D.J.; Fini, J.M.; Nelson, L.E. Space-division multiplexing in optical fibres. *Nat. Photonics* **2013**, *7*, 354–362. [CrossRef]
13. Ellis, A.D.; Suibhne, N.M.; Saad, D.; Payne, D.N. Communication networks beyond the capacity crunch. *Philos. Trans. R. Soc. A Math. Phys. Eng. Sci.* **2016**, *374*, 20150191. [CrossRef]
14. Turitsyn, K.S.; Turitsyn, S.K. Nonlinear communication channels with capacity above the linear Shannon limit. *Opt. Lett.* **2012**, *37*, 3600–3602. [CrossRef]
15. Sakaguchi, J.; Awaji, Y.; Wada, N.; Kanno, A.; Kawanishi, T.; Hayashi, T.; Taru, T.; Kobayashi, T.; Watanabe, M. Space Division Multiplexed Transmission of 109-Tb/s Data Signals Using Homogeneous Seven-Core Fiber. *J. Light. Technol.* **2012**, *30*, 658–665. [CrossRef]
16. Nakajima, K.; Sillard, P.; Richardson, D.; Li, M.J.; Essiambre, R.J.; Matsuo, S. Transmission media for an SDM-based optical communication system. *IEEE Commun. Mag.* **2015**, *53*, 44–51. [CrossRef]
17. Mizuno, T.; Takara, H.; Shibahara, K.; Sano, A.; Miyamoto, Y. Dense space division multiplexed transmission over multicore and multimode fiber for long-haul transport systems. *J. Light. Technol.* **2016**, *34*, 1484–1493. [CrossRef]
18. Wang, J.; Padgett, M.J.; Ramachandran, S.; Lavery, M.P.; Huang, H.; Yue, Y.; Yan, Y.; Bozinovic, N.; Golowich, S.E.; Willner, A.E. Multimode Communications Using Orbital Angular Momentum. In *Optical Fiber Telecommunications*; Elsevier: Amsterdam, The Netherlands, 2013. [CrossRef]
19. Berdagué, S.; Facq, P. Mode division multiplexing in optical fibers. *Appl. Opt.* **1982**, *21*, 1950–1955. [CrossRef] [PubMed]
20. Bozinovic, N.; Yue, Y.; Ren, Y.; Tur, M.; Kristensen, P.; Huang, H.; Willner, A.E.; Ramachandran, S. Terabit-scale orbital angular momentum mode division multiplexing in fibers. *Science* **2013**, *340*, 1545–1548. [CrossRef]
21. Qu, Z.; Djordjevic, I.B. 500 Gb/s free-space optical transmission over strong atmospheric turbulence channels. *Opt. Lett.* **2016**, *41*, 3285. [CrossRef]
22. Willner, A.E.; Zhao, Z.; Liu, C.; Zhang, R.; Song, H.; Pang, K.; Manukyan, K.; Song, H.; Su, X.; Xie, G.; et al. Perspectives on advances in high-capacity, free-space communications using multiplexing of orbital-angular-momentum beams. *APL Photonics* **2021**, *6*, 030901. [CrossRef]
23. Zhao, Z.; Zhang, R.; Song, H.; Pang, K.; Almaiman, A.; Zhou, H.; Song, H.; Liu, C.; Hu, N.; Su, X.; et al. Modal coupling and crosstalk due to turbulence and divergence on free space THz links using multiple orbital angular momentum beams. *Sci. Rep.* **2021**, *11*, 2110. [CrossRef] [PubMed]
24. Khonina, S.; Kazanskiy, N.; Soifer, V. Optical Vortices in a Fiber: Mode Division Multiplexing and Multimode Self-Imaging. In *Recent Progress in Optical Fiber Research*; InTech: London, UK, 2012. [CrossRef]
25. Wang, J.; Yang, J.Y.; Fazal, I.M.; Ahmed, N.; Yan, Y.; Huang, H.; Ren, Y.; Yue, Y.; Dolinar, S.; Tur, M.; et al. Terabit free-space data transmission employing orbital angular momentum multiplexing. *Nat. Photonics* **2012**, *6*, 488–496. [CrossRef]
26. Zhao, Y.; Askarpour, A.N.; Sun, L.; Shi, J.; Li, X.; Alù, A. Chirality detection of enantiomers using twisted optical metamaterials. *Nat. Commun.* **2017**, *8*, 1–8. [CrossRef]
27. Tamburini, F.; Anzolin, G.; Umbriaco, G.; Bianchini, A.; Barbieri, C. Overcoming the Rayleigh criterion limit with optical vortices. *Phys. Rev. Lett.* **2006**, *97*, 163903. [CrossRef] [PubMed]
28. Kozawa, Y.; Matsunaga, D.; Sato, S. Superresolution imaging via superoscillation focusing of a radially polarized beam. *Optica* **2018**, *5*, 86–92. [CrossRef]
29. Fu, H.; Wang, S.; Chang, H.; You, Y. A high resolution and large range fiber Bragg grating temperature sensor with vortex beams. *Opt. Fiber Technol.* **2020**, *60*, 102369. [CrossRef]
30. Yue, Z.; Ren, H.; Wei, S.; Lin, J.; Gu, M. Angular-momentum nanometrology in an ultrathin plasmonic topological insulator film. *Nat. Commun.* **2018**, *9*, 1–7. [CrossRef]
31. Fang, J.; Zhou, C.; Mou, Z.; Wang, S.; Yu, J.; Yang, Y.; Gbur, G.J.; Teng, S.; Cai, Y. High order plasmonic vortex generation based on spiral nanoslits. *New J. Phys.* **2021**, *23*, 033013. [CrossRef]
32. Liu, E.; Yan, B.; Zhou, H.; Liu, Y.; Liu, G.; Liu, J. OAM mode-excited surface plasmon resonance for refractive index sensing based on a photonic quasi-crystal fiber. *J. Opt. Soc. Am. B* **2021**, *38*, F16–F22. [CrossRef]
33. Vahala, K.J. Optical microcavities. *Nature* **2003**, *424*, 839–846. [CrossRef]
34. Cai, X.; Wang, J.; Strain, M.J.; Johnson-Morris, B.; Zhu, J.; Sorel, M.; O'Brien, J.L.; Thompson, M.G.; Yu, S. Integrated Compact Optical Vortex Beam Emitters. *Science* **2012**, *338*, 363–366. [CrossRef]
35. Strain, M.J.; Cai, X.; Wang, J.; Zhu, J.; Phillips, D.B.; Chen, L.; Lopez-Garcia, M.; O'Brien, J.L.; Thompson, M.G.; Sorel, M.; et al. Fast electrical switching of orbital angular momentum modes using ultra-compact integrated vortex emitters. *Nat. Commun.* **2014**, *5*, 4856. [CrossRef]
36. Kim, Y.; Han, J.H.; Ahn, D.; Kim, S. Heterogeneously-Integrated Optical Phase Shifters for Next-Generation Modulators and Switches on a Silicon Photonics Platform: A Review. *Micromachines* **2021**, *12*, 625. [CrossRef] [PubMed]

37. Li, R.; Feng, X.; Zhang, D.; Cui, K.; Liu, F.; Huang, Y. Radially Polarized Orbital Angular Momentum Beam Emitter Based on Shallow-Ridge Silicon Microring Cavity. *IEEE Photonics J.* **2014**, *6*, 1–10. [CrossRef]
38. Soref, R. Tutorial: Integrated-photonic switching structures. *APL Photonics* **2018**, *3*, 021101. [CrossRef]
39. Soref, R.; Bennett, B. Electrooptical effects in silicon. *IEEE J. Quantum Electron.* **1987**, *23*, 123–129. [CrossRef]
40. Nedeljkovic, M.; Soref, R.; Mashanovich, G.Z. Free-Carrier Electrorefraction and Electroabsorption Modulation Predictions for Silicon Over the 1–14-μm Infrared Wavelength Range. *IEEE Photonics J.* **2011**, *3*, 1171–1180. [CrossRef]
41. Bogaerts, W.; De Heyn, P.; Van Vaerenbergh, T.; De Vos, K.; Kumar Selvaraja, S.; Claes, T.; Dumon, P.; Bienstman, P.; Van Thourhout, D.; Baets, R. Silicon microring resonators. *Laser Photonics Rev.* **2012**, *6*, 47–73. [CrossRef]
42. Hao, B.; Leger, J. Experimental measurement of longitudinal component in the vicinity of focused radially polarized beam. *Opt. Express* **2007**, *15*, 3550. [CrossRef]
43. Lerman, G.M.; Levy, U. Effect of radial polarization and apodization on spot size under tight focusing conditions. *Opt. Express* **2008**, *16*, 4567. [CrossRef] [PubMed]
44. Khonina, S.N.; Degtyarev, S.A. Analysis of the formation of a longitudinally polarized optical needle by a lens and axicon under tightly focused conditions. *J. Opt. Technol.* **2016**, *83*, 197. [CrossRef]
45. Helseth, L. Optical vortices in focal regions. *Opt. Commun.* **2004**, *229*, 85–91. [CrossRef]
46. Rashid, M.; Maragò, O.M.; Jones, P.H. Focusing of high order cylindrical vector beams. *J. Opt. A Pure Appl. Opt.* **2009**, *11*, 065204. [CrossRef]
47. Khonina, S.N. Vortex beams with high-order cylindrical polarization: Features of focal distributions. *Appl. Phys. B* **2019**, *125*, 100. [CrossRef]
48. Almazov, A.A.; Khonina, S.N.; Kotlyar, V.V. Using phase diffraction optical elements to shape and select laser beams consisting of a superposition of an arbitrary number of angular harmonics. *J. Opt. Technol.* **2005**, *72*, 391. [CrossRef]
49. Khonina, S.N.; Podlipnov, V.V.; Karpeev, S.V.; Ustinov, A.V.; Volotovsky, S.G.; Ganchevskaya, S.V. Spectral control of the orbital angular momentum of a laser beam based on 3D properties of spiral phase plates fabricated for an infrared wavelength. *Opt. Express* **2020**, *28*, 18407. [CrossRef] [PubMed]
50. Lavery, M.P.J.; Speirits, F.C.; Barnett, S.M.; Padgett, M.J. Detection of a Spinning Object Using Light's Orbital Angular Momentum. *Science* **2013**, *341*, 537–540. [CrossRef]
51. Cvijetic, N.; Milione, G.; Ip, E.; Wang, T. Detecting Lateral Motion using Light's Orbital Angular Momentum. *Sci. Rep.* **2015**, *5*, 15422. [CrossRef] [PubMed]
52. Ren, H.; Wang, X.; Li, C.; He, C.; Wang, Y.; Pan, A.; Maier, S.A. Orbital-Angular-Momentum-Controlled Hybrid Nanowire Circuit. *Nano Lett.* **2021**, *21*, 6220–6227. [CrossRef]
53. Li, S.; Ding, Y.; Guan, X.; Tan, H.; Nong, Z.; Wang, L.; Liu, L.; Zhou, L.; Yang, C.; Yvind, K.; et al. Compact high-efficiency vortex beam emitter based on a silicon photonics micro-ring. *Opt. Lett.* **2018**, *43*, 1319. [CrossRef] [PubMed]

sensors

MDPI

*Article*

# Twisted Few-Mode Optical Fiber with Improved Height of Quasi-Step Refractive Index Profile

**Anton V. Bourdine [1,2,3,4], Vladimir V. Demidov [2], Artem A. Kuznetsov [5,*], Alexander A. Vasilets [5,6], Egishe V. Ter-Nersesyants [2], Alexander V. Khokhlov [2], Alexandra S. Matrosova [2,7], Grigori A. Pchelkin [2,8], Michael V. Dashkov [1], Elena S. Zaitseva [1], Azat R. Gizatulin [9], Ivan K. Meshkov [9], Airat Zh. Sakhabutdinov [5], Eugeniy V. Dmitriev [10], Oleg G. Morozov [5], Vladimir A. Burdin [1], Konstantin V. Dukelskii [2,4,7], Yaseera Ismail [11], Francesco Petruccione [11,12], Ghanshyam Singh [13], Manish Tiwari [14] and Juan Yin [15]**

1 Department of Communication Lines, Povolzhskiy State University of Telecommunications and Informatics, 23, Lev Tolstoy Street, Samara 443010, Russia; bourdine@yandex.ru (A.V.B.); mvd.srttc@gmail.com (M.V.D.); zaytzewa@inbox.ru (E.S.Z.); burdin@psati.ru (V.A.B.)

2 JSC "Scientific Production Association State Optical Institute Named after Vavilov S.I.", 36/1, Babushkin Street, St. Petersburg 192171, Russia; demidov@goi.ru (V.V.D.); ter@goi.ru (E.V.T.-N.); khokhlov@goi.ru (A.V.K.); a.pasishnik@gmail.com (A.S.M.); beegrig@mail.ru (G.A.P.); kdukel@mail.ru (K.V.D.)

3 "OptoFiber Lab" LLC, Skolkovo Innovation Center, 7, Nobel Street, Moscow 143026, Russia

4 Department of Photonics and Communication Links, Saint Petersburg State University of Telecommunications Named after M.A. Bonch-Bruevich, 22, Bolshevikov Avenue, St. Petersburg 193232, Russia

5 Department of Radiophotonics and Microwave Technologies, Kazan National Research Technical University Named after A.N. Tupolev-KAI, 10, Karl Marx Street, Kazan 420111, Russia; a.vasilets@mail.ru (A.A.V.); azhsakhabutdinov@kai.ru (A.Z.S.); microoil@mail.ru (O.G.M.)

6 Faculty of Management and Engineering Business, Kazan Innovative University Named after V.G. Timiryasov (IEML), 42, Moskovskaya Street, Kazan 420111, Russia

7 Faculty of Photonics and Optical Information, School of Photonics, ITMO University, Bldg. A, 49, Kronverksky Alley, St. Petersburg 197101, Russia

8 Institute of Physics, Nanotechnology and Telecommunications, Peter the Great St. Petersburg Polytechnic University, Bldg. II, 29, Politekhnicheskaya Street, St. Petersburg 194064, Russia

9 Department of Telecommunication Systems, Ufa State Aviation Technical University, 12, Karl Marx Street, Ufa 450000, Russia; azat_poincare@mail.ru (A.R.G.); mik.ivan@bk.ru (I.K.M.)

10 Department of Quantum Communication Systems Research, Radio Research and Development Institute, 16, Kazakova Street, Moscow 105064, Russia; dmitriev@niir.ru

11 Quantum Research Group, School of Chemistry and Physics, University of KwaZulu-Natal, Durban 4001, South Africa; ismaily@ukzn.ac.za (Y.I.); petruccione@ukzn.ac.za (F.P.)

12 National Institute for Theoretical Physics, University of KwaZulu-Natal, Durban 4001, South Africa

13 Department of Electronics and Communication Engineering, Malaviya National Institute of Technology (MNIT) Jaipur, J.L.N Road, Jaipur 302017, Rajasthan, India; gsingh.ece@mnit.ac.in

14 Department of Electronics and Communication Engineering, School of Electrical and Electronics & Commu-nication Engineering, Manipal University Jaipur, Jaipur-Ajmer Expressway, Jaipur 303007, Rajasthan, India; manish.tiwari@jaipur.manipal.edu

15 Division of Quantum Physics and Quantum Information, University of Science and Technology of China, 99, Xiupu Road, Pudong District, Shanghai 200093, China; yinjuan@ustc.edu.cn

* Correspondence: artem.a.kuznetsov@bk.ru; Tel.: +7-919-6425689

check for updates

**Citation:** Bourdine, A.V.; Demidov, V.V.; Kuznetsov, A.A.; Vasilets, A.A.; Ter-Nersesyants, E.V.; Khokhlov, A.V.; Matrosova, A.S.; Pchelkin, G.A.; Dashkov, M.V.; Zaitseva, E.S.; et al. Twisted Few-Mode Optical Fiber with Improved Height of Quasi-Step Refractive Index Profile. *Sensors* **2022**, 22, 3124. https://doi.org/10.3390/s22093124

Academic Editor: Francesco Fienga

Received: 14 February 2022
Accepted: 13 April 2022
Published: 19 April 2022



**Abstract:** This work presents designed and fabricated silica few-mode optical fiber (FMF) with induced twisting 10 and 66 revolutions per meter, core diameter 11 μm, typical "telecommunication" cladding diameter 125 μm, improved height of quasi-step refractive index profile and numerical aperture 0.22. Proposed FMF supports 4 guided modes over "C"-band. We discussed selection of specified optical fiber parameters to provide desired limited mode number over mentioned wavelength range. Some results of tests, performed with pilot samples of manufactured FMF, are represented, including experimentally measured spectral responses of laser-excited optical signals, that comprise researches and analysis of few-mode effects, occurring after fiber Bragg grating writing.

**Keywords:** twisted optical fiber; laser beam profile; differential mode delay; laser-based few-mode optical signal transmission; fiber Bragg grating; few-mode effects

## 1. Introduction

Twisted optical fibers have been known since the early 1980s: here, the concept of fiber spinning was firstly originally introduced in the work [1]. The fabrication technique of twisted optical fibers is based on rotation of preform during the fiber drawing [1] or directly spinning of the drawn optical fiber [2]. Twisted single mode optical fibers are usually considered as fibers with reduced polarization mode dispersion (PMD) [1–4], while induced chirality (including twisting) over multimode optical fibers is declared as the method for differential mode delay (DMD), decreasing with total bandwidth improvement [5,6].

Nowadays, twisted optical fibers (both with typical coaxial geometry (core, bounded by intermediate and/or outer solid cladding) and microstructured/photonic crystal optical fibers) are considered as alternative unique fiber optic elements with great potentiality for various applications in fiber optic sensors [7–10].

At the same time, many recently published works demonstrated new effects, occurring in fiber Bragg gratings (FBGs), written in FMFs as well as in multimode optical fibers (MMFs), under laser-source excitation. A lot of many recently published works demonstrated few-mode operation of these conventional, tilted or slanted FBGs on MMFs and FMFs in vibration, temperature, deformation, displacement, bending, etc., fiber optic sensors [11–20].

A few-mode regime adds a new other dimension to the space of parameters: it is associated with guided modes of particular order, in which a limited number (from two to a few dozen) transfer the most part of optical signal power over tested optical fiber. We suppose that twisted FMF with recorded FBG can be considered as a new complicated fiber optic element with unique features and great potentiality for application in fiber optic sensors.

This work is focused on design and fabrication of twisted few-mode optical fiber (FMF) with specified limited (3…4) guided modes, supporting over "C"-band. Therefore, at the first stage, by using commercially available software with rigorous numerical finite element method, technological/geometrical parameters were specified to provide the desired few-mode regime operation of designed FMF.

We performed adaptation of the conventional technique for drawing optical fibers to fabricate designed FMF with induced twisting under small as well as large number of revolutions per meter. Some results of theoretical and experimental researches, performed for pilot samples of manufactured FMF, are represented, including experimentally measured spectral responses of laser-excited optical signals, that comprise researches and analysis of few-mode effects, occurring after fiber Bragg grating recording.

## 2. Design of FMF: Selection of Geometrical Parameters to Provide Desired Limited Number of Guided Modes

At the first stage, we utilized rigorous numerical finite element method, used in COMSOL Multiphysics software, to analyze a preliminary designed set of specified step-index optical fibers with the same typical "telecommunication" cladding diameter 125 μm, but differing by combination of core diameter and numerical aperture (e.g., height of refractive index profile). Here, for each designed optical fiber sample, guided modes (which satisfy to the cut-off condition) were defined, and their effective refractive indexes were computed. The main criterion was focused on providing limited (3…4) transversal guided modes' propagation over FMF under laser source excitation at the wavelength $\lambda = 1550$ nm.

We performed analysis of FMF by an earlier developed and successfully verified method [21–23], based on the numerical solution of linear Maxwell equation system, written for a homogeneous isotropic dielectric in the absence of free charges and currents and reduced to wave equations for the vectors of electric ($E$) and magnetic (B) fields [24]:

$$\nabla \times \left( \frac{1}{\mu} \nabla \times E \right) - k_0^2 \varepsilon E = 0, \quad (1)$$

where $k_0$ is wave number ($k_0 = 2\pi/\lambda$); $\varepsilon$ is the dielectric permeability ($\varepsilon = n^2$, $n$ is refractive index); $\mu$ is magnetic permeability.

By taking into account satisfaction to the perfectly matched layer (PML) conditions, Equation (1) was transformed to the following form [25]:

$$\nabla \times \left( \frac{1}{\mu'} \nabla \times \frac{1}{[S]} E \right) - k_0^2 \varepsilon' \frac{1}{[S]} E = 0, \quad (2)$$

where $\varepsilon'$ and $\mu'$ are modified dielectric and magnetic permeability; $[S]$ is matrix of PML layer coefficients.

Solution of Equation (2) is equation of electromagnetic wave, propagating along $z$-axis of optical fiber [24]:

$$E(z,\ t) = E_0 \exp\left[ j\left( \omega t - \frac{\omega}{c} n_{\text{eff}} z \right) \right], \quad (3)$$

where $E_0$ is amplitude of electric field strength; $\omega$ is circular frequency; $c$ is light speed in vacuum; $t$ is time.

Effective refractive index $n_{\text{eff}}$ is defined by numerical solution of Equation (3) and related transversal mode is identified (in terms of linear polarized modes $LP_{lm}$, e.g., azimuthal and radial orders $l$ and $m$) by comparison computed radial mode field distribution with predefined field patterns of known order modes $LP_{lm}$, which are exact solutions for scalar wave equation, written for model optical fiber with ideal step index or unbounded parabolic refractive index profiles and described by Bessel or Laguerre-Gaussian functions [26,27].

Table 1 shows results of optical fiber analysis, performed by a rigorous finite element numerical method in COMSOL Multiphysics software under wavelength λ = 1550 nm. We considered ideal step-index refractive index profile, the same cladding diameter 125 μm (that corresponds to conventional telecommunication optical fibers), three various core diameters (8.3, 10.0 and 11.0 μm) and six values of numerical aperture $NA$ (0.14, 0.16, 0.18, 0.20, 0.22 and 0.24—it corresponds to approximately the difference between core and cladding refractive indexes 0.02). We start from the core diameter 8.3 μm as the typical value for standard single mode optical fibers (SMFs) of ITU-T Rec. G.652 [28]. It was supposed that even the weak improvement of refractive index profile height, in comparison with ratified SMF, may provide desired few-mode regime with 3…4 transversal guided mode propagation at λ = 1550 nm. However, results of computation showed that following increasing both core diameter and numerical aperture (e.g., refractive index profile height) is required).

**Table 1.** Results of optical fiber analysis, performed by rigorous numerical method: step-index optical fibers under various combinations of core diameter and numerical aperture (λ = 1550 nm).

| № | Core Diameter, μm | Cladding Diameter, μm | Numerical Aperture $NA$ | Mode Composition | $n_{\text{eff}}$ | $\Delta n_{\text{eff}}$ |
|---|---|---|---|---|---|---|
| 1 | 8.3 | 125 | 0.14 | $LP_{01}$ | 1.460478 | – |
| 2 | 8.3 | 125 | 0.16 | $LP_{01}$<br>$LP_{11}$ | 1.462210<br>1.457688 | 0.004522 |
| 3 | 8.3 | 125 | 0.18 | $LP_{01}$<br>$LP_{11}$ | 1.464263<br>1.459082 | 0.005181 |
| 4 | 8.3 | 125 | 0.20 | $LP_{01}$<br>$LP_{11}$ | 1.466624<br>1.460940 | 0.005684 |

**Table 1.** *Cont.*

| № | Core Diameter, μm | Cladding Diameter, μm | Numerical Aperture *NA* | Mode Composition | $n_{eff}$ | $\Delta n_{eff}$ |
|---|---|---|---|---|---|---|
| 5 | 8.3 | 125 | 0.22 | $LP_{01}$<br>$LP_{11}$ | 1.469284<br>1.463199 | 0.006085 |
| 6 | 8.3 | 125 | 0.24 | $LP_{01}$<br>$LP_{11}$<br>$LP_{21}$<br>$LP_{02}$ | 1.472237<br>1.465821<br>1.458136<br>1.457159 | 0.006416<br>0.014101<br>0.015078 |
| 7 | 10 | 125 | 0.14 | $LP_{01}$<br>$LP_{11}$ | 1.461181<br>1.457847 | 0.003334 |
| 8 | 10 | 125 | 0.16 | $LP_{01}$<br>$LP_{11}$ | 1.463027<br>1.459219 | 0.003808 |
| 9 | 10 | 125 | 0.18 | $LP_{01}$<br>$LP_{11}$ | 1.465177<br>1.461012 | 0.004165 |
| 10 | 10 | 125 | 0.20 | $LP_{01}$<br>$LP_{11}$<br>$LP_{21}$<br>$LP_{02}$ | 1.467622<br>1.463176<br>1.457854<br>1.457139 | 0.004446<br>0.009768<br>0.010483 |
| 11 | 10 | 125 | 0.22 | $LP_{01}$<br>$LP_{11}$<br>$LP_{21}$<br>$LP_{02}$ | 1.470355<br>1.465682<br>1.459875<br>1.458498 | 0.004673<br>0.010480<br>0.011857 |
| 12 | 10 | 125 | 0.24 | $LP_{01}$<br>$LP_{11}$<br>$LP_{21}$<br>$LP_{02}$ | 1.473371<br>1.468510<br>1.462343<br>1.460621 | 0.004861<br>0.011028<br>0.012750 |
| 13 | 11 | 125 | 0.14 | $LP_{01}$<br>$LP_{11}$ | 1.461499<br>1.458455 | 0.003044 |
| 14 | 11 | 125 | 0.16 | $LP_{01}$<br>$LP_{11}$ | 1.463387<br>1.459994 | 0.003393 |
| 15 | 11 | 125 | 0.18 | $LP_{01}$<br>$LP_{11}$<br>$LP_{21}$<br>$LP_{02}$ | 1.465572<br>1.461914<br>1.457563<br>1.457061 | 0.003658<br>0.008009<br>0.008511 |
| 16 | 11 | 125 | 0.20 | $LP_{01}$<br>$LP_{11}$<br>$LP_{21}$<br>$LP_{02}$ | 1.468048<br>1.464179<br>1.459378<br>1.458241 | 0.003869<br>0.008670<br>0.009807 |
| 17 | 11 | 125 | 0.22 | $LP_{01}$<br>$LP_{11}$<br>$LP_{21}$<br>$LP_{02}$ | 1.470808<br>1.466767<br>1.461642<br>1.460198 | 0.004041<br>0.009166<br>0.010610 |
| 18 | 11 | 125 | 0.24 | $LP_{01}$<br>$LP_{11}$<br>$LP_{21}$<br>$LP_{02}$<br>$LP_{31}$ | 1.473847<br>1.469664<br>1.464288<br>1.462650<br>1.458078 | 0.004183<br>0.009559<br>0.011197<br>0.015769 |

For example, combination of the SMF "nominal" core diameter 8.3 μm and maximal (from the researched range) numerical aperture value $NA$ = 0.24 provides technical satisfaction of the cut-off condition for desired 4 modes—the fundamental $LP_{01}$ and higher order modes $LP_{02}$, $LP_{11}$, $LP_{21}$. However, the last two modes $LP_{02}$ and $LP_{21}$ are unacceptably instable to propagation over long distances due to their field concentration in the cladding:

here, optical confinement factor $P_{co}$ (e.g., mode power, transferred over core) for both aforementioned modes is inadmissibly low ($P_{co} < 0.5$). Therefore, we conclude that none of the researched combinations of core diameter 8.3 μm and 6 tested numerical aperture values $NA = 0.14 \ldots 0.24$ do not provide desired 4-mode operation at wavelength $\lambda = 1550$ nm. The same matter corresponds to core diameter 10.0 μm and $NA = 0.20$: here also, higher-order modes $LP_{02}$ and $LP_{21}$ satisfy cut-off condition under unacceptable low optical confinement factor $P_{co} < 0.5$, while desired 4-mode operation is achieved for numerical aperture range $NA = 0.22 \ldots 0.24$. Following improvement of core diameter up to 11 μm showed the best results for $NA = 0.20$ and $NA = 0.22$: all 4 modes satisfy the cut-off condition under the required optical confinement factor $P_{co} > 0.5$. Lower $NA = 0.18$ led to inappropriate low $P_{co} < 0.5$ for the same last two higher-order modes, while increased $NA = 0.24$ provides satisfaction of the cut-off condition for 5th mode $LP_{31}$.

Therefore, according to computation results, we selected the following configuration for fabricated FMF: core diameter 11 μm, typical "telecommunication" cladding diameter 125 μm, numerical aperture $NA = 0.22$.

## 3. Pilot FMF 11/125 with Improved Height of Quasi-Step Refractive Index Profile and Induced Twisting

According to the aforementioned technological parameters, a preform of the desired FMF 11/125 with the numerical aperture $NA = 0.22$ was prepared by conventional modified chemical vapor deposition (MCVD) method [29]. Figure 1 presents measured refractive index profile with improved height of FMF fabricated preform.

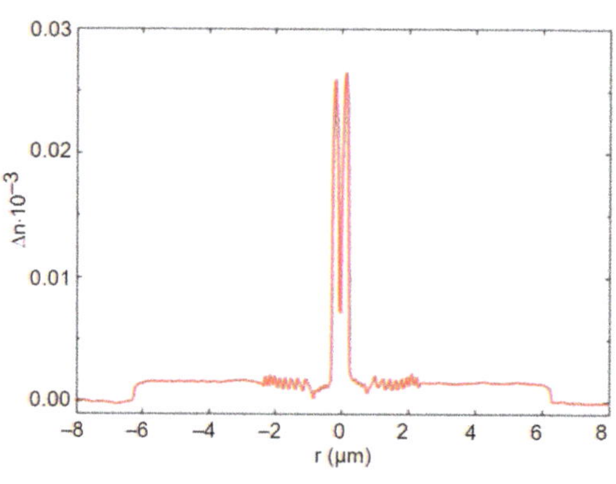


**Figure 1.** Refractive index profile of pilot FMF preform (measured by refractometer Photon Kinetics P101).

The general form of fabricated preform refractive index profile is quasi-step. Moreover, there is a dip of refractive index in the core center, which is typical for MCVD technique: it is caused by highly volatile $GeO_2$ dopant diffusion during support tube collapse. Here, the absolute height of the profile reaches ~0.27, while dip is ~0.08. As a result, to correctly evaluate refractive index profile height, we computed the area of the central (core) part and further estimated the effective height of the profile. For researched prepared FMF 11/125 preform, this parameter was ~0.018, that is equivalent to numerical aperture $NA = 0.22$.

We performed some modifications of the drawing tower to induce twisting on FMF during its drawing. The detailed description of modification is represented in the earlier published work [30]. Usually, preform is fixed in a mechanical chuck of the feed unit, which inputs preform to the heat space of a high temperature furnace. Preform is kept in a stationary position and redrawn without rotation. To induce desired twisting over manufactured FMF, we integrated the stepper motor to the feed unit, which continuously rotates preform under the set speed and adds a new rotation function to the drawing system. The minimal motor rotation speed is 20 revolutions per minute, while the maximal is 200.

Therefore, under slow drawing speed 2. . . 3 m/min (that is usually used for manufacturing special or experimental optical fibers), it induces twisting with 10 and 66 revolutions per meter (rpm), respectively.

Figure 2 shows an image of the end-face of fabricated pilot sample of twisted FMF 11/125 with numerical aperture $NA = 0.22$, drawn from the aforementioned manufactured preform. Figure 3 presents near field laser beam profile (operating wavelength $\lambda = 1550$ nm) after propagation over the fabricated FMF 11/125 by CCD camera.

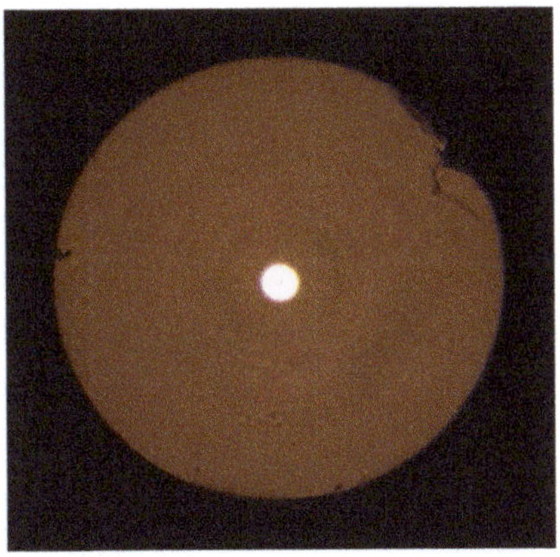

**Figure 2.** Image of the end-face of fabricated pilot 4-mode FMF 11/125 with numerical aperture $NA = 0.22$ (high-resolution optical microscope Nikon Eclipse N-U).

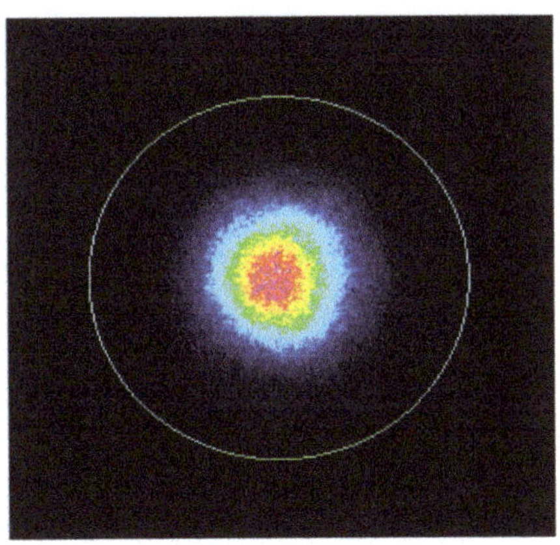

**Figure 3.** Near field laser beam profile (operating wavelength $\lambda = 1550$ nm), measured after propagation over pilot sample of FMF 11/125 by CCD camera DataRay WinCamD-LCM-C-TE.

We measured both 10 rpm and 66 rpm pilot sample 50 m length FMF 11/125 attenuation $\alpha(\lambda)$ by cutback method over wavelength band $\lambda = 900$–1700 nm by using a halogen lamp (OSRAM 64642 HLX) as a light source, programmable monochromator (ANDO), germanium photodiode (wavelength range 900. . . 1700 nm), optical amplifier (eLockIn) and optical power meter (and ANDO AQ-1135E).

Measured attenuation curves $\alpha(\lambda)$ contain typical resonance "water" peaks with strong loss due to simplified and rapid technique for FMF preform fabrication without hydroxyl ($OH^-$) dopants extraction: here, we just focused on twisted FMF pilot sample length manufacturing with specified geometry parameters, which should provide desired few-mode operation by low widening core diameter and strong improvement of refractive profile height and did not pay attention to attenuation reduction. In the same way, increased attenuation (in comparison with commercially available silica optical fibers [28]) at the central regions of the "C"- and "O"-bands, which reaches almost $\alpha = 7\ldots8$ dB/km, was expected due to intentional excluding (to reduce reagent consumption and also to simplify preform fabrication process) of typical operation of Fluorine (F) doping to the core region, which helps to decrease $GeO_2$ dopant unwanted influence on attenuation increasing.

Figure 4 demonstrates that attenuation curve $\alpha(\lambda)$ over "flat" regions between the resonance peaks for FMF with twisting 66 rpm being lower in comparison with 10 rpm twisted FMF. It may be explained by more smoothing of refractive index profile typical MCVD technological defects under more rapid twisting of preform during optical fiber drawing.

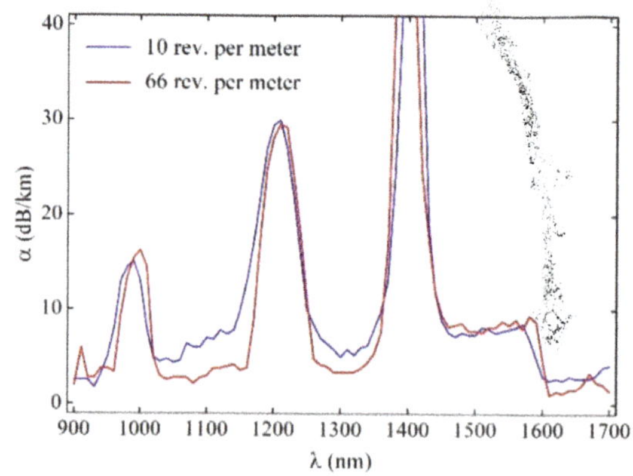


**Figure 4.** Attenuation of manufactured 50 m length FMF 11/125 samples with induced twisting 10 and 66 rpm.

## 4. Dispersion Parameters of Guided Modes, Propagating in the Pilot Sample FMF 11/125 with Improved Height of Quasi-Step Refractive Index Profile

During the next stage, we computed spectral characteristics of dispersion parameters of guided modes, satisfying the cut-off condition for fabricated pilot sample of FMF 11/125. For this purpose, it was proposed to utilize an earlier on developed simple and fast approximate method, which is a modification of the Gaussian approximation, extended to the case for estimation of the transmission parameters of arbitrary order modes, propagating in a weakly guiding optical fiber with an arbitrary axially symmetric refractive index profile [31], with following, optionally (in appropriate case), accuracy improvement by rigorous numerical method of mixed finite elements [32]. This extended modification of the Gaussian approximation (EMGA) is based on combination of the stratification method [26] and "classical" Gaussian approximation [27]. Stratification method provides ability to represent complicated form of researched optical fiber refractive index profile with high detailing and corresponding technological defects (including local refractive index fluctuations), in spite of the most approximate methods, which typically utilize one or a set of smooth functions. Proposed approach significantly reduces computational error during direct calculation of transmission parameters of guided modes in optical fiber with large core diameter (in comparison with single mode optical fibers) and complicated form of refractive index profile [31,32]. Here, only one variational parameter—normalized mode field radius $R_0$—should be determined as a result of characteristic equation solution, while $R_0$ within the "classical" Gaussian approximation is the basis and it completely defines all desired guided mode transmission parameters. According to Gaussian approximation, radial mode field distribution is represented by a well-known approximating expression, based on Laguerre-Gaussian functions [27], that corresponds to exact solution of scalar wave equation, written for weakly guiding optical fiber with an ideal inbounded parabolic refractive index profile. This permits to derive and write analytical expressions for variational expression and characteristic equation in the form of finite nested sums, and further, their first and second derivations—mode delay and chromatic dispersion parameter. Therefore, developed approximate method EMGA does not require high computational resources (even during higher-order mode dispersion parameter estimation) and provides low (less than 1% [26,27]) computational error.

Figure 5 shows an equivalent quasi-step refractive index profile of the analyzed FMF 11/125 with a numerical aperture $NA = 0.22$, restored by report of measurements, performed for drawn optical fiber.

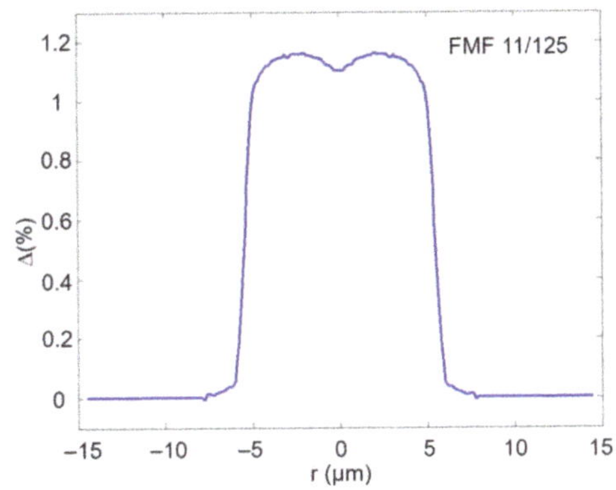


**Figure 5.** Equivalent quasi-step refractive index profile with improved height, restored by measurement report data.

At the first stage, we computed optical confinement factor for modes, propagating in a mentioned above FMF 11/125, over wavelength range λ = 700. . . 1700 nm. The results are presented in the form of a diagram in Figure 6.

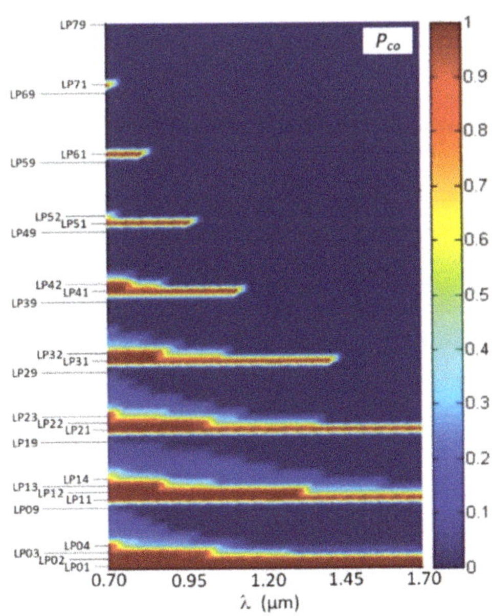


**Figure 6.** Diagram of the optical confinement factor distribution between modes of FMF 11/125 over the wavelength range λ = 700. . . 1700 nm.

According to computational results, desired 4-mode optical signal transmission is provided by researched FMF 11/125 over band λ = 1450...1700 nm. Generally, 38 $LP_{lm}$ modes with $l$ = 0...7 azimuthal and $m$ = 1...9 radial orders nominally satisfy the cut-off condition at the least researched wavelength range bound λ = 700 nm. However, only for 19 modes with also $l$ = 0...7, but $m$ = 1...4, orders optical confinement factor as more $P_{co} \geq 0.5$ for the same wavelength.

At the central region of the "O"-band (λ = 1300 nm), researched FMF 11/125 supports 6 guided modes that satisfy the cut-off condition under the optical confinement factor value more $P_{co} \geq 0.5$: they are listed in Section 2—$LP_{01}$, $LP_{11}$, $LP_{21}$, $LP_{02}$ modes and two additional higher-order modes $LP_{12}$, $LP_{31}$. We computed spectral curves of dispersion parameters for those 6 aforementioned guided modes. Results are represented in Figure 7a with spectral characteristics of mode delay and Figure 7b with chromatic dispersion coefficient. Analysis of mode delay curves show that DMD reaches 18.35 ns/km over λ = 1300 nm wavelength region, while near λ = 1550 nm DMD decreases down to 14.93 ns/km due to "suppression" of two higher-order modes $LP_{12}$ and $LP_{31}$.

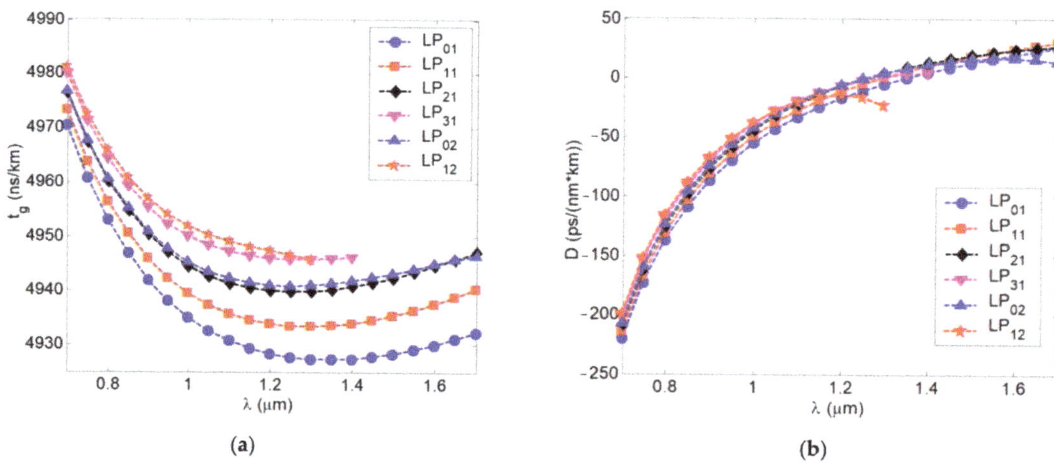


**Figure 7.** Spectral characteristics of guided mode dispersion parameters: (**a**) mode delay; (**b**) chromatic dispersion coefficient.

By comparing spectral characteristics of the chromatic dispersion coefficient for the fundamental and higher-order modes, computed curves are generally similar to spectral characteristic of chromatic dispersion coefficient for standard telecommunication single mode optical fiber (SMF) of ITU-T Rec. G.652 [28]. Here, zero dispersion wavelength of both the fundamental and higher-order guided modes corresponds to wavelength range λ = 1300...1350 nm. Maximal deviation of this parameter $D$ between higher-order guided modes was 27.09 ps/(nm·km) at λ = 1300 nm and 4.97 ps/(nm·km) at λ = 1550 nm.

### 5. Experimental Research of Spectral Responses of FBG, Written over Twisted FMF 11/125 with Improved Height of Quasi-Step Refractive Index Profile

Two FBG samples were written on the short (less 1.5 m) segments of SMF (Rec. ITU-T G.652 [28]) and fabricated pilot sample of twisted FMF 11/125 by Lloyd ineterferometric setup workstation under the same mask (with the same grating period), providing expected Bragg wavelength about $\lambda_B \approx 1550$ nm. We performed preliminary measurements of both FBG spectral responses under propagation of optical signal, generated by continuous emission (CE) wideband laser diode (LD) with operating wavelength λ = 1550 nm and pigtailed by SMFs. Conventional setup was utilized for FBG spectral response measurement by optical spectrum analyzer (OSA) with fiber optic circulator (CIR), also pigtailed by SMFs.

The described above scheme for testing of FBG, written on FMF, is shown in Figure 8. Both tested FBGs were jointed to SMF pigtail by fusion splicer and further connected to circulator via corresponding fiber optic adapter.

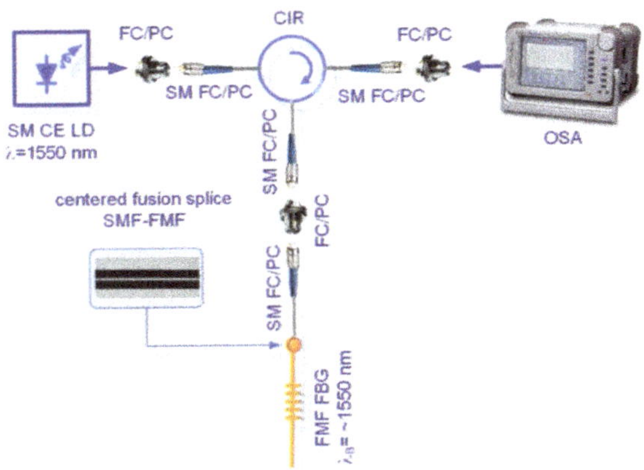


**Figure 8.** Conventional setup for reflected FBG spectral response measurement: testing of FBG, written on FMF under laser-based few-mode operation.

Results of measurements—OSA software screenshots—are represented in Figure 9a,b. Comparison of two measured spectral responses show that detected Bragg wavelength of FBG, written on FMF 11/125, is higher up to 16.46 nm ($\lambda_B$ = 1567.50 nm), than for FBG on SMF ($\lambda_B$ = 1551.04 nm). This suggests that effective refractive index for the fundamental mode $LP_{01}$ of FMF is somewhat higher, in comparison with the fundamental mode $LP_{01}$ of SMF, in approximately 1%. Spectral response of FBG on FMF contains main and periphery peaks. It may be considered as superposition of several modes, corresponding to transversal mode components, that led to response widening and confirms desired few-mode regime of FMF operation at the central wavelength of "C"-band ($\lambda$ = 1550 nm).

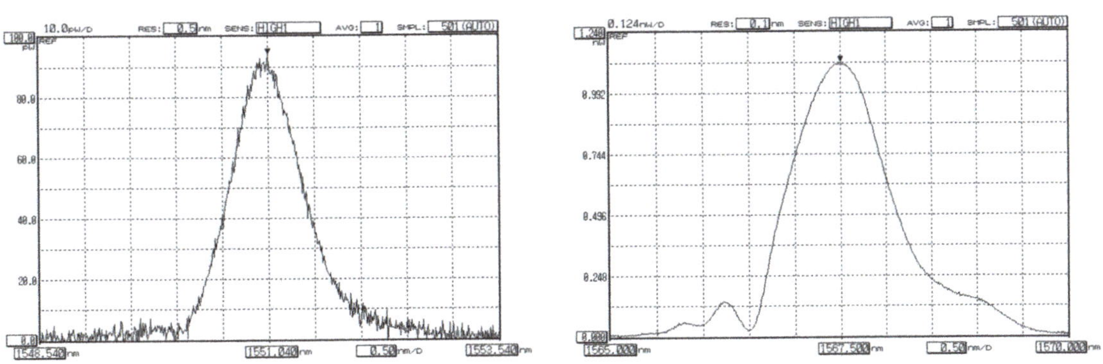

(**a**) (**b**)

**Figure 9.** Spectral responses of FBG, excited by laser-source (CE LD, $\lambda$ = 1550 nm): (**a**) FBG on SMF; (**b**) FBG on FMF 11/125.

The next set of tests was concerned with research of FBG Bragg wavelength $\lambda_B$ shifting sensitivity to the temperature action with the following comparison. We placed sequentially both FBGs to the thermostat and discretely varied temperature from +40 °C up to +120 °C with a step of 20 °C. Here, Bragg wavelength $\lambda_B$ under the least bound temperature +40 °C was considered as the reference value for the following estimation of $\lambda_B$ shifting under the temperature increasing. Results are represented in Figure 10. Both dependences are highly linear, while the slope for FBG on FMF is somewhat higher (approximately on 5%).

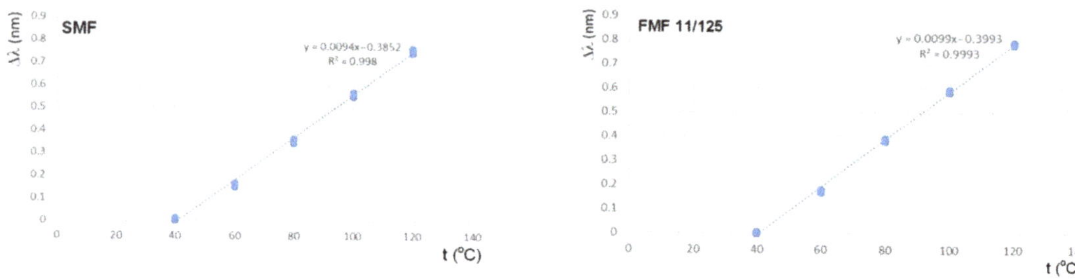


(a) (b)

**Figure 10.** FBG Bragg wavelength $\lambda_B$ shifting sensitivity to the temperature action: (**a**) FBG on SMF; (**b**) FBG on FMF 11/125.

By analogy with the previous measurements, we performed test series which focused on research and comparison sensitivity of FBGs on SMF and FMF to mechanical action. For these researches, we placed FBGs to the precision translation stage, which provides tensile and following with precision particular elongation of researched optical fiber segment with written FBG over the range 100. . . 250 µm with a step of 50 µm. Here, Bragg wavelength under the unstrained state was considered as the reference value for the following estimation $\lambda_B$ shifting $\Delta\lambda$ under the described mechanical action. Results are represented in Figure 11. Both dependences are also highly linear, while the slope for FBG on FMF is somewhat higher (approximately on 2.5%).

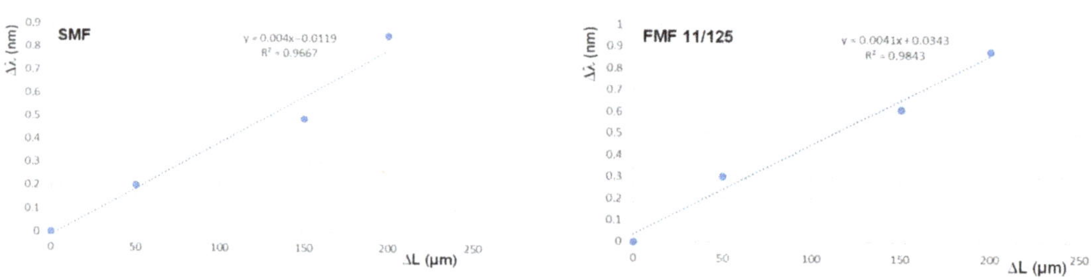


(a) (b)

**Figure 11.** FBG Bragg wavelength $\lambda_B$ shifting sensitivity to the mechanical action: (**a**) FBG on SMF; (**b**) FBG on FMF 11/125.

The next test series was concerned with researches of few-mode effects, occurring during laser-excited optical signal propagation over FBG, written in FMF 11/125, under some various stress actions. Here, we utilized a "direct" FBG spectral response measurement scheme without fiber optic circulator (Figure 12). We tested sample FBG on FMF which was pigtailed by using fusion splicer by short SMF pigtails with length not more than 140 mm to avoid conversion of higher-order FMF guided modes to leakage/cladding modes in SMF pigtail [33].

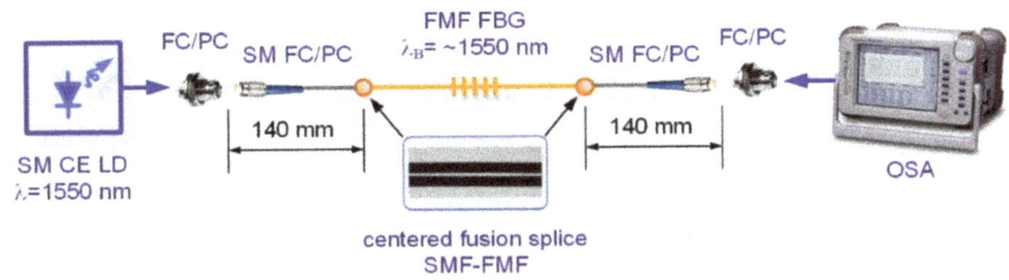


**Figure 12.** Conventional setup for direct FBG spectral response measurement: testing of FBG, written on FMF under laser-based few-mode operation.

Spectral response, measured for FBG on FMF 11/125 in an unperturbed state (that would be further considered as the reference) is presented in Figure 13. Here, 3 peaks (1 main peak (1567.24 nm) and 2 periphery peaks (1566.65 nm and 1567.96)) could be seen quite distinctly. During the next tests, we measured spectral responses under forming FMF loop with radius 15 mm before, on and after FBG. Results are shown in Figure 14. As expected, in all cases, Bragg wavelength shifting was detected. However, response smoothing as well as periphery peak dropout were noticed. Here, $\lambda_B$ shifted down to $\Delta\lambda$ = 0.24 nm under loop before and after FBG, while the loop on FBG $\lambda_B$ became longer up to 0.08 nm in comparison with the reference response main peak value. Second test series was concerned with spectral response measurements after placing loops with radius 86 and 63 mm over researched segment of FMF 11/125 with written FBG. Results are demonstrated in Figure 15. Here again, response smoothing and periphery peak dropout are noticed under the same Bragg wavelength shifting down to 0.16 and 0.20 nm.

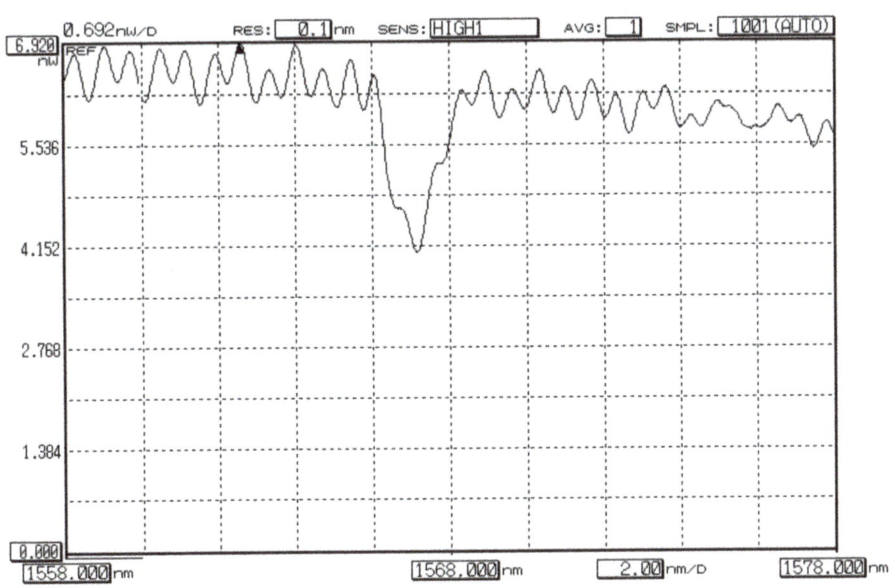


**Figure 13.** Reference spectral response of unperturbed FBG, written on FMF 11/125.

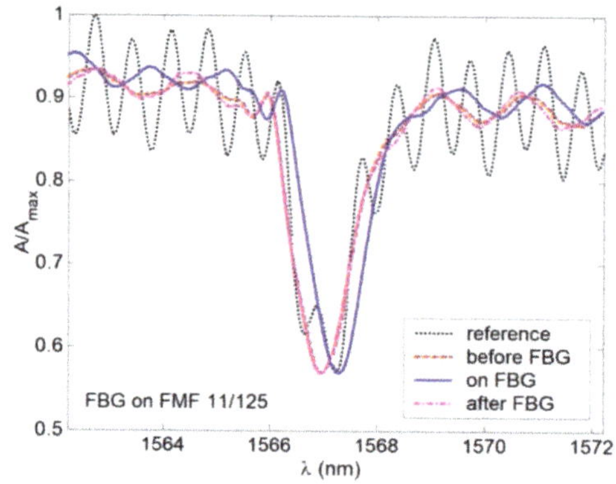


**Figure 14.** Spectral responses under 15 mm loop before, after and on the FBG, written on FMF 11/125.

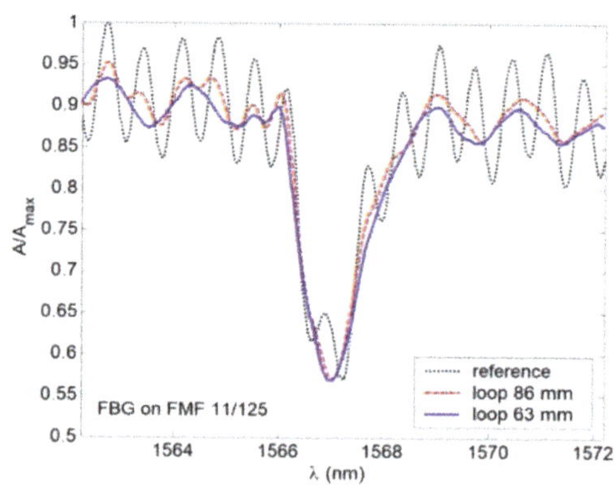


**Figure 15.** Spectral responses under 86 and 63 mm loops, placed over FMF 11/125 with written FBG.

## 6. Conclusions

This work is devoted to the design and fabrication, as well as experimental and theoretical researches of the parameters of FMF 11/125 with induced twisting and improved height of a quasi-step refractive index profile, which provides 4-mode operation over "C"-band. Based on the series of simulation of described optical fiber, we selected specified technological parameters to support the desired 4 guided modes over the mentioned above "C"-band: core diameter 11 μm, cladding diameter 125 μm and numerical aperture $NA = 0.22$.

Successfully fabricated pilot sample lengths of the described above FMF 11/125 with induced twisting of 10 and 66 rpm are presented. Results of measured attenuation showed expected increased loss $\alpha = 7 \ldots 8$ dB/km over "C"- and "O"-bands, explained by an intentionally simplified technique for FMF preform manufacturing by excluding typical F doping to optical fiber preform core region, which helps to decrease $GeO_2$ dopant unwanted influence on attenuation increasing.

We performed analysis of designed and fabricated FMF 11/125: here, data from the measurement report were utilized to restore the real form of quasi-step refractive index profile. Orders of guided modes, satisfying the cut-off condition, were defined over researched wavelength band $\lambda = 700 \ldots 1700$ nm (4…6 guided modes were localized over "C"- and "O"-band, respectively). Spectral characteristics of dispersion parameters (mode delay and chromatic dispersion coefficient) for defined guided modes were computed for the mentioned above researched wavelength range. Analysis of the mode delay curves showed that at the wavelength $\lambda = 1300$ nm, DMD reaches 18.35 ns/km, while at the wavelength λ = 1550 nm, it reduces down to 14.93 ns/km. By comparing spectral characteristics of the chromatic dispersion coefficient for the fundamental and higher-order modes, computed curves are generally similar to spectral characteristics of chromatic dispersion coefficient for conventional SMF (ITU-T Rec. G.652): here, zero dispersion wavelength of both the fundamental and higher-order guided modes corresponds to wavelength range λ = 1300. . . 1350 nm.

FBG was written in the sample of fabricated FMF 11/125 segment and test series were performed to research a few-mode effects, occurring during laser-excited optical signal propagation over FMF with written FBG, both unperturbed and under the temperature or mechanical actions. Main and periphery peaks were localized on the spectral response of unperturbed FMF FBG, while under the stress besides the expected Bragg wavelength shifting, spectral response smoothed and periphery peaks dropped out. Results of performed theoretical and experimental researches showed a good potentiality for utilization of designed and fabricated twisted FMF 11/125 in various applications of selected order guided mode management as well as in fiber optic sensors.

**Author Contributions:** Conceptualization, A.V.B., V.V.D., G.S., M.T. and F.P.; methodology, A.V.B., V.A.B., K.V.D., Y.I., O.G.M., F.P. and J.Y.; development of technique for twisted optical fiber fabrication, modification of drawing tower, twisted optical fiber fabrication, V.V.D., K.V.D., A.V.K., A.S.M., G.A.P. and E.V.T.-N.; measurements of geometry parameters, laser beam profile, attenuation, V.V.D., A.V.K., A.S.M., G.A.P. and E.V.T.-N.; writing FBG, measurements of FBG spectral responses, A.V.B., M.V.D., A.A.K., A.Z.S., E.S.Z., A.R.G., I.K.M. and E.V.D.; writing—review and editing, A.V.B., V.V.D., A.A.V. and A.A.K.; visualization, A.V.B., V.V.D., M.V.D., A.A.K., A.A.V., E.V.D., A.R.G. and I.K.M.; supervision, A.V.B.; project administration, A.V.B. All authors have read and agreed to the published version of the manuscript.

**Funding:** This research was funded by RFBR, DST, NSFC and NRF according to the research project 19-57-80016 BRICS_t.

**Institutional Review Board Statement:** Not applicable.

**Informed Consent Statement:** Not applicable.

**Conflicts of Interest:** The authors declare no conflict of interest.

## References

1. Barlow, A.J.; Ramskov-Hansen, J.J.; Payne, D.N. Birefringence and polarization mode dispersion in spun single-mode fibers. *Appl. Opt.* **1981**, *20*, 2962–2968. [CrossRef] [PubMed]
2. Hart, A.C.; Huff, R.G.; Walker, K.L. Method of Making a Fiber Having Low Polarization Mode Dispersion Due to a Permanent Spin. U.S. Patent 5,298,047, 29 March 1994.
3. Blaszyk, P.E.; Christoff, W.R.; Gallagher, D.E.; Hawk, R.M.; Kiefer, W.J. Method and Apparatus for Introducing Controlled Spin in Optical Fibers. U.S. Patent 6,324,872 B1, 4 December 2001.
4. Li, M.-J.; Chen, X.; Nolan, D.A. Fiber spinning for reducing polarization mode dispersion in single-mode fibers: Theory and applications. In Proceedings of the Optical Transmission Systems and Equipment for WDM Networking II, Orlando, FL, USA, 7–11 September 2003; Volume 5247, pp. 97–110.
5. DiGiovanni, D.J.; Golowich, S.E.; Jones, S.L.; Reed, W.A. Method of Making an Improved Multimode Optical Fiber and Fiber Made by Method. U.S. Patent 2001/0019652, 6 September 2001.
6. DiGiovanny, D.J.; DiMarcello, F.V.; Jiang, X.L.; Oulundsen, G.E.; Pandit, S.P. Multimode Optical Fiber with Increased Bandwidth. U.S. Patent 2004/0228590 A1, 18 November 2004.
7. Bohnert, K.; Gabus, P.; Kostovic, J.; Brandle, H. Optical fiber sensors for the electric power industry. *Opt. Lasers Eng.* **2005**, *43*, 511–526. [CrossRef]
8. Kopp, V.I.; Churikov, V.M.; Singer, J.; Neugroschl, D. Chiral fiber sensors. In Proceedings of the Fiber Optic Sensors and Applications VII, Orlando, FL, USA, 5–9 April 2010; Volume 7677.
9. Kopp, V.I.; Genack, A.Z. Adding twist. *Nat. Photonics* **2011**, *5*, 470–472. [CrossRef]
10. Wong, G.K.L.; Beravat, R.; Russell, P.S.J. Helically twisted photonic crystal fibres. *Philos. Trans. R. Soc. A Math. Phys. Eng. Sci.* **2017**, *375*, 20150440.
11. Liu, T.; Wang, D.; Raenaei, R.; Cheng, X.; Zhang, L.; Bennion, I. A low-cost multimode fiber Bragg grating sensor system. In Proceedings of the Advanced Sensor Systems and Applications II, Beijing, China, 8–11 November 2004; Volume 5634, pp. 54–61.

12. Sang Xi Yu Ch Yan, B. Bragg gratings in multimode optical fibres and their applications. *J. Optoelectron. Adv. Mater.* **2006**, *8*, 1616–1621.
13. Zhao, C.-L.; Li, Z.; Demokan, M.S.; Yang, X.; Jin, W.; Lu, C. Studies on strain and temperature characteristics of a slanted multimode fiber Bragg grating and its application in multiwavelength fiber Raman ring laser. *IEEE J. Lightwave Technol.* **2006**, *24*, 2394–2400. [CrossRef]
14. Mizunami, T.; Djambova, T.V.; Niiho Ts Gupta, S. Bragg gratings in multimode and few-mode optical fibers. *IEEE J. Lightwave Technol.* **2006**, *18*, 230–235. [CrossRef]
15. Yong, Z.; Zhan, C.; Lee, J.; Yin, S.; Ruffin, P. Multiple parameter vector bending and high temperature sensors based on asymmetric multimode fiber Bragg gratings inscribed by an infrared femtosecond laser. *Opt. Lett.* **2006**, *31*, 1794–1796. [CrossRef]
16. An, J.; Liu, T.; Jin, Y. Fiber optic vibration sensor based on the tilted fiber Bragg grating. *Adv. Mater. Sci. Eng.* **2013**, *2013*, 545013. [CrossRef]
17. Li, D.; Gong Yu Wu, Y. Tilted fiber Bragg grating in graded-index multimode fiber and its sensing characteristics. *Photonic Sens.* **2013**, *3*, 112–117. [CrossRef]
18. Fang, S.; Li, B.; Song, D.; Zhang, J.; Sun, W.; Yuan, L. A smart graded-index multimode fiber based sensor unit for multi-parameter sensing applications. *Opt. Photonics J.* **2013**, *3*, 265–267. [CrossRef]
19. Guo, T.; Guan, B.-O.; Albert, J. Polarimetric multi-mode tilted fiber grating sensors. *Opt. Express* **2014**, *22*, 7330–7336. [CrossRef] [PubMed]
20. Schmid, M.J.; Muller, M.S. Measuring Bragg gratings in multimode optical fibers. *Opt. Express* **2015**, *23*, 8087–8094. [CrossRef] [PubMed]
21. Gatchin, Y.; Demidov, V.; Dukelskii, K.; Ter-Nersesyants, E. Quasi-single-mode fibers with an increased core size based on non-hexagonal microstructures. *Proc. Telecommun. Univ.* **2017**, *3*, 37–42. [CrossRef]
22. Demidov, V.V.; Dukelskii, K.V.; Leonov, S.O.; Matrosova, A.S. Nonlinear optical transformations of picosecond laser pulses in multimode microstructured fibers with moderate nonlinearity. *Proc. Telecommun. Univ.* **2018**, *4*, 61–66. [CrossRef]
23. Ananiev, V.A.; Demidov, V.V.; Leonov, S.O.; Nikonorov, N.V. Hollow antiresonant fibers with a large effective area of the mode field for operation in the near and mid-IR spectral regions. *Proc. Telecommun. Univ.* **2019**, *5*, 6–14. [CrossRef]
24. Agrawal, G.P. *Nonlinear Fiber Optics*; Academic Press: Burlington, UK, 2012.
25. Olszewski, J.; Szpulak, M.; Urbanczyk, W. Effect of coupling between fundamental and cladding modes on bending losses in photonic crystal fibers. *Opt. Express* **2005**, *13*, 6015–6022. [CrossRef]
26. Adams, M.J. *An Introduction to Optical Waveguides*; John Wiley & Sons Ltd.: New York, NY, USA, 1981.
27. Snyder, A.W.; Love, J. *Optical Waveguide Theory*; Chapman & Hall: London, UK, 1983.
28. ITU. *Optical Fibres, Cables and Systems*; ITU-T Manual; ITU: Geneva, Switzerland, 2009.
29. Guenther, B.; Steel, D. *Encyclopedia of Modern Optics*, 2nd ed.; Elsevier Science: Amsterdam, The Netherlands, 2018; Volume 1.
30. Bourdine, A.V.; Barashkin, A.Y.; Burdin, V.A.; Dashkov, M.V.; Demidov, V.A.; Dukelskii, K.V.; Evtushenko, A.S.; Ismail, Y.; Khokhlov, A.V.; Kuznetsov, A.A.; et al. Twisted silica microstructured optical fiber with equiangular spiral six-ray geometry. *Fibers* **2021**, *9*, 27. [CrossRef]
31. Bourdine, A.V. Modeling and simulation of piecewise regular multimode fiber links operating in a few-mode regime. *Adv. Opt. Technol.* **2013**, *2013*, 469389. [CrossRef]
32. Bourdine, A.V.; Delmukhametov, O.R. Calculation of transmission parameters of the launched higher-order modes based on the combination of a modified Gaussian approximation and a finite element method. *Telecommun. Radio Eng.* **2013**, *72*, 111–123. [CrossRef]
33. Vasil'ev, S.A.; Dianov, E.M.; Medvedkov, O.I.; Protopopov, V.N.; Constantini, D.M.; Iocco, A.; Limberger, H.G.; Salathe, R.P. Properties of cladding modes excited in a fiber optical waveguide by refractive index gratings. *Quantum Electron.* **1999**, *26*, 65–68. [CrossRef]

MDPI AG
Grosspeteranlage 5
4052 Basel
Switzerland
Tel.: +41 61 683 77 34

*Sensors* Editorial Office
E-mail: sensors@mdpi.com
www.mdpi.com/journal/sensors

MDPI

Academic Open
Access Publishing

mdpi.com

ISBN 978-3-7258-1639-2

www.ingramcontent.com/pod-product-compliance
Lightning Source LLC
Chambersburg PA
CBHW061508210926
51850CB00008B/1857
* 9 7 8 3 7 2 5 8 1 6 4 0 8 *